CASED HOLE AND PRODUCTION
LOG EVALUATION

CASED HOLE
AND PRODUCTION
LOG EVALUATION

James J. Smolen, Ph.D.

Copyright © 1996 by
PennWell Publishing Company
1421 South Sheridan / P.O. Box 1260
Tulsa, OK 74101

Library of Congress Cataloging-in-Publication Data
Smolen, Jim.
    Cased hole and production log evaluation / Jim Smolen.
        p.   cm.
    Includes index.
    ISBN 0-87814-456-X
    1. Oil well logging.   I. Title
TN871.35.S547   1995   95-46489
622'.1828--dc20   CIP

Printed in the United States of America

4 5 6   04 03

Contents

Preface

The topic of cased hole logging encompasses a very broad spectrum of applications. These include formation evaluation for shows of hydrocarbons, assurance of well integrity, and the mapping of fluid movement downhole. For these applications, many technologies have been and continue to be developed. This book is an attempt to provide a somewhat complete and grand overview of such technologies and examples of their application.

The text is presented to maximize the reader's physical sense of how tools work and why they work. The interpretation models are presented in a concise and logical manner. The discussion focuses on the basic computational techniques and on higher order effects where such effects may be useful for new and novel measurements.

The present state of the industry comprises the range from basic analog technologies to highly sophisticated computer modeling and imaging techniques. This book covers the full spectrum, with emphasis on the tools and their uses. This book is up-to-date with discussion of the newest tools available today. Since this book is published only in black and white, many of the new full color imaging presentations cannot be shown, and the interested reader is advised to request examples from the service companies. Older tools which may no longer be used, and which often occur in well files, are also covered.

Certain types of cased hole logs are not discussed in this book. These include the gravel pack evaluation logs and the so-called stuck point or free point indicators. Also, nonlogging wireline apparatus, such as perforating equipment, casing cutters, and the like are not covered.

The author would like to thank the wireline service companies for their assistance and contributions to this book. They have been extremely helpful in proofreading various sections and supplying technical information requested. The wireline business is very competitive. The author has attempted to avoid any bias or preference among service companies or their hardware, and has tried to show examples originating from numerous service company sources. The names of the companies are used throughout the book. The author would also like to thank his many friends at major oil companies for their comments and assistance.

Certain people deserve special thanks, for they have been pivotal in the author's path of life leading to this document. First and most recently, the author would like to thank Fred Bradburn of Shell Internationale Petroleum Maatschappij B.V., Group Training, in the Netherlands. Fred reviewed training materials used in a course and provided strong initial prodding to write this book. The author would like to thank the late Bob Kudrle of Schlumberger. Bob many years ago provided the vision necessary for me to remain on the technical course which ultimately led to this book. Dr. Antoni K. Oppenheim, of the University of California at Berkeley, deserves special thanks. He, like a surrogate father, provided me a direction when my long term objectives appeared to vacillate day to day. And, of course, my family's patience through this period is greatly appreciated.

It would be the author's greatest delight for this book to open the reader's minds to new ideas and applications for cased hole logging techniques not previously used in their areas. As production declines in many producing provinces, these techniques hold the key to stemming the decline and restoring production. Just as one would not have surgery without a proper diagnosis, the cased hole logs are the diagnostic tools for producing reservoirs and wells. Their maintenance and workover should not be done without the use of cased hole logs.

Chapter 1

INTRODUCTION

BACKGROUND

Why run cased hole logs? They are run to assess well integrity, improve reservoir management, and scan the well for bypassed production before plugging and abandoning. These services are also key diagnostic tools for workover planning and operations.

In new wells, cased hole logs are run to establish that the primary cement job has been accomplished properly. Is there cement over the interval to be completed? Is such cement adequate to provide a reasonable assurance of zonal isolation? Is the well integrity adequate to complete the well as planned? Is a squeeze cement job needed and where should it be done?

In most older producing wells, cased hole logs are all that can be run. Traditional reasons include diagnosing the illness of old wells, wells which may now produce excessive amounts of water. How can the water production be reduced and hydrocarbon production increased? Cased hole logs shed light on the sources of water and hydrocarbon production downhole and offer hope that such production restorations may be possible.

Wells need not be sick or exhibit production changes to be candidates for cased hole logs. Indeed, logs are frequently run in casing to monitor the reservoir so that changes in production can be anticipated and planned for. Such monitoring is commonly done to detect movement of water-oil or gas-oil contacts during production in individual wells and thereby help manage overall reservoir production among a number of wells. Saturation changes at unanticipated locations may dictate selectively perforating new intervals to assure proper reservoir depletion.

Cased hole logs are extremely important in secondary or tertiary recovery programs. Typically water (or other fluid) is injected into certain wells and produced along with hydrocarbons at other wells in the field. Injection and production profiles are run to detect anomalous injection/production, i.e., is the flow going in or coming out near the top or bottom of the formation? What must be done to assure that the flood front is uniform and prevent operations from simply circulating water between wells?

Sometimes, old wells are logged for the purpose of detecting bypassed producible zones, typically uphole from current or recently produced zones. This occurs when old wells do not have a good suite of open hole logs available, or such bypassed production was not detectable using older logging or interpretation techniques. Perhaps such upper zones are gas and were not of interest at the time of initial completion. Locating such bypassed production will provide new reserves to a company from assets thought to be depleted and worthless.

Due to the increased use of Logging While Drilling (LWD) operations, cased hole logs are frequently used as supplemental sources of information on formation lithology and hydrocarbon saturation. If the day comes where wells will be routinely drilled using coiled tubing units with turbine drill bits, cased hole logs will then have come of age. It is anticipated that such operations would utilize the coil tubing as the production casing string and virtually no open hole logs would be available. The only logging information would be that obtainable through casing.

1

Cased hole logs are the diagnostic tools for efficient reservoir management and production. Problems must be properly defined before they can be corrected. Virtually no workover operation or step taken to optimize or improve production would not be helped in some way by cased hole logs. This text will examine a wide variety of logging tools and types of services. The previous examples are only a tiny fraction of the applications of such services. This text is designed to highlight the operation and application of logging tools. Emphasis is on the physical sense of what each tool does and how it does it. It is hoped that this approach will provide the reader with the kind of information which stimulates creative thinking in applying these logs to solve both traditional as well as new problems in oilfield operations.

CATEGORIES OF CASED HOLE LOGS

For purposes of this text, cased hole logs may be placed into the following four categories:

1. Formation Evaluation
2. Wellbore Integrity
3. Fluid Movement During Production/Injection
4. Other

A description of each category follows:

Formation Evaluation

Logging tools in this category are designed to evaluate formation properties. Included are formation shale content, clay type, and vertical definition of zones which are clean and shale free. Logging services in this category are also capable of determining the type of rock (sand, lime, etc.), the type of hydrocarbon, i.e., gas or oil, and its saturation. Other information available includes mechanical properties of the rock, to mineralogy, its permeability, skin damage, pressure, natural fractures, and even samples of formation fluids.

Wellbore Integrity

This category of logs includes the wide variety of logs to evaluate the cement sheath around the casing. Cement top location, fraction of annular fill, and cement compressive strength can be measured. This information provides some assurance of hydraulic isolation. Casing condition in terms of depth and extent of damage may also be evaluated. Certain tools even discriminate damage on the inner wall from that on the outer surface of the casing.

Fluid Movement During Production/Injection

This category includes tools which detect channels behind pipe in both injection and production wells. Such tools furthermore detect zones of fluid injection, location of pumped-in materials such as fracture fluid or proppant, and even can directionally detect the orientation of certain injected particulates. Flow profiles in both injection and production may be evaluated along with the contributions of each phase of produced fluid on a zone by zone basis. Combined with pressure information, these contributions may be the basis for determining a zone by zone inflow performance relationship (IPR).

Other

This category is a catch-all for services whose application or environment may be unusual. For example, the gravel pack logs are designed to evaluate the presence of gravel outside of a wire wrapped or slotted liner, a condition which is neither in casing nor out of it. Another grouping in this catch-all category are the stuck point or free point indicator tools. These are typically not closely related to reservoir management and are of a more immediate operational concern. These surveys are not covered in this text.

Operational Considerations

Cased hole logging tools are typically run on wireline. This line may be either armored electrical cable or "slick line." Units using electrical cable transmit data to the surface and data is gathered in real time. Most cased hole tools are run on a cable with a single conductor, i.e., a monocable, although some services may be run on multiconductor cable. With slick line units, data is gathered and recorded downhole and retrieved when the tool string is brought to the surface.

In highly deviated and horizontal wells, various tricks must be done to get the tools into and moving across the interval of interest. Such operations are typically done with coiled tubing units where the electric wireline is inside of the coiled tubing to which the tool string is rigidly attached. The tubing is used to push or pull the tool across the interval to be logged. Units without electric cable and using downhole recorders are also available. Alternatively, pump down techniques in which the tool string with wireline attached is pumped down through tubing to the end of the horizontal section, then pulled back. Needless to say, this technique would only work using a nuclear tool which sees through the tubing.

In most cases tool diameter is important since it is the deciding factor between "through tubing" or "in casing" operations. The former requires little well preparation and is least expensive in terms of lost production and other required operations. The latter typically is done at the time of completion or later. After completion and production, such operations require shutting in the well and removing the tubing and packer at great expense.

PURPOSE OF THIS BOOK

This book is written to present in one place a grand overview of virtually all of the cased hole logging tools available today. This is increasingly important since certain types of equipment are often used for cross discipline purposes. Most notable are the pulsed neutron capture tools which are primarily used for porosity, water saturation, and gas detection. In recent years these have been used for water flow measurement, gas entry detection, and phase holdup evaluation. Such comments, however, also apply to the natural gamma ray, compensated neutron, ultrasonic, and other tools. A basic understanding of such cross-applications hopefully will provide the basis for creative solutions to complex downhole diagnostic problems.

This text attempts to provide the reader with a physical sense of what the tools do, how they do it, and why they do what they do. Modern tools as well as certain older generation tools are reviewed since all are likely to be present in well files. A wide variety of applications if not discussed in the text will be shown in the log examples at the end of each chapter.

CHAPTER 2

THE CASED HOLE LOGGING JOB

THE WELL

Typical Well Profile

Most wells are cased over a large interval to protect the shallow fresh water sands, to prevent collapse of the formations into the borehole, to maintain pressure control over widely varying pressured zones encountered during drilling, and to allow for a controlled completion. A typical casing string is illustrated on Figure 2.1.[1] The conductor pipe is designed to be of a large diameter and to prevent the collapse of surface formation and stabilize the wellsite. The surface casing string protects those zones containing fresh water. The oil string

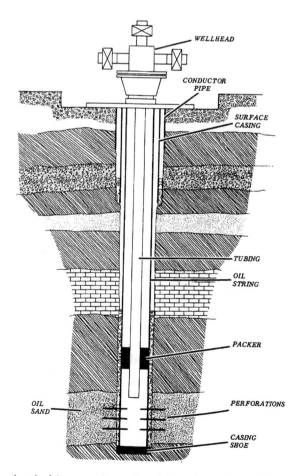

Figure 2.1 Typical casing/tubing configuration of a producing well (Courtesy Petroleum Extension Service (PETEX), The University of Texas at Austin, Ref. 1)

is set typically over the zones to be completed. There may be a number of intermediate casing strings which are necessary due to formation pressure considerations or wear on casing as deeper drilling is being done. Deeper casing strings may be set as liners, which do not return to the surface, but instead are hung from near the bottom of the larger string above. Notice in the figure that casing strings are not necessarily cemented all the way back to the surface. Tubing and packer are typically set just above the completed zone and the tubing is run to the surface. Produced fluids flow through the tubing to the wellhead at the surface, and then to separators, flow lines, storage tanks, and the like. The tubing contains the pressure of the flowing fluids and protects the casing from corrosion and wear.

Natural Completions

Natural completions are those in which the pressure downhole is sufficient to drive the produced fluid to the surface without the use of pumping equipment. The following are typical well configurations for natural completions. Each type may pose some unique problems for acquiring downhole information using logging tools. This is especially the case when dealing with deviated and horizontal wells.

Perforated Cased Hole Completion.   This is the completion shown on Figure 2.1. Primary production comes from a zone which is cased and cemented to isolate it from neighboring zones. That zone, which is hydrocarbon bearing, is perforated to establish communication between it and the wellbore. Sometimes, the oil string is set across more than one zone bearing hydrocarbons. If more than one zone is completed, i.e., perforated, the production from such zones is said to be mixed or "commingled."

Open Hole or "Barefoot" Completion.   This type of completion occurs when casing is set to just above the zone or zones bearing hydrocarbons. After setting casing, further drilling is done through the hydrocarbon zone(s). Rather than casing and cementing this interval, the hole remains open and the formation fluids are allowed to flow into it from such zone(s).

Slotted Liners or Casing Completion.   Such a completion is in fact an open hole completion with a slotted pipe set in the producing interval to prevent the hole from collapsing. The slotted pipe is not cemented into place.

Gravel Packed Completions.   This type of completion is typically done to prevent unconsolidated formation material from being produced into the wellbore. If produced, such sand production damages downhole and surface equipment, and ultimately fills the producing interval with sand and kills the well. Usually casing is cemented in place and perforated. Then, a wire wrapped screen is set inside the casing. The annulus between the screen and perforated casing is filled with a gravel which is sized to hold back the formation sand and yet not pass through the wire wrapped screen.

Dual or Multiple Completions.   These completions typically have two or more tubing strings in the same well. These tubing strings are each designed to produce different zones and are set up with special packers capable of passing two or more tubing strings. Alternatively, there are "tubingless" completions in which there are no packers and the individual small tubing strings are cemented in place and each perforated over different intervals.

Pumping Wells

Pumping wells are those in which pumping equipment is used to reduce the pressure down hole and bring the produced fluid to the surface. These techniques are used when formation pressures are not adequate to force produced fluids to the surface.

Rod Pumped Wells.   Rod pumped wells are probably the most common type of pumping well, especially among wells with low flowrates. These use a horse's head and walking beam arrangement. The motion of the horse's head causes the up and down movement of the sucker rods which, in turn, operates a pump at the end of tubing downhole. A special wellhead is required to log such wells and the logging tool is run down the tubing-casing annulus.

Electric Submersible Pumps.   These are subsurface hydraulic pumps located at the bottom of tubing and powered by electrical or hydraulic energy. These generally cannot be logged unless a "Y" tool is situated at the end of the tubing string. This tool allows passage of the logging tool even as the pump works.

Gas Lift.   Gas lift is a method wherein gas is pumped into the tubing string, thereby removing the heavier fluids present and reducing the hydrostatic pressure on the formation. This type of well can be logged much like a naturally completed well.

THE LOGGING JOB

General Information Required

Each cased hole logging job is a unique adventure. The first thing to consider may seem obvious, but "why are we running this log?" should be asked. To answer this question, we need to know exactly what the problem is and what we need to know to resolve the relevant issues. When the well is acting up don't just guess at what the problem is. Use all other information which you have available. Consider open hole log data regarding adjacent zones, reservoir conditions, production history, any sudden changes in the well's performance, recent workover operations, offset wells in the same reservoir, and the like.

Avoid fishing expeditions. Production logs that run without a theory to confirm are likely to be ill thought out and end in failure to get the necessary information to properly understand the problem downhole. Establish a theory which the logs are to confirm and upon which you can act. If there are other possibilities, be sure to run sensors necessary to confirm or refute those also.

Much detailed information is necessary before running cased hole logs. While each type of logging job may have unique requirements, the following list will provide guidelines regarding the kinds of detailed information likely to be needed.

1. A complete and accurate well sketch should be available showing all sizes and weights of tubing and casing, all restrictions and other trinkets down hole, and their depths. Such a sketch should also show the deviation angle.
2. Anticipate downhole conditions. What is the temperature and pressure? Is H_2S present? Is gas present downhole? The logging tools or even cable may not be rated or suitable for the conditions to be encountered and you may require special equipment. Be sure to work with the service company closely on such matters.
3. What is the well head pressure? What special pressure equipment is required? Will weights be needed to run the tool into the hole?
4. How long is the tool string with all of the necessary sensors? Consider riser pipe length limitations, need for weights, and the longer the tool the less likely it will pass through tubing which may be coiled up above the packer.
5. Is the well head configuration known to the service company so that they can flange up without undue delay?
6. Subject to safety considerations, it is usually preferable to run the tool downhole with the well flowing. To do this, you may have to consider the flowrates anticipated downhole and in the tubing. If you must shut in, the well will have to be restabilized with the tool in the hole.
7. Is a dummy run required? It is not uncommon for tubing to coil up above the packer enough that the tool cannot pass. Such a dummy run is done using weights prior to risking the expensive logging tools in the hole.
8. Will the operations be daylight only or will they also proceed at night?

Can the Log Be Run through Tubing?

Refer to Figure 2.1. The last piece of piping installed in the well is the tubing string. This string is necessary if the well is to be producing while logging. Some logs are run through tubing, some not. This may or may not be a problem depending upon when in the life of the well the logging job occurs.

If the tubing has not yet been installed and the casing not perforated, then it is convenient to run certain surveys. For example, this is the ideal time to run a cement evaluation type of log such as a cement bond log (CBL). Also, logs for depth correlation such as gamma ray and collar locator are run during this time.

Once the tubing has been installed and the well is producing, then it is most desirable to run logs through tubing. Such surveys are characterized by small diameter tools, typically 1 11/16 in. (4.3 cm). After appropriate rigging up at the surface, the tools can be run directly down into the well through the tubing and measurements made across the producing interval. Operations performed in this manner do not necessarily require shutting in the well, although the flow may be stopped or reduced when running in. Therefore, there is no or little loss of production. Surveys run for monitoring saturation changes in the formation or for locating water or other production sources are most often run through tubing.

Sometimes the logging tool is too large to run through tubing. This may occur if a casing inspection tool is required or a larger diameter carbon/oxygen tool is required to evaluate formation water saturations. To run a larger diameter tool, the tubing must be removed and the well killed. This results in a loss of production, excessive rig time costs, and the risk that production will not resume at its earlier rates. So, when considering running cased hole logs, it is important to determine the maximum tool size that can be run and to try to select those surveys which can be run through tubing if such are available to do the job.

RIGGING UP FOR THE JOB—NATURALLY PRODUCING WELLS

Rigging Up at the Wellsite

The equipment necessary at the wellsite for a typical cased hole logging job through tubing in a naturally producing well is shown on Figure 2.2.[3] The numbered descriptions correspond to the figure.

1. The logging truck. The cable, winch, surface computers, and logging personnel are in this truck.
2. The mast truck. This unit has a mast which folds or tele-scopes up to the position shown and back for movement off location.
3. The wellhead with valves and flowlines connected to it.
4. Lubricator or riser pipe. This pipe is used to store the tool before running into the hole. The lubricator is mounted atop the wellhead and the pressure in the lubricator is equalized to that of the wellhead before logging. Note that a number of riser pipe sections may be connected to accommodate longer tool strings.
5. Cable. This cable is usually a single conductor (monocable) armored cable. The cable is wound onto the winch of the logging truck for storage.
6. Pressure bleed-off hose to relieve pressure from the riser pipe after the logging job.
7. Grease line to main grease seal.
8. Grease pump and reservoir for grease seal.
9. Grease seal. These seals assure hydraulic seal around the cable even when running the cable in and out of the hole.
10. Instrument truck. This unit typically may or may not be needed depending on the services run. Most modern trucks are fully contained and this unit is usually not necessary.
11. Pressure being released from the lubricator through the bleed-off hose.
12. Upper sheave wheel. Note also the lower sheave wheel chained to the wellhead.

PRODUCTION LOGGING
WELLSITE SET – UP

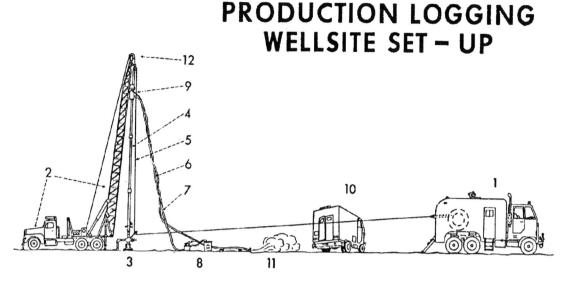

Figure 2.2 Production logging wellsite set-up (Courtesy Canadian Well Logging Society, Ref. 3)

Pressure Control Equipment

Figure 2.3 shows a detailed diagram of the typical pressure control equipment mounted atop the wellhead.[4] It includes a hydraulically operated blowout preventer and a tool trap to catch the tool if it is inadvertently pulled off in the riser pipe. The length can be adjusted prior to rigging up by using multiple sections of riser pipe and the tool placed inside prior to connecting this equipment to the wellhead. The grease seals are shown above the riser pipe sections. They are comprised of seal tubes which closely fit the cable diameter. Grease is injected in the annulus between the cable and seal tubes and thereby provides a seal even when the cable is moving in and out of the well. The hydraulically actuated packing gland is used to contain pressure when the cable is stopped, but is not used when the tool is moving. The configuration of the pressure control equipment and wellhead is such to allow the tool to pass directly into the tubing string and be run downhole.

Such equipment is typically rated up to 15,000 psi (103 MPa) in increments of 5,000 psi (34.3 MPa). Each set of equipment must also be periodically tested to pressures of 1.5x or greater than their rated pressure. Special equipment is necessary for H_2S work. Consult with the service company prior to the job.

The Logging Truck

A schematic of a typical logging truck is shown on Figure 2.4.[5] The operator views and controls the winch and cable from the air conditioned operator's compartment. The computer and data processing system is located here also. The cable passes through the measuring and spooling system for measurement of depth and cable tension. Both the back and top of the truck are open to allow full view of the rig area. The figure is otherwise self explanatory.

"Slickline" Operations

Some cased hole logs are run on a "slickline" unit. This is a wireline unit where the data is recorded downhole and not sent to the surface. Typically, the data is recorded mechanically by scratching a surface which moves as a function of time or digitally recording data on magnetic media for playback at the surface. Pressure gauges for pressure transient analysis are typically run on slickline. The wire used in such units is a nonconducting and non-armored piano wire, ranging in diameter from about 0.06 to 0.1 in. (0.15-0.25 cm), much smaller than conventional electrical wireline. Due to its light weight and short tools, the riser pipe typically also serves as the mast.

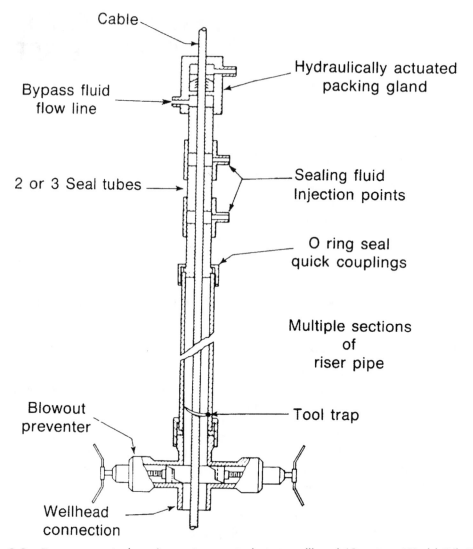

Figure 2.3   Pressure control equipment mounted atop wellhead (Courtesy World Oil, Ref. 4)

RIGGING UP—PUMPING WELLS

Typical Set Up for a Pumping Well

For a typical pumping well consisting of a horse's head and walking beam arrangement, along with sucker rods and pump at the bottom of tubing, the typical logging set up is shown on Figure 2.5.[6] Here the truck has its own mast unit and is quite close to the wellhead. This is not usually the case when dealing with high pressure, but pumping wells generally have little or no pressure at the surface. Before rigging up, the well must typically be prepared by removing the tubing anchor and setting the top of tubing about 50 ft (15 m) above the interval to be logged. Production should be allowed to stabilize for a few days prior to logging. The logging is done through the annulus between the tubing and casing.

A special wellhead for logging through the annulus is shown on Figure 2.6.[7] It features an access opening for logging tools adjacent to the polished rod. The lubricator is mounted to the access by a swivel coupling. The logging tool is lowered down the annulus to below the bottom of tubing and the producing interval is logged. When returning to the surface, there is a danger of the cable wrapping around the tubing and preventing retrieval of the logging tool. If this occurs, a pulling unit is needed to raise the wellhead to allow the service company personnel to manually unwrap the tool and allow its removal.

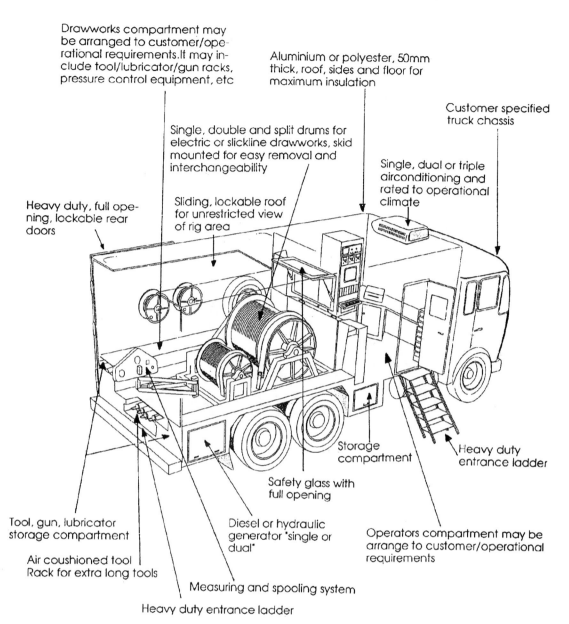

Drawworks compartment may be arranged to customer/operational requirements.It may include tool/lubricator/gun racks, pressure control equipment, etc

Aluminium or polyester, 50mm thick, roof, sides and floor for maximum insulation

Customer specified truck chassis

Single, double and split drums for electric or slickline drawworks, skid mounted for easy removal and interchangeability

Single, dual or triple airconditioning and rated to operational climate

Heavy duty, full opening, lockable rear doors

Sliding, lockable roof for unrestricted view of rig area

Tool, gun, lubricator storage compartment

Air coushioned tool Rack for extra long tools

Diesel or hydraulic generator 'single or dual'

Safety glass with full opening

Storage compartment

Heavy duty entrance ladder

Operators compartment may be arrange to customer/operational requirements

Measuring and spooling system

Heavy duty entrance ladder

Figure 2.4 Wireline Logging Truck (Courtesy Sodesep, Ref. 5)

RIGGING UP—HIGH DEVIATION ANGLE AND HORIZONTAL WELLS

Typical Set Up for Coil Tubing Logging Operations

When a wellbore becomes excessively deviated, say over about 60°, great difficulty is usually encountered in running the tool into the hole. This is a result of friction, poor hole conditions, and the like. Tool strings having rollers, sinker bars, special nose guides, and rigidized cable have been used to help the tool slide downhole. Where tool strings are long, the tool may not be able to negotiate the curvature of the borehole, and hence knuckle joints are used to enhance tool flexibility. In open holes, it is common practice to run logging tools on the end of the drill string and mate up later with the tools by means of a wireline "wet connection." The cable comes out of the drill string through a side entry sub, and the well is logged one joint or stand of pipe at a time.

The use of coil tubing to convey logging tools began in the mid to late 1980's. Coil tubing equipment had been used prior to this time for workover operations. It was very useful to spot fluids downhole and frequently used for squeeze cementing. For logging purposes, the cable is threaded through the coil tubing and then a cablehead for the logging

Figure 2.5 Production logging set-up for a pumping well (Courtesy Cardinal Surveys, Ref. 6)

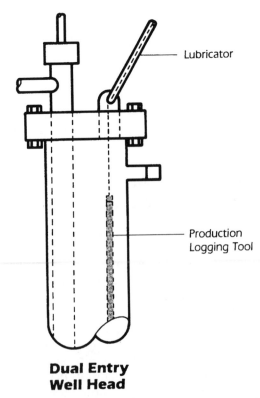

Lubricator

Production
Logging Tool

**Dual Entry
Well Head**

Figure 2.6 Dual entry well head for a rod-pumped well (Copyright SPE, Ref. 7)

tool is rigidly affixed to the end of the coil tubing. When the logging tool is connected, the coil tubing, which is relatively flexible, can be pushed downhole to the interval to be logged. Logging speeds can be controlled and, if needed, fluids can be pumped down through the coil tubing to the logging tool.

Figure 2.7 shows a typical job set up for a coil tubing logging operation.[8] The coil tubing containing the wireline is fed through an injector head and the logging tools are then pushed downhole.[9,10] Tubing is typically 1.25 or 1.5 in. (3.2 - 3.8 cm) diameter with wall thickness slightly less than 0.10 in. (0.25 cm). The maximum depth for such a rig is about 15,000 ft. (5,000 m). Logging may be done at speeds of up to 200 ft/min (65 m/min) and the log may be run in either the up or down direction. At about 2,500 ft. (800 m) horizontally, the coil tubing may begin to act like a piece of spaghetti and it cannot be pushed farther. Larger coil tubing sizes may extend this reach.

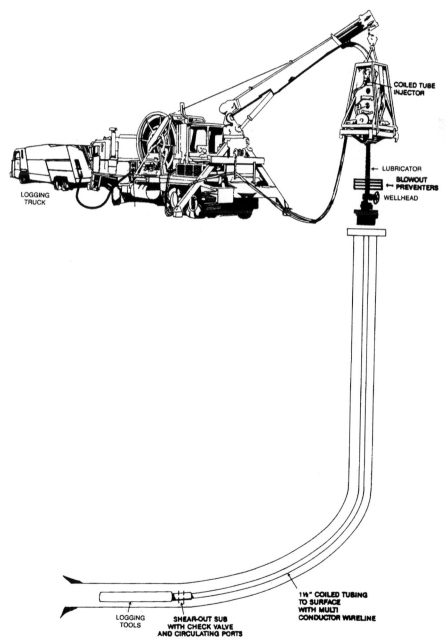

Figure 2.7 Coil tubing logging in a highly deviated well (Courtesy Canadian Institute of Mining, Metallurgy & Petroleum, CIM, Ref. 8)

The coil tubing, although somewhat flexible, may fail due to buckling or tension during the pushing and pulling of the tool downhole. As a result, computer modeling for the job is advisable prior to the job to assure that the planned job is within the physical capability of the coil tubing.

Pump Down Operations

Pump down operations are much less common than the coiled tubing operations. For this kind of operation, conventional tubing is run into the hole across the horizontal interval to be logged. The tools are typically reduced to shorter sections and joined by flexible knuckle joints. The tool string is pulled downhole by a pump down sub which has swab cups engaging the inside wall of the tubing. Upon reaching bottom, the pump pressure is increased to open a bypass around the swab cups. The tool can now be turned on and pulled back along the horizontal section to be logged. Needless to say, this technique only works with nuclear or other tools whose depth of investigation is beyond wall of the tubing.[11]

POINTS TO REMEMBER

- Prepare a complete well sketch showing all tubing, casing, and equipment downhole.
- What is the minimum diameter restriction?
- What is downhole temperature and pressure?
- Review production history, open hole, and reservoir information to develop theory of problem to be solved.
- Don't fish. Develop a theory which the logging job is to provide for information on which to act.
- Using PVT or other data, determine flow rates and phases present downhole.
- Run in with well flowing, subject to safety considerations.
- Is H_2S present?
- Determine wellhead connection information.
- What wellhead pressure is expected?
- Is operation to be done in daylight only?
- In pumping wells, tubing is to be set 50 ft (16 m) above the producing zone.
- Stabilize production before running into pumping or any well.
- Slim line tools may be required for annular logging on pumping wells.
- Dual wellhead for pumping wells to be logged through annulus.
- Tool cable wrap may occur when pulling out of annular logging job.
- Wrap occurs in less than 10% of logging runs. Wrap occurs in top 40 ft (12 m) from surface.
- To unwrap, use pulling unit and unwrap cable by hand.
- Tool opening of dual head must face away from walking beam.
- Difficulty in getting tool downhole is typically encountered at angles as low as 40-60° deviation.
- Weights, rollers, nose guides, and rigidized cable are used to help tool downhole.
- Coil tubing with cable inside of tubing is effective at pushing logging tools across intervals of interest, even in horizontal wells.
- Coil tubing logging allows tool speed to be controlled in both up and down direction.
- Fluids can be pumped through coil tubing when logging.
- Tools with downhole recorders can be mounted on coil tubing.
- Coil tubing is limited to about 15,000 ft (4,500 m).
- Pump down tool are available. Special tools and tubulars are required.

References

1. Petroleum Extension Service, *Introduction of Oilwell Service and Workover,* The University of Texas at Austin, Austin, Texas, 1971.
2. Schlumberger, *Wireline Services Catalog,* Document SMP-7004, Houston, Texas, 1991.
3. Connolly, E.T., "Interpretation and Recognition of Calibration and Recording Abnormalities on Production Logs," *CWLS Journal,* pp. 89-122, December, 1970.
4. Snyder, R.E., and Suman, G.O., "High Pressure Well Completions Handbook," *World Oil,* Houston, Texas, 1979.
5. Sodesep, *Wireline Products Catalog,* Vitrolles, France, 1994.
6. Cardinal Services Company, *Services Catalog,* Midland, Texas, 1984.
7. Hammack, G.W., Myers, B.D., and Barcenas, G.H., *Production Logging Through the Annulus of Rod-Pumped Wells to Obtain Flow Profiles,* Paper SPE 6042, Richardson, Texas 1976.
8. Lohuis, G., Lancaster, G., and Redmond, S., *The Expanded Use of Coil Tubing in Both Completion and Workover Operations,* Paper No. 88-39-49, 39th Annual Meeting of the Petroleum Society of CIM, Calgary, Canada, June, 1988.
9. Fertl, W.H., and Nice, S.B., *Well Logging in Extended-Reach and Horizontal Boreholes,* Paper OTC 5828, 20th Annual Offshore Technology Conference, Houston, Texas, May, 1988.
10. Brown, E., Thomas, R., and Milne, A., "The Challenge of Completing and Stimulating Horizontal Wells," *Schlumberger Oilfield Review,* pp. 52-63, July, 1990.
11. Noblett, B.R., and Gallagher, M.G., *Utilizing Pumpdown Pulsed Neutron Logs in Horizontal Wellbores to Evaluate Fractured Carbonate Reservoir,* 34th Annual SPWLA Symposium, June, 1993.

CHAPTER 3

LOGGING TOOLS AND THE CASED HOLE ENVIRONMENT

THE CASED HOLE ENVIRONMENT

Overview of the Completed Interval

The cased hole environment in which the logging tools must operate is shown on Figure 3.1. In this schematic, there are four zones, A, B, C, and D, which are porous and permeable and which can produce some type of fluid. These zones are separated from each other by impermeable shale layers. Casing is run and cement is circulated to create the situation shown. The operator of this well is interested in producing zones A, B, and D. The well is then completed by perforating into each of the desired productive zones. This process establishes communication between the completed zones and the common wellbore. The production is said to be "commingled."

As often occurs, things do not work out exactly as planned. Zone A is producing, but zone B is taking some of the production which would ordinarily be produced to the surface. Zone B must be at a lower pressure than the wellbore fluid and is called a "thief" zone. Zone D appears to be producing, but some, none, or all of that production is coming from zone C, depending on the pressure of each zone and the wellbore. In this case, the cement is intended to isolate all of the zones along the wellbore by filling the casing to formation annulus. Here, there is a deficiency of such fill and fluid produced from C flows through a "channel" up to zone D. If zone C is gas bearing, then the fluids in the borehole vary from completion fluid (salt water) at the bottom of the well, oil between zones A to D, and bubbling gas through oil above zone D.

What Can Logging Tools Do?

Even if no open hole information is available, and the well is cased, logs can be run to help properly complete or remedy this well. If no information is known about the formation, there are formation evaluation logs available. A gamma ray log may be used to discriminate the shaley intervals from clean sands. A compensated neutron log evaluates the porosity of zones A to D. Pulsed neutron capture or carbon/oxygen logs can identify the fluids in the pore space. A cased hole formation tester determines formation pressures and possibly taking samples would have indicated low pressure in zone B. As a result, the completion of zone B may have been delayed or a "dual" completion well configuration used.

The well integrity could have been assessed prior to completion. The cement bond type logs are available to evaluate the cement annular fill. This log would have indicated the likelihood of a channel between zones C and D and the need to squeeze cement into that interval. Even though this is probably a new well, casing inspection logs could be helpful to assure that the casing is in good condition and the perforations are located at the proper depths.

The so called fluid movement or "production logs" like the spinner flowmeter would indicate that both A and D are producing and that B is thieving. The fluid identification

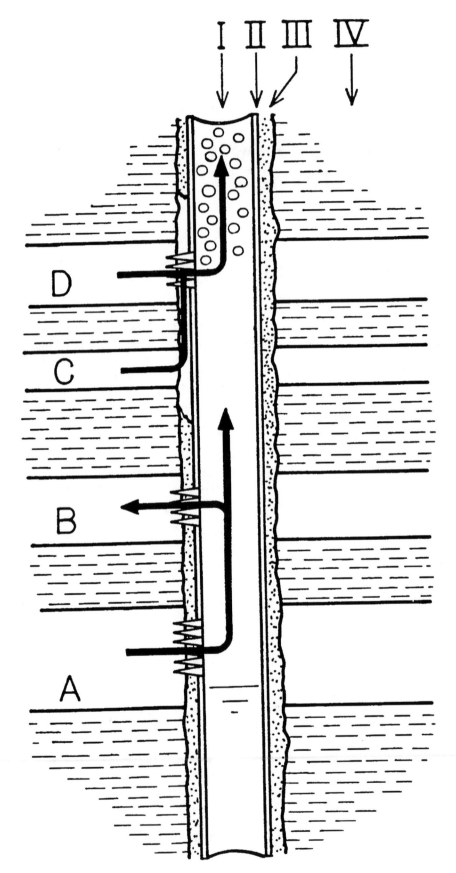

Figure 3.1 Cased hole logging environment

tools would indicate that completion fluid is at the bottom and the gas in entering at zone D. Temperature and noise logs would also be helpful to detect the channel between zones C and D.

Cased hole logs are the diagnostic tools necessary to provide an accurate picture of what a well can do, whether it is ready to do it, and if it is doing as predicted after completion. The prior discussion elaborated on three overall applications for cased hole logs. These are:

1. Formation Evaluation
2. Wellbore Integrity
3. Fluid Movement During Production/Injection

The remaining chapters of this text focus on the differing technologies available to each of these applications.

Regions of Investigation of Cased Hole Logging Tools

While overall applications of tools are a handy categorization, it is not suggestive of cross discipline applications and misses important details of the logging services. However, another type of classification is better suited to categorize the technologies and hint at other new applications. Referring to Figure 3.1, Roman numerals I, II, III, and IV are indicated at the top. These numerals point to regions of investigation in which tools are primarily designed to operate. For example, the production logging tools measure the region inside casing, region I. Casing inspection devices operate in region II. Cement evaluation logs inspect region III. The various formation evaluation services are measuring parameters in region IV.

The importance of this type of categorization is realized when one considers that tools are usually designed primarily for a single objective. For example, pulsed neutron capture logs are primarily designed for evaluation of formation water saturation. However, this tool may be affected by borehole, casing, and annular conditions. As these tools improved over the years, it was realized that there is information about the other regions in the acquired data. Early on it was realized that wellbore water movement could be detected by oxygen activation. Examples emerged where gas channels were detected. And today it appears that a whole new round of pulsed neutron equipment is being developed solely for such secondary information. The point is that if you have a log in the well file, it may have added information on it regarding its secondary regions which has not been considered or even understood.

Figure 3.2 is a table which lists the generic names of virtually all of the types of logging tools run today. On the left are the application categories. At the top are the regions of investigation. The black dot under a region indicates that the tool is primarily designed to make measurements in that region. A hollow dot under another region indicates that the tool is or may be affected by another region not of primary interest. For example, the pulsed neutron capture is primarily evaluating the formation but may be affected by wellbore fluids, casing, or the annulus.

FACTORS AFFECTING LOG RESPONSE

The Borehole Environment

Detailed discussion of borehole environmental factors affecting tool response is contained in the respective tool chapter. However, a few observations will immediately confirm that given an open hole and equivalent cased hole measurement, the open hole is always better, other things being equal. (Refer to Figure 3.1.) A cased hole logging tool, if it is to evaluate the formation, must look through the casing and cement. These cause an increased standoff from the formation, a smaller signal from the formation, plus signals from the casing and cement which must be compensated for and removed. The proper compensation is difficult since the cement thickness is, in fact, often not known with any accuracy. Furthermore, the cement may have pockets of water, mud, oil, or gas, which may affect the

	SURVEY	REGION OF INVESTIGATION I	II	III	IV
FLUID FLOW	TEMPERATURE	●	○	○	○
	DIFFERENTIAL TEMPERATURE	●	○	○	○
	NOISE (STATIONARY)	●	○	○	
	NOISE (CONTINUOUS)	●	○	○	
	RADIOACTIVE TRACER	●	○	○	○
	OXYGEN ACTIVATION WATER FLOW	●	○	○	
	CONTINUOUS SPINNER	●	○		
	FLOW DIVERTING SPINNER	●			
	HORIZONTAL SPINNER	●	○		
	FLUID IDENTIFICATION	●	○		
	FLUID SAMPLER	●			
WELL INTEGRITY	BOW SPRING CALIPER		●		
	MULTIFINGER CALIPER		●		
	ELECTROMAGNETIC (PAD TYPE)		●		
	ELECTROMAGNETIC (PHASE SHIFT)		●		○
	ACOUSTIC PULSE ECHO SURVEY	○	●		
	BOREHOLE VIDEO CAMERA	○	●		
	CASING POTENTIAL SURVEY	○	●	○	○
	COLLAR LOCATOR		●		
	ACOUSTIC BOND LOG	○	○	●	○
	PULSE ECHO BOND LOG	○	○	●	○
	PAD TYPE BOND LOG	○	○	●	○
	RADIAL DIFFERENTIAL TEMPERATURE	○	○	●	○
FORMATION EVALUATION	GAMMA RAY AND SPECTRAL GR	○	○	○	●
	DIRECTIONAL GR (ROTASCAN)	○	○	○	●
	CHLORINE LOG	○	○	○	●
	NEUTRON-COMPENSATED NEUTRON LOG	○	○	○	●
	PULSED NEUTRON CAPTURE	○	○	○	●
	CARBON/OXYGEN (INDUCED GR)	○	○	○	●
	DENSITY	○	○	○	●
	ACOUSTIC	○	○	○	●
	GRAVIMETER				●
	PRESSURE	○			●
	FORMATION TESTER			○	●

Figure 3.2 Generic names of cased hole logging services

tool's response. Certain tools are seriously affected by borehole fluids as well, especially gas, which is present in the figure. There is, however, one bit of good news. Unless channeling is present, there should be no formation invasion provided the log is run adequately long after circulating cement.

Other factors of the environment relate to tool ratings of temperature and pressure. Each service company's equipment is rated to slightly different limits and they should be closely consulted as temperature and/or pressure become high. Hostile high temperature equipment is available for some types of tools. The presence of H_2S requires special attention and knowledge of it is critical not only to the measurement but to the safety of the personnel on the job site.

Reservoir Fluids

When dealing with production logs, it is often necessary to compare downhole production with that indicated on the surface. This cannot be done without an understanding of how fluids behave when coming from the formation to the surface. Fluid conversions are discussed in most reservoir engineering texts and are not discussed here except in a very general sense.

Oil downhole usually has a substantial amount of gas held in solution. (See Figure 3.3.) If the oil is produced as a liquid downhole, and its flow rate is measured at, say, 1,000 units per day, this will not in general match the measured oil production at the surface, which

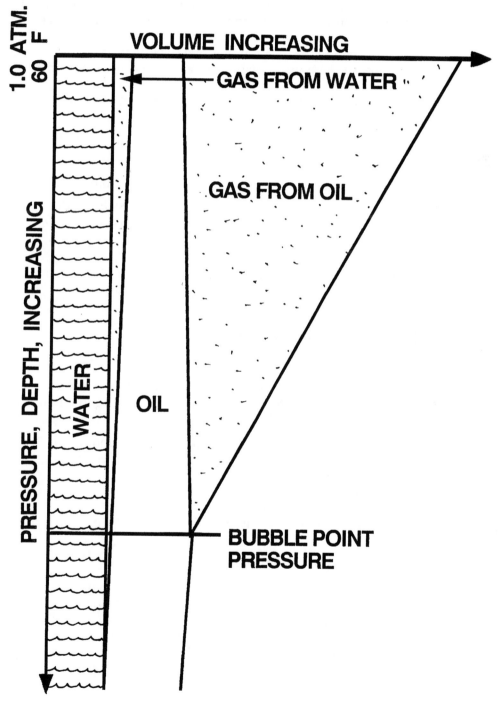

Figure 3.3 Change of fluid volumes in traveling from downhole to surface

may be 750 or 800 units per day. What happened to the oil? At downhole conditions, the oil contains gas in solution. When the oil comes to the surface, its pressure is reduced to below the bubble point and the gas begins to come out of solution. This solution gas accounts for the missing oil.

If the wellbore pressure is greater than the oil's bubble point pressure, only the oil phase is produced and flows in the wellbore. The oil is said to be "undersaturated." If the wellbore pressure is less than the bubble point pressure, then gas bubbles emerge from the oil and are present downhole. The oil is said to be "saturated," since it can hold no more gas at these conditions. As the pressure and temperature are decreased further, more and more gas comes out of solution until surface conditions are reached.

Water also has gas in solution. However, the gas volume associated with water is usually quite small. So, if 1,000 units per day of water are measured downhole, the corresponding surface production is, for all intents and purposes, the same—certainly within the accuracy of the measurements that we are making.

If gas is present in the reservoir, it is said to be "free gas" and not associated with oil production. Special charts are available for free gas to convert downhole to surface production and vice-versa.

Certain volume ratios are frequently used to convert downhole volumes to standard conditions of one atmosphere and 60°F (15.5°C). These are:

$$\text{GAS:} \quad B_g = \frac{\text{Gas Volume At Reservoir Conditions}}{\text{Gas Volume At Standard Conditions}}$$

$B_g \ll 1$ and typically its reciprocal, $1/B_g$, is used

$$\text{OIL:} \quad B_o = \frac{\text{Oil Volume At Reservoir Conditions}}{\text{Oil Volume At Standard Conditions}}$$

$B_o > 1$ typically, and is due to loss of solution gas

For accurate values of these volume ratios, consult PVT data on reservoir fluids or consult the algorithms and charts in reservoir engineering texts.[1,2]

TYPICAL PRODUCTION PROBLEMS

Poor Initial Performance

Poor initial performance is indicated by the well's failure to meet the producibility expectations for it. Most of the following problems can be diagnosed using cased hole logs. Such performance may be caused by:

1. Damaged zone causing significant skin effect around the wellbore
2. Plugged or ineffective perforations
3. Poor depth control during completion
4. Improperly sized tubing
5. Drawing down below bubble point
6. Thief zones created by commingling high and low pressured zones (see Figure 3.1)

Changes in or Unanticipated Phase Production

Changes in phases produced typically refers to increases in water or gas production. Unanticipated fluid production can mean changes from or unexpected initial production of an undesired phase. Most of these problems can be diagnosed with cased hole logs. Indeed, monitoring of fluid contacts allows the operator to anticipate changes in production. Some causes of this condition are indicated below:

1. Encroachment of water-oil contact (WOC) or gas-oil contact (GOC), Figure 3.4, A.
2. Water or gas coning, Figure 3.4, B.
3. Fingering through high permeability zones, Figure 3.5.

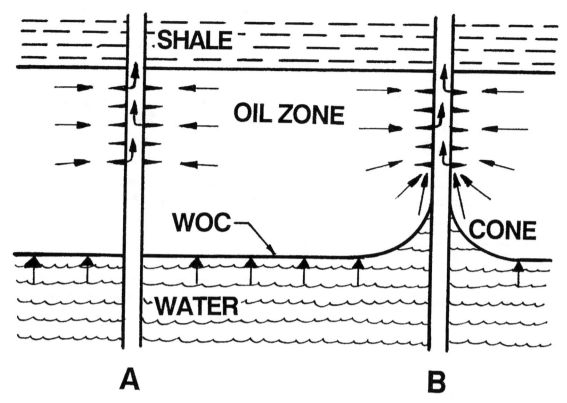

Figure 3.4   A. Flat water-oil contact (WOC), and B. Coning

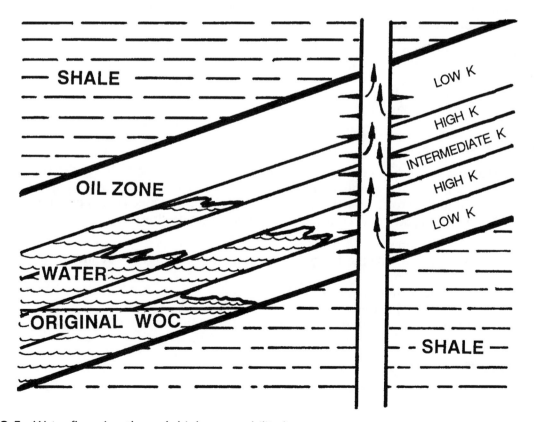

Figure 3.5   Water fingering through high permeability layers

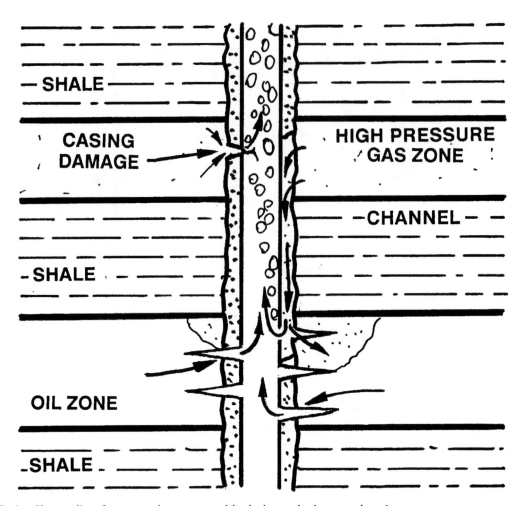

Figure 3.6 Channeling from nearby zone and leak through damaged casing

  4. Channeling from nearby zone, Figure 3.6.
  5. Casing damage, Figure 3.6.

Decline in Production

Declines in production are gradual and usually result from a slowly deteriorating condition. Some of these problems may be difficult to diagnose with logs, but monitoring of the well from time to time with logs may shed light on the mechanism causing the problem. The following are some contributors to production decline:

  1. Reservoir pressure decreases.
  2. Near wellbore permeability loss due to particulate movement and the like.
  3. Relative permeability effects.
  4. Fracture closure.
  5. Restriction due to sand fill.
  6. Increase in hydrostatic head due to increased production of heavy phase.

Miscellaneous Problems

This category really encompasses an uncountable number of problems. The following are just a few.

  1. Effectiveness of fracture, acid, or squeeze cement job.
  2. Sand production.
  3. Tubing, casing, or packer leaks.

POINTS TO REMEMBER

- When multiple zones are completed to the same casing string, they are said to be commingled.
- Low pressure zones which take fluid otherwise going to the surface are "thief" zones.
- Regions of poor cement through which fluids flow are "channels."
- "Production logs" are those used to assess fluid movement.
- There are three overall applications for cased hole logs:
    1. Formation Evaluation
    2. Wellbore Integrity
    3. Fluid Movement During Production/Injection
- Logging tools may be classified as to their region of investigation.
- Other things being equal, a log run in open hole is always better than that same log run through casing.
- Temperature, pressure, and H_2S are considerations important to logging tool selection.
- Oil volume flow rate downhole is significantly larger that it is at the surface due to the escape of solution gas.
- Water flow volume down hole is approximately the same as at the surface.
- Gas produced from a gas reservoir is free gas and undergoes a substantial volume change from downhole to surface.
- Production problems can be grouped into the following categories:
    1. Poor initial performance
    2. Changes in or unanticipated phase production
    3. Decline in production
    4. Miscellaneous

References

1. Amyx, J.W., Bass, D.M. Jr., and Whiting, R.L., *Petroleum Reservoir Engineering,* McGraw-Hill Book Company, New York, 1960.
2. Schlumberger, "Fluid Conversions in Production Log Interpretation," Document C-11952, Houston, Texas, 1986.

CHAPTER 4

FORMATION EVALUATION: NATURAL GAMMA RAY LOGGING

CONVENTIONAL GAMMA RAY LOGGING

Overview

The standard gamma ray (GR) tool is usually included on a modern cased hole logging tool string. This tool contains no source and responds only to gamma ray emissions from the downhole environment. It is run to tie-in with open hole logs (depth control), to provide information regarding shale, and for numerous other applications relating to naturally occurring or intentionally placed radioactive materials downhole. Detecting intentionally placed materials is discussed in a later section entitled "Radioactive Tracer Logging." This section focuses on applications arising from naturally occurring sources downhole. Gamma ray count rates are generally displayed on the left hand track of the standard American Petroleum Institute (API) log presentation.

Sources of Gamma Radiation in Earth Formations

Naturally occurring gamma rays in earth formations arise primarily from three basic radioactive chemical sources.[1,2] These are potassium 40 (K^{40}), uranium 238 (U^{238}) and daughter elements, and thorium 232 (Th^{232}) and daughter elements. Daughter elements are intermediates in the decay sequence prior to arriving at a stable isotope. Conventional gamma ray logging measures the total number of gamma rays at a point downhole regardless of source or energy. High gamma ray counts are typically associated with shales or clays since the potassium and thorium series elements tend to concentrate there. Clean formations tend to have fairly low gamma ray readings. Uranium is typically found in source beds and its salts are soluble.

Types of Gamma Ray Detectors and Tools

Two main types of gamma ray detectors have been commonly used in the oilfield. The least sensitive is the Geiger Mueller detector which measures incident gamma rays by gas ionization. This type tends to be rugged and unaffected by high temperatures. It, however, lacks the sensitivity of the newer scintillation detectors and hence is not often used by the major wireline companies today. The scintillation detector is typically a sodium iodide crystal coupled with a photomultiplier tube to detect tiny flashes of light associated with penetrations of the crystal by gamma rays. These flashes are converted to electrical impulses, counted, and that count rate is presented on the log. The count rate is displayed in API units which are calibrated to standard cores at the American Petroleum Institute Gamma Ray Test Pit facility at the University of Houston.[3] An API unit is defined as 1/200 of the difference between two of the test zones.

Gamma ray tools are typically available in all sizes from as small as 1.0 in. (25.4 mm) diameter for through tubing applications to larger sizes. They are rated to at least 350°F

(177°C) and 15,000 psi (103.4 MPa) for both temperature and pressure, respectively. Check with the service company if you are near to or expect to exceed these parameters. The gamma ray tools may be run with any fluid, including gas, in the well bore.

APPLICATIONS OF THE GAMMA RAY MEASUREMENT

The Gamma Ray-Collar Locator for Depth Control

The casing collar locator (CCL) is a magnetic device which is sensitive to the increased metal at a casing collar. It is almost universally run with cased hole logs and is the primary depth control log, although a gamma ray alone may be used for depth control in some cases. However, the CCL can only be effective if it is properly tied in to the open hole information. The key to accomplish such a tie-in lies in comparing nuclear logs from both the open and cased hole. Most any nuclear log can be used, but the gamma ray and sometimes the neutron logs are most common. Since the casing is somewhat transparent to formation gamma rays (this is really not the case since formation gamma ray counts are reduced and downscattered in energy by the cement and casing), the open and cased hole gamma ray logs should look alike.

The correlation of open and cased hole gamma ray logs is shown on Figure 4.1. After the well is cased, the collar locator is run with the gamma ray. The tool configuration is shown at the right. In this schematic, the CCL is "X" distance above the primary sensor, the gamma ray. If modern computerized equipment is used, the tool string is entered and the distance between the CCL and the GR is automatically corrected to where all sensors are presented at the correct depth. It is, of course, critically important that the correct tool string be entered into the computer for proper depth corrections. In older logs likely to be found in many well files, a depth correction must be made manually. When the GR is recorded on correct depth and a collar is detected, that collar is located "X" distance above the depth of the GR. In this example the correction to the collars is shown by the diagonal lines to the right of the collar log and corrected collar depths are noted in the depth track. It is highly advisable that a short or "pup" joint be put in the casing string to assure that no uncertainty arises due to each joint of pipe being nearly the same length. Once this tie-in is accomplished, the logger can ascertain his position with respect to the formation with a high degree of accuracy.

Bed Definition

Because the gamma ray looks into the formation, it is useful for bed definition purposes. Within a well it is used to define potential zones which are relatively clean and shale free. Such zones are characterized by low gamma ray count rates while shales would exhibit significantly higher count rates. See for example the gamma ray log of Figure 4.1, which shows three clean intervals or beds among four shaley sections. These clean zones are typically further analyzed for possible hydrocarbon content. On an inter-well basis, the gamma ray is useful to correlate beds between wells and picking marker beds, i.e., distinctive beds which appear throughout or over a portion of a geologic area. Such correlations are useful to determine geologic structure.

Evaluation of Shale Volume from the Gamma Ray Log

It is extremely important to note that after a well is in production, the gamma ray log may change due to the build-up of radioactive salts over time. These salt deposits or scales are often referred to as "Naturally Occurring Radioactive Materials" or NORM. As a result, if an analysis is to be done for shale volume, the initial open hole gamma ray log should be used if possible. As we will see later, spectral gamma ray logs run during the productive life of a well may also be used for this purpose.

If the cased hole gamma ray has not been affected by such NORM, or if the open hole gamma ray is used, higher readings tend to be associated with shales while lower readings are associated with clean intervals. A gamma ray log is shown on Figure 4.2. The gamma

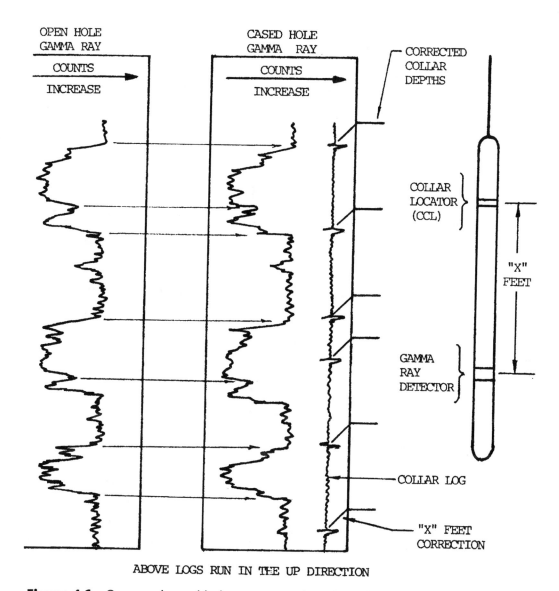

Figure 4.1 Open and cased hole gamma ray logs tie in the CCL to formation depths

ray counts are increasing to the right and it is scaled from 0 to 100 API units. Scanning the log, it is apparent that a number of intervals tend to have similar "clean" readings while others appear to have similar "shale" readings, i.e., the log appears to vary within a corridor bounded by the average clean and shale values. A clean and shale line may be drawn defining this corridor as shown in Figure 4.2. The clean line is on 30.0 API units and the shale line is on 81.0 API units. The shale line is assumed to be 100% shale, while the clean line corresponds to formations which are shale free.

Shale volume at a point "i" is computed by first evaluating the relative deflection X_i at that point using the equation

$$X_i = \frac{GR_i - GR_{CL}}{GR_{SH} - GR_{CL}}$$

This relative deflection may be used as a shale volume in some areas, but usually overestimates it. As a result, a number of correlations have been developed to improve the estimate of shale volume, V_{SH}.[1] A number of such correlations are shown on Figure 4.3.[1] All have a concave character and V_{SH} is always less than the relative deflection X. For accurate shale volumes, empirically developed curves such as those shown on Figure 4.3 should be developed for individual areas. The equations for the shale volume correlations are shown below on Figure 4.3.

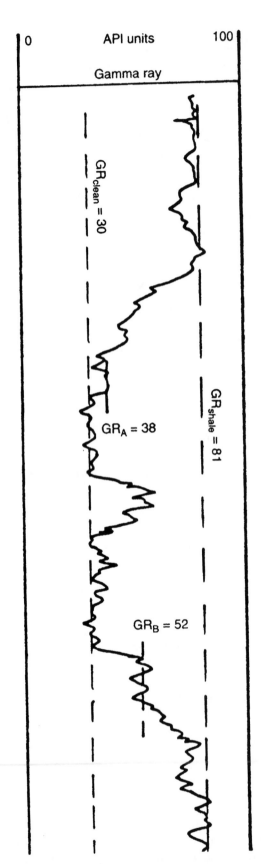

$$X = \frac{GR - GR_{clean}}{GR_{shale} - GR_{clean}}$$

Shale volume computation:

Point A

$$X = \frac{38 - 30}{81 - 30} = \frac{8}{51} = .16$$

$$V_{SH_A} = .05$$

Point B

$$X = \frac{52 - 30}{81 - 30} = \frac{22}{51} = .43$$

$$V_{SH_B} = .20$$

Figure 4.2 Technique to determine shale volume from gamma ray

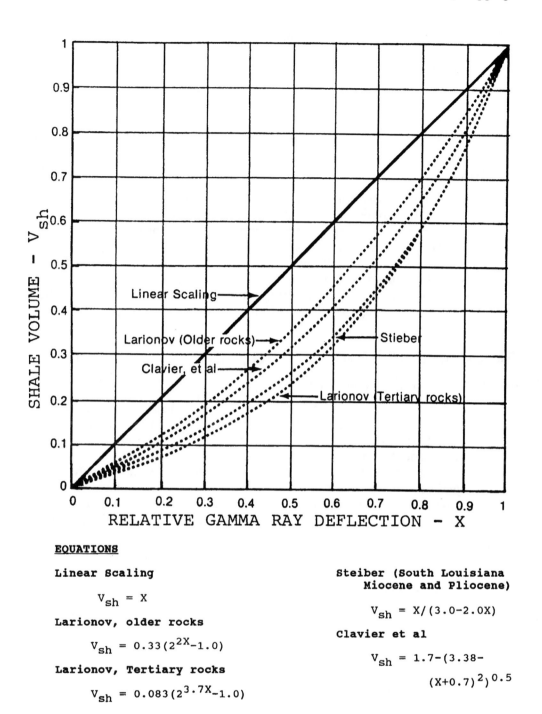

EQUATIONS

Linear Scaling

$$V_{sh} = X$$

Larionov, older rocks

$$V_{sh} = 0.33(2^{2X} - 1.0)$$

Larionov, Tertiary rocks

$$V_{sh} = 0.083(2^{3.7X} - 1.0)$$

Steiber (South Louisiana Miocene and Pliocene)

$$V_{sh} = X/(3.0 - 2.0X)$$

Clavier et al

$$V_{sh} = 1.7 - (3.38 - (X+0.7)^2)^{0.5}$$

Figure 4.3 Shale volume correlation curves (Courtesy Western Atlas and World Oil, Ref. 1)

Note that for shale volume, the relative deflection between the clean and shale line is used as a starting point. Hence, the absolute values of the GR log readings are irrelevant and the log can arbitrarily be scaled from 0 to 100. This is convenient if quick computations of shale volume are to be made by hand.

To compute shale volumes from the gamma ray, refer again to Figure 4.2. Shale volume will be computed for the points marked A and B. For point A, GR_A equals 38, and for point B, GR_B equals 52. The relative deflection for point A, X_A, is given by

$$X_A = \frac{38 - 30}{81 - 30} = \frac{8}{51} = .16$$

Putting X_A = .16 into the shale correlation chart of Figure 4.2 yields a value of V_{shA} = .05, i.e., 5% of the bulk formation volume is shale. For point B, GR_B equals 52 and the relative deflection, X_B, becomes

$$X_B = \frac{52 - 30}{81 - 30} = \frac{22}{51} = .43$$

Putting X_B = .43 into the shale correlation chart of Figure 4.2 yields a value of V_{shB} = .20.

"Hot" Zones

The presence of "hot" or highly radioactive zones is sometimes detected by a gamma ray survey. Hot zones may indicate a highly radioactive bed. Hot zones may also develop over time in a well. Producing wells often show increasing count rates over time in or above the producing interval. This is a result of radioactive salts precipitating out of solution as formation fluids, especially water, flow toward lower pressures. Perforations, channels, and even formations may become "hot." Such scale also builds up in casing and tubing and is referred to as naturally occurring radioactive material or NORM. Changes in the gamma ray over time are often used as indicators of water movement.

NATURAL GAMMA RAY SPECTROMETRY MEASUREMENTS

Gamma Ray Emission Spectra

The main sources of gamma rays downhole are the uranium 238 (U^{238}) and the thorium 232 (Th^{232}) series, including their daughter elements, and potassium 40.[4] Unlike conventional gamma ray tools, the natural gamma ray spectrometry tools evaluate the energy spectrum of the incident gamma radiation.[5,6] Figure 4.4[4] shows the natural gamma ray energy spectrum of potassium along with the uranium and thorium series. Clearly, the distribution of energies is fortunate, since the Th and K peaks can be easily isolated. Indeed, earlier tools utilized three to five gates to perform such evaluation. Newer tools tend to have either 100 or 256 gates from which groups may be looked at to better isolate Th, U, and K.

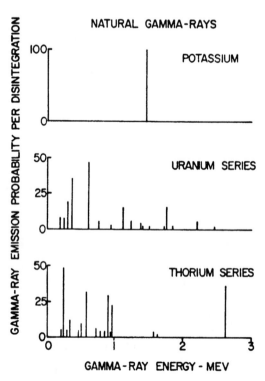

Figure 4.4 Gamma ray energy spectra for Potassium, Uranium, and Thorium (Courtesy Schlumberger, Ref. 4)

Characteristics of K, Th, and U Downhole

Why are such discriminations important? It turns out that these elements are generally found in unique and specific environments downhole. The following Table 4.1 highlights some of these differences.[2]

Tools and Service Companies

The service companies that run spectral natural gamma ray equipment and the trade name of the service is listed below. These tools are typically at least 3 3/8 in. (85.7 mm) in diameter. They are rated to at least 300°F (149°C) and 20,000 psi (137.9 MPa). Check with the service company if you plan to operate anywhere near to or exceed these conditions. The trade names are listed below on Table 4.2. The * indicates that that name is a mark of the company under which it is listed. Note that "Western Atlas" refers to Western Atlas International, Inc.

A smaller 1 11/16 in. (42.8mm) diameter family of tools is available, but these are primarily designed to detect much stronger radioactive materials placed downhole as part of some workover or stimulation operation. Since these tools have a smaller diameter, they provide fewer counts and are not generally used for detecting natural gamma rays. The tools/service names are listed in Table 4.3 below.

This group of tools is further discussed in the section on radioactive tracer logging.

Log Presentation

Figure 4.5 is a typical gamma ray spectral log presentation. This is a Schlumberger NGS log.[7] The right-hand tracks show Thorium (THOR) and Uranium (URAN) scaled in parts per million (ppm) with Potassium (POTA) scaled in percent concentration (%). The left track shows the conventional gamma ray (total counts) as SGR and the corrected gamma ray (CGR) with the uranium subtracted out. This is the basic presentation used by all service companies. Each has its own special variations which will show up on examples which follow.

TABLE 4.1

THORIUM, Th	Insoluble in water
	Associated with shale and heavy minerals
URANIUM, U	Not generally related to shale
	Salts are soluble, both in water and, to a lesser extent, in oils
	Found in organic source beds
POTASSIUM, K	Associated with typical shales
	May be present in drilling muds/workover fluids

TABLE 4.2

SCHLUMBERGER	NGS*—Natural Gamma Ray Spectrometry Log*
WESTERN ATLAS	Spectralog®*
HALLIBURTON ENERGY SERVICE	CSNG*—Compensated Spectral Natural Gamma Log*
	SGR*—Spectral Gamma Ray Log*

TABLE 4.3

SCHLUMBERGER	MTT*—Multiple Tracer Tool*
WESTERN ATLAS	PRISM*—Precision Radioactive Isotope Spectral Measurement*
HALLIBURTON ENERGY SERVICE	TracerScan*

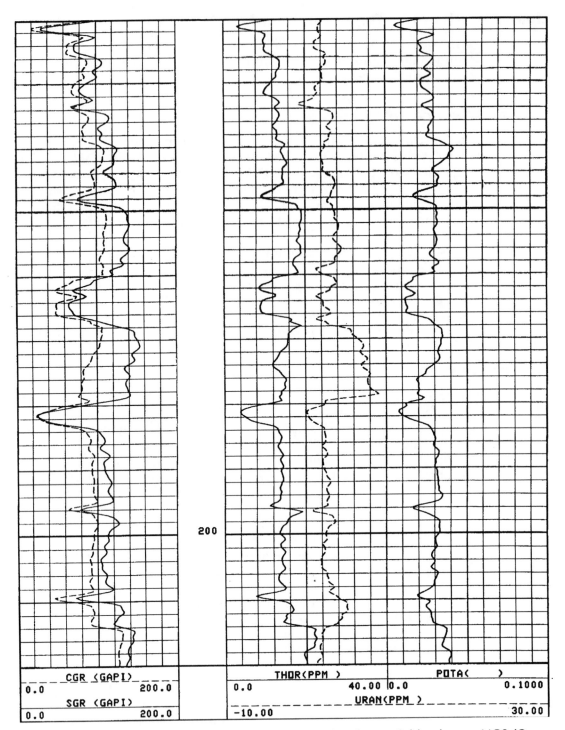

CGR (GAPI)		THOR(PPM)		POTA( )	
0.0	200.0	0.0	40.00	0.0	0.1000
SGR (GAPI)			URAN(PPM)		
0.0	200.0	-10.00			30.00

Figure 4.5 Typical gamma ray spectral log presentation, here a Schlumberger NGS (Courtesy Schlumberger, Ref. 7)

APPLICATIONS OF SPECTRAL GAMMA RAY MEASUREMENTS

Improved Shale Volume Computations

Uranium, as noted earlier, is soluble and not necessarily associated with shales. Hence, uranium counts may cause the gamma ray to read too high and therefore overestimate shale volume. To correct this problem in areas where it has been observed, the relative gamma ray deflection, X, discussed earlier may now be based on the gamma ray curve with the uranium subtracted out, i.e., Th+K, or even Th or K alone. The Th+K, the Th, or the K curve is treated just like the GR curve in that a corridor is defined between the clean and shale values, and X becomes

$$X_i = \frac{(Th + K)_i - (Th + K)_{cL}}{(Th + K)_{SH} - (Th + K)_{cL}}$$

or

$$X_i = \frac{Th_i - Th_{cL}}{Th_{SH} - Th_{cL}}$$

or

$$X_i = \frac{K_i - K_{cL}}{K_{SH} - K_{cL}}$$

If a relationship between X and V_{sh} is known for these isotopes, then X is used to evaluate shale volume as before.

Mineral and Clay Identification and Volumes

Mineral identification is a very complex art and numerous ratios between Th, K, and U may be taken along with counts to evaluate mineralogy, igneous rock, marine or continental depositional environments, organic shale/source beds, and the like downhole. Such ratios may be presented on the logs if requested. Details of such analysis techniques abound in the literature and are beyond the scope of this text. The chart of Figure 4.6 shows how mineralogy may be related to the Th/K ratio and concentration.[8] Note that each line emanating from the origin corresponds to a constant Th/K ratio. For example, if Th = 10.6 ppm and K = 3.9%, then the ratio Th/K = 10.6/3.9 = 2.7, and the clay mineral is indicated to be illite.

It frequently occurs that shale and clay are mixed up and treated as the same thing. This is not the case. Shale is a rock, having a matrix and porosity. Clay is a mineral and hence has no porosity. It is the matrix component of the shale.

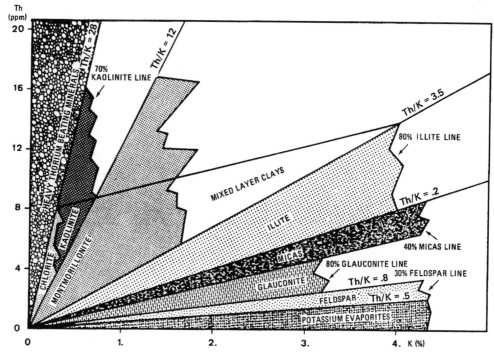

Figure 4.6 Clay and mineralogy identification from Th and K (Courtesy Shlumberger, Ref. 8)

Logs Showing Spectral Gamma Ray Applications

The following annotated log examples highlight the applications listed below:

- Source rock/false shale having production potential
- Location of water producing perforations
- Radioactive scaling within reservoir rock
- Effect of muds/wellbore fluids containing potassium
- Natural fracture identification
- CSNG* detects changes in wellbore hardware

Source Rock/False Shale with Production Potential (Figure 4.7). The total counts gamma ray log shows what at first glance appears to be two shale sections above and below the relatively clean and tight Buda limestone. The Del Rio shale below 4,218 ft (1,285.6 m)

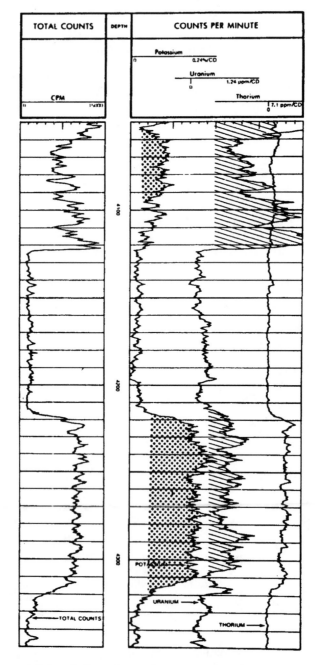

Figure 4.7 Source rock/false shale with production potential (Courtesy Western Atlas and Copyright SPE, Ref. 9)

is a true shale. The upper "shale" zone above 4,122 ft (1,256.4 m) is, in fact, the Eagle Ford shale, an organic rich source rock with some production potential. This zone has high uranium counts while the shale indicating Th and K are low.[9]

Location of Water Producing Perforations (Figure 4.8). The total counts gamma ray exhibits high peaks over perforated intervals in an old well. The uranium curve shows these peaks to be associated with uranium salts (radiobarite, $BaRaSO_4$) deposited over an otherwise clean section. These peaks are interpreted to be points of water entry. Comparison of a current gamma ray log with a base open hole gamma ray would yield the same conclusion.[10]

Radioactive Scaling within Reservoir Rock (Figure 4.9). This example shows a base GR followed by a later Spectralog. Increased total counts below XX670 over time are clearly evident while the spectral information shows this increase to be associated with uranium salts. This pay zone was produced from structurally higher offset wells. The increased U counts show that the reservoir has flowed and is now watered out.[10]

Effect of Muds/Wellbore Fluids Containing Potassium (Figure 4.10). This log (open hole) exhibits a somewhat excessive potassium percentage for the area where the log was run. These excessive potassium counts are attributed to potash in the mud.[10]

Spectral Gamma Ray Tool Detects Natural Fractures (Figure 4.11). This example shows a ratty response across a Mississippian limestone section overlain by Pennsylvanian shale. The rattiness is attributed to uranium count variations and is interpreted to be natural fractures. Over geologic periods of time, water migrates through the fracture system, leaving salt deposits in the fractures which are now detected as uranium count spikes. Note that there is no way to discriminate open from filled fractures.[10]

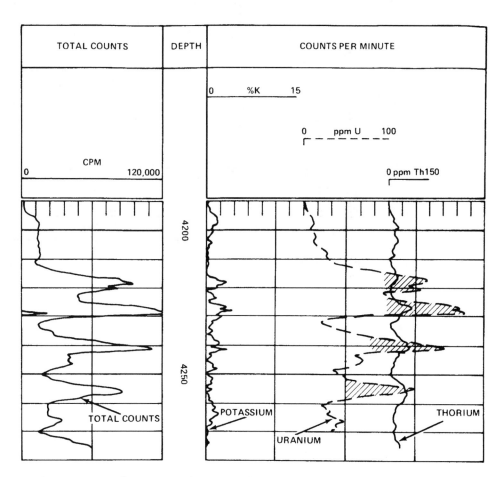

Figure 4.8 Location of water producing perforations (Courtesy Western Atlas and SPWLA, Ref. 10)

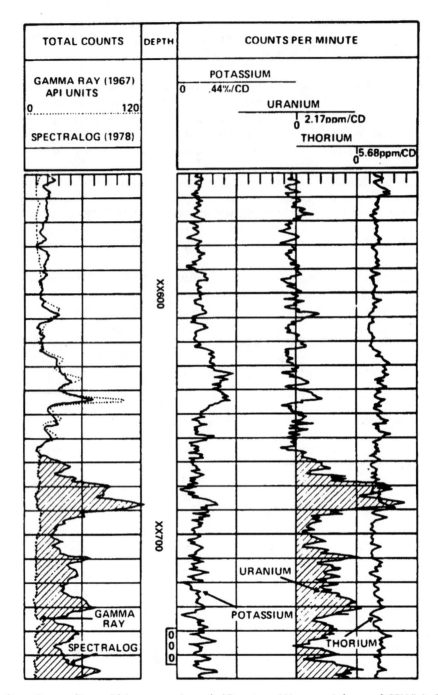

Figure 4.9 Radioactive scaling within reservoir rock (Courtesy Western Atlas and SPWLA, Ref. 10)

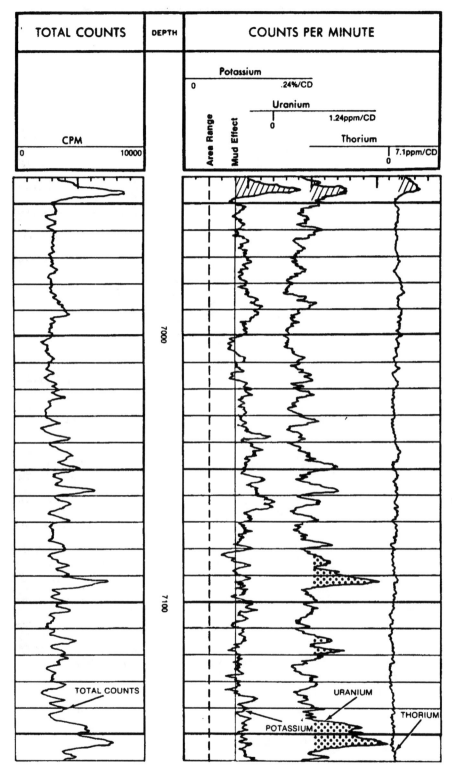

Figure 4.10 Effect of Muds/wellbore fluids containing potassium (Courtesy Western Atlas and SPWLA, Ref. 10)

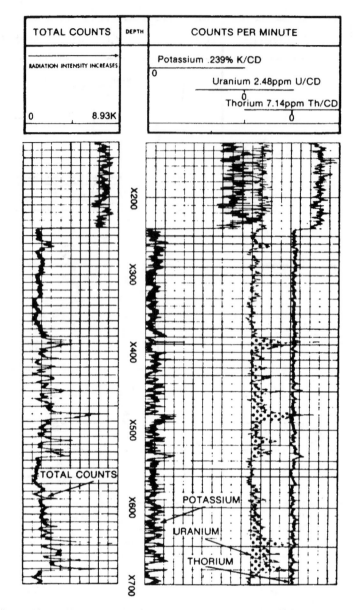

TOTAL COUNTS	DEPTH	COUNTS PER MINUTE

RADIATION INTENSITY INCREASES

0   8.93K

Potassium .239% K/CD
0
Uranium 2.48ppm U/CD
0
Thorium 7.14ppm Th/CD
0

X200
X300
X400
X500

TOTAL COUNTS

X600

POTASSIUM

URANIUM

X700

THORIUM

Figure 4.11 Spectral gamma ray tool detects natural fractures (Courtesy Western Atlas and SPWLA, Ref. 10)

Halliburton CSNG Measures Casing Condition (Figure 4.12). The primary CSNG presentation above shows a "RATIO CASING" measurement. This is a ratio of counts in selected windows which are optimized for casing in cased holes. This curve shows a clear ability to respond to casing and has been used as a casing wear indicator. The CSNG also has a quality suite of logs (not shown) to indicate the standard deviation of the spectral measurements and other quality related parameters.[11]

POINTS TO REMEMBER

- The conventional gamma ray log responds to all sources of gamma ray emissions downhole regardless of energy.
- The gamma ray equipment contains no radioactive material.
- The primary sources of naturally occurring gamma rays downhole are potassium 40, thorium 232, and uranium 238.
- Th and K tend to accumulate in shales while U is soluble and not necessarily associated with shales.

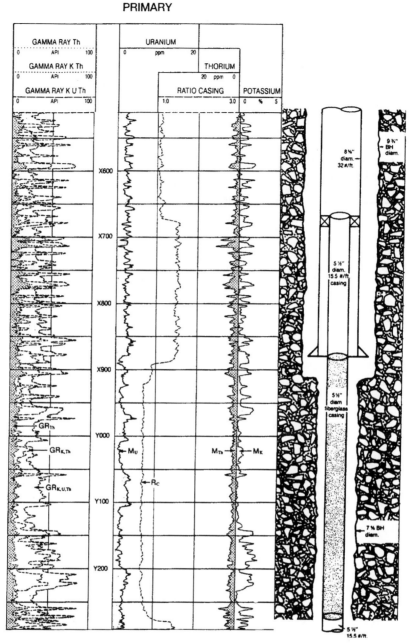

Figure 4.12 Halliburton CSNG measures casing condition (Courtesy Halliburton Energy Services, Ref. 11)

- The GR-CCL is the primary tie-in for depth control in casing.
- Shales exhibit higher gamma ray counts than clean intervals.
- Shale volumes may be computed if the appropriate correlation curve is known.
- The GR log may change in time and therefore the initial open hole GR is best for shale volume calculations.
- Change in the GR over time may indicate water movement.
- Spectral gamma ray tools measure Th, K, and U individually.
- Th, K, or Th+K may be better shale volume indicators than total gamma ray counts.
- Th/K and other ratios and counts are useful for mineral and source bed identification, depositional environment, and the like.
- Excessive U counts may indicate fluid movement (usually water entries) or natural fractures.

References

1. Fertl, W. H., "Gamma Ray Spectral Logging: A New Evaluation Frontier," *World Oil,* Series from March to November, 1983.
2. Gearhart Industries, Interpretation of the Spectral Gamma Ray, Publication No. G-1963, Fort Worth, Texas, 1986.
3. Scott, H.D., "Analysis of Samples from the API K-U-Th Logging Calibration Facility," SPWLA 30th Annual Logging Symposium, June 11-14, 1989.
4. Serra, O., Baldwin, J. L., and Quirein, J. A., "Theory and Practical Application of Natural Gamma-Ray Spectroscopy," 21st Annual SPWLA Symposium, Lafayette, Louisiana, July 8-11, 1980.
5. Dresser Atlas, "Spectralog," Publication No. 3334, Houston, Texas, December, 1980.
6. Smith, H.D. Jr., Robbins, C.A., Arnold, D.M., Gadeken, L.L., and Deaton, J.G., "A Multi-Function Compensated Spectral Natural Gamma Ray Logging System," SPE Paper 12050, 1983.
7. Schlumberger, "Well Evaluation Developments—Continental Europe," 1982.
8. Schlumberger, "Log Interpretation Charts," Publication SMP-7006, Houston, Texas, 1994.
9. Fertl, W.H., Stapp, W.L., Vaello, D.B., and Vercellino, W.C. "Spectral Gamma-Ray Logging in the Texas Austin Chalk Trend," JPT, pp. 481-488, March, 1980.
10. Fertl, W.H., Gamma Ray Spectral Data Assists in Complex Formation Evaluation," SPWLA 6th European Evaluations Symposium, London, England, March, 1979.
11. Gadeken, L.L., Arnold, D.M., and Smith, H.D.Jr., "Applications of the Compensated Spectral Natural Gamma Tool," SPWLA 25th Annual Symposium, New Orleans, Louisiana, June, 1984.

CHAPTER 5

FORMATION EVALUATION: PULSED NEUTRON CAPTURE LOGS

EQUIPMENT AND APPLICATIONS

Overview

Pulsed neutron capture (PNC) logs are the most important devices for evaluation of formations through casing. These tools are small diameter 1 11/16 in.(42.9mm) or less. They are designed for through tubing operations and are often run without shutting in production from the well being logged. These tools do not contain chemical neutron sources. Instead, they are electronically pulsed and emit bursts of neutrons periodically. They do, however, contain tritium as part of the sealed neutron tube from which the neutrons are generated.[1]

The pulsed neutron capture logs are primarily used for measurement of water saturation, porosity, and presence of gas in the formation.[2-5] Many new applications have recently emerged, some relating to measurements typically associated with fluid movement downhole. This latter group is discussed only briefly here, but more extensively in a later section covering oxygen activation and other fluid detection applications.

Industry Versions of Pulsed Neutron Capture Tools

There are a number of industry versions of this type of tool currently and historically available. Unlike many other services, there is no central calibration standard to which all tools must conform. As a result, it is important when comparing one log to another to appreciate that differences might exist between them. These differences will arise in the values of the primary measurement and in the variety of unique parameters measured by each tool. A listing of the main tools available follows on Table 5.1. The * indicates that the name and mnemonics are marks of the company under which that tool is listed.

Other pulsed neutron capture tools may be available, but these listed are or have been commercially available over a wide geographical area. The Dual Burst TDT is sometimes referred to as the TDT-P.

Applications of the PNC measurement

The applications listed below for measurements by PNC tools are by no means exhaustive of the possibilities.

1. Porosity
2. Water saturation
3. Gas detection
4. Location and monitoring of gas/oil and water/oil contacts
5. Correlation with open hole resistivity logs
6. Shale indicator

TABLE 5.1

SCHLUMBERGER
  TDT-K*, TDT-M*, Thermal Decay Time tool*
  Dual Burst Thermal Decay Time tool*, Dual Burst TDT*
  RST-A*, RST-B*; Reservoir Saturation Tool*

WESTERN ATLAS
  NLL*; Neutron Lifetime Log*
  PDK-100®*; Pulsed Decay (DK) with 100 counting gates

HALLIBURTON ENERGY SERVICES
  TMD*; Thermal Multigate Decay tool*
  TMD-L*; Thermal Multigate Decay-Lithology tool*

COMPUTALOG
  PND®-S; Pulsed Neutron Decay-Spectrum tool*
  (PND® is registered in the U.S. Patent and Trademark Office)

7. Evaluate changes in saturation due to zones watering out
8. Measure residual oil saturation (ROS)
9. Locate and select zones for recompletion

The above are the formation evaluation applications. The applications regarding fluid flow are listed below, but are discussed in detail in another chapter:

1. Waterflow in and near the wellbore by oxygen activation
2. Measure water holdup
3. Detect gas entry by inelastic count rate
4. Detect substantial gas channels
5. Locate zones of acid injection
6. Detect natural fractures with the gamma ray log
7. Detect channels by Boron log technique
8. Locate injected water breakthrough when run with carbon/oxygen (C/O) log
9. Silicon activation for gravel pack evaluation

Originally designed for formation evaluation, the PNC logs have flowered in recent years with new and unexpected applications.

PRINCIPLE OF OPERATION

Tool Configuration

The typical PNC tool configuration is shown on Figure 5.1. The pulsed source or minitron emits a burst of 14 million electron volts (mev) neutrons periodically at about 1,000 microsecond intervals. These neutrons interact with the formation causing gamma ray emissions which may be detected at the two detectors. There is a near (N) or short spaced (SS) detector and a far (F) or long spaced (LS) detector. The near is about 1 foot (30.5 cm) and the far is about 2 feet (61 cm) above the source. This spacing will vary among the service companies and among the tool models.

Gamma Ray Detectors

The gamma ray (GR) detectors are typically sodium iodide (NaI) detectors, similar to those used for conventional gamma ray logs. The Atlas Wireline PDK-100 tool uses a Bismuth Germinate (BGO) crystal. While this is a more efficient crystal at detecting gamma rays, it must be kept cool with a Dewar Flask arrangement within the tool housing. The Schlumberger RST uses a Gadolinium Oxyorthosilicate (GSO) crystal. These latter two crystals are significantly more sensitive than the NaI crystal. Measurements are made of the changes

TABLE 5.2 Capture Cross Sections of Common Downhole Materials, Σ, C.U.

WATER (200°F (93.3°C), 5,000 psi (34MPa))*			
Fresh (0 ppm)	22.2 c.u.	150,000 ppm	77.0 c.u.
50,000 ppm	38.0 "	200,000 ppm	98.0 "
100,000 ppm	58.0 "	250,000 ppm	120.0 "

HYDROCARBONS**	
Crude Oil (Dead, Stock Tank)	22.0 c.u.
Reservoir Oil (See Fig. 5.3)	21.0 "
Gas at Reservoir Conditions (See Fig. 5.3)	< 10.0 "

FORMATION MATRIX	
Sandstone	6–13 c.u.
Limestone	6–14 "
Dolomite	6–12 "
Anhydrite	13–21 "
Shale	25–50 "

PURE MINERALS			
Quartz (SiO_2)	4.36 c.u.	Calcite ($CaCO_3$)	7.48 c.u.
Dolomite ($CaMg(CO_3)_2$)	4.78 "	Anhydrite ($CaSO_4$)	12.30 "
Gypsum ($CaSO_4 2H_2O$)	19.40 "	Coal	1–2 "
Halite (NaCl)	762.36 "	Iron (Fe)	214.90 "
Water (H_2O)	22.2 "	Boron (B)	760. "

* ppm equivalent NaCl; Capture cross section of water is slightly dependent on temperature and pressure—see service company chart books for exact values.

** Approximate values. Actual values for oil depend on American Petroleum Institute (API) gravity and gas oil ratio. Actual values for gas depend on gas gravity and reservoir temperature and pressure. (See chart of Figure 5.6.)

movable and connate water. A somewhat different model, the "Dual Water" model, is discussed in a later section of this chapter.

Each formation constituent of Figure 5.4 has shown with it its fraction of the total volume. It turns out that the bulk capture cross section seen by the PNC tool, Σ_{LOG}, is a convenient linear combination of the contributions of each constituent. Therefore, Σ_{LOG} is given by

$$\Sigma_{LOG} = \underset{\text{MATRIX}}{(1 - V_{SH} - \phi_e)\Sigma_M} + \underset{\text{HYDROCARBON}}{\phi_e(1 - S_W)\Sigma_H} + \underset{\text{WATER}}{\phi_e S_W \Sigma_W} + \underset{\text{SHALE}}{V_{SH}\Sigma_{SH}} \qquad \text{(Equation 5.2)}$$

where

Σ_{LOG} = Bulk capture cross section of the formation

Σ_H = Capture cross section of the hydrocarbon

Σ_M = Capture cross section of the matrix

Σ_{SH} = Capture cross section of the shale

Σ_W = Capture cross section of the water

V_{SH} = Fractional shale volume (0 to 1)

ϕ_e = Effective porosity (0 to 1)

S_W = Water saturation (0 to 1)

Theoretically, Σ_{LOG} would equal the formation capture cross section, Σ_F but this may not be so since tool responses may vary among the service companies and tool types.

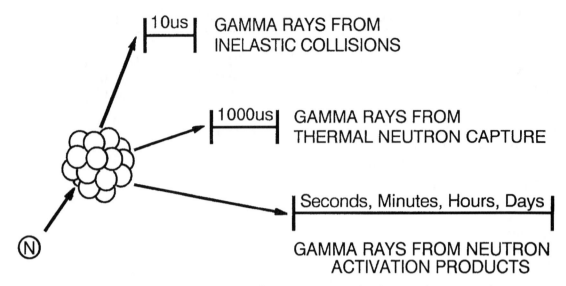

Figure 5.2 Time frames for gamma rays from neutron activation products

Capture Cross Sections

The thermal neutrons are captured by various formation materials at differing rates. The measure of the probability of the capture of a thermal neutron by formation materials is the capture cross section in capture units (c.u.). The capture unit is equal to 10^{21} Barns/cm^3 or 10^{-3}cm^{-1}. The higher the capture cross section, the more likely a capture event and gamma ray emission will occur. The symbol representing the capture cross section is sigma, Σ. The capture cross sections of some important formation constituents are shown on Table 5.2.[6]

The formation materials capture cross section table indicates that fresh water at 22.2 c.u. and reservoir oil at 21 c.u. are very close and would be difficult to discriminate from each other. Salty waters, having high NaCl ppm content, are needed to discriminate water from oil using capture measurements, with the greater the salinity, the better the discrimination. Fresh water looks like oil to capture tools. Furthermore, the large overlap among sandstone, limestone, and dolomite indicates that the capture measurement is not useful for matrix type discrimination. Note however that shales have fairly high values and are readily discriminated downhole.[7,8]

Hydrocarbons, both oil and gas, may vary widely in capture cross section downhole, depending on their characteristics and reservoir conditions. The chart of Figure 5.3 may be used to obtain more accurate values.[6] Note that the capture cross section for gas is given for methane. For other heavier hydrocarbon gases, the following equation may be used to correct for gases based on their gravity:

$$\Sigma_{GAS} = \Sigma_{METHANE} (.23 + 1.4\ \gamma_g) \qquad \text{(Equation 5.1)}$$

where γ_g is the gas gravity.

Waters also vary depending on reservoir temperature and pressure, but such variations downhole are not large relative to the data of Table 5.2.

FORMATION MODEL AND LOG RESPONSE

Formation Model

The simplest formation model for analyzing PNC logs is that shown on Figure 5.4.[9] The formation consists of the rock matrix, effective porosity filled with water and a hydrocarbon, and shale.[9,10] Note that the effective porosity is used in this model and it specifically excludes porosity associated with the shale. Shale porosity is impermeable, contained within the shale rock, and not relevant to saturation calculations. The water includes both

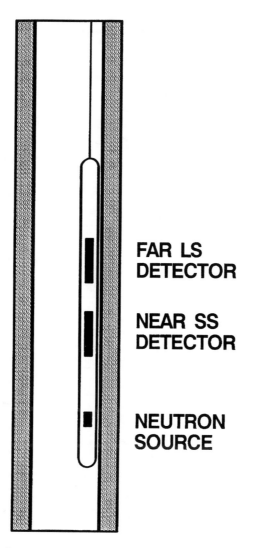

Figure 5.1 Typical pulsed neutron capture, PNC, tool configuration

in the rate of counts over the time period between bursts. Up until about 1993, no tools made any sort of spectral measurement of the incident gamma ray energies, although it is clear that all new tools will have this capability.

Neutron Interactions

The 14 mev neutrons emitted by the source interact with the borehole and formation environment following the burst. After collision with these neutrons, atoms from the environment emit gamma rays of distinct energies at characteristic times depending upon their atomic number. Within the first tens of microseconds, high energy inelastic collisions occur. Gamma rays emitted in this period are important for carbon/oxygen measurements, but of no interest for capture logging. (See Figure 5.2.) From this time on to about 1,000 microseconds or longer, the neutrons are slowed and become low energy "thermal" neutrons which are "captured." A capture event occurs upon collision with certain nuclei in the environment and emission of a gamma ray. The rate of such capture is a result of thermal neutron collisions mainly with hydrogen and chlorine and is of prime importance in PNC logging. Finally, some of the instabilities created by neutron collisions may take numerous seconds, minutes, or longer to return to normal. Gamma rays emitted at such times are of no use in capture logging, but are important for oxygen activation water movement logs or silicon activation gravel pack logs.

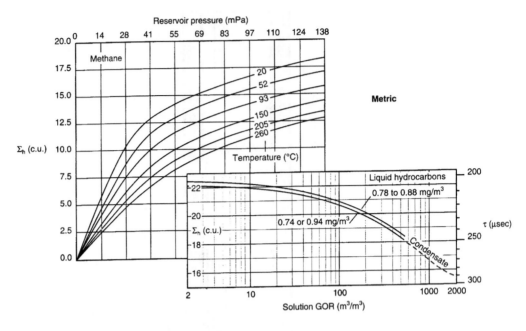

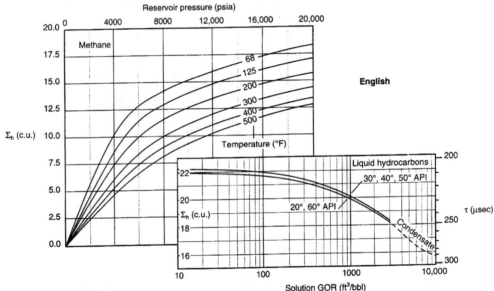

Figure 5.3 Capture cross sections of hydrocarbons at reservoir conditions (Courtesy Schlumberger, Ref. 6)

Log Response to Water and Hydrocarbon Zones

With this model, it is instructive to compute the PNC response to water, oil, and gas zones under ideal conditions. PNC logs are well suited to Texas and Louisiana Gulf Coast reservoirs with porosities in the range of 30 percent and salinities in the neighborhood of 100,000 ppm, but are not well suited to conditions of fresher waters and lower porosities. As a rule of thumb, a minimum of 15 percent porosity and 50,000 ppm NaCl formation water salinity are required for "quantitative" evaluation.

Consider a clean shale free Texas Gulf Coast zone with a water, oil, and gas interval, bound above and below by shale. The gamma ray of Figure 5.5 shows such an interval. Notice that the PNC log has a shale response of 40 c.u. in this hypothetical example. Equation 5.2 may be simplified for a clean zone by setting $V_{sh}=0$.

$$\Sigma_{LOG} = (1-\phi_e)\Sigma_M + \phi_e(1-S_W)\Sigma_H + \phi_e S_W \Sigma_W \qquad \text{(Equation 5.3)}$$

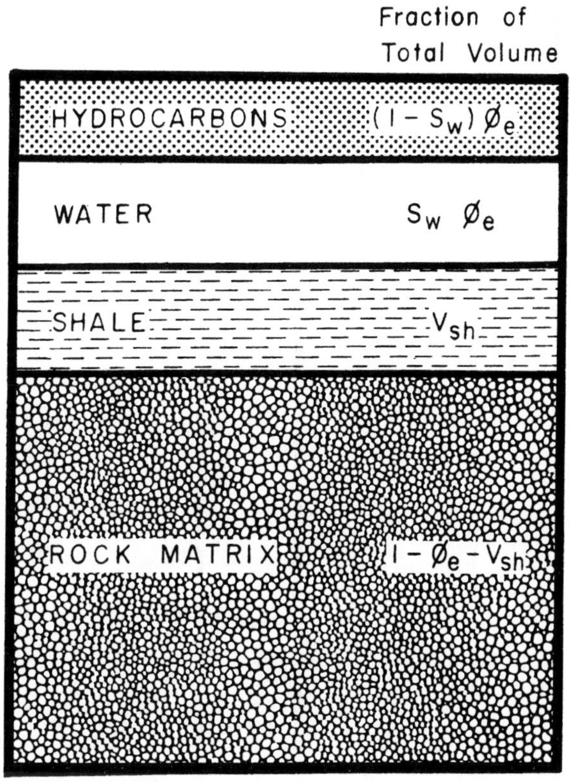

Fraction of
Total Volume

Figure 5.4 Formation model for PNC analysis (Courtesy Schlumberger, Ref. 9)

Applying this equation for Gulf Coast conditions of $\phi_e = .30$ $\Sigma_M = 10.0$ c.u., $\Sigma_H = 21.0$ c.u., and $\Sigma_w = 58.0$ c.u.(100,000 ppm NaCl), water ($S_w = 1$), oil ($S_w = .20$) and gas ($S_w = .20$) zone responses are calculated as follows:

Water Zone—$S_w = 1.0$

$$\Sigma_{LOG} = (1.0 - .30) \times 10.0 + .30 \times (1.0 - 1.0) \times 21.0 + .30 \times 1.0 \times 58.0 =$$

$$7.0 \text{ c.u.} + 0.0 \text{ c.u.} + 17.4 \text{ c.u.} = 24.4 \text{ c.u.}$$

$$\text{matrix} \quad \text{hydrocarbon} \quad \text{water}$$

Oil Zone—$S_w = .20$

$$\Sigma_{LOG} = (1.0 - .30) \times 10.0 + .30 \times (1.0 - .20) \times 21.0 + .30 \times .20 \times 58.0 =$$

$$7.0 \text{ c.u.} + 5.04 \text{ c.u.} + 3.48 \text{ c.u.} = 15.5 \text{ c.u.}$$

Gas Zone—$S_w = .20$, $\Sigma_g = 8.0$ c.u.

$$\Sigma_{LOG} = (1.0 - .30) \times 10.0 + .30 \times (1.0 - .20) \times 8.0 + .30 \times .20 \times 58.0 =$$

$$7.0 \text{ c.u.} + 1.92 \text{ c.u.} + 3.48 \text{ c.u.} = 12.4 \text{ c.u.}$$

A PNC log run over these intervals would look like the heavy capture cross section curve of Figure 5.5. The water, oil, and gas zones are clearly identifiable over this interval. The water-oil contact (WOC) and gas-oil contact (GOC) can easily be found. The bed boundaries at the shales stand out.

The typical PNC tool is said to have an error bar (plus or minus one standard deviation of measured capture cross section in c.u.) of +/–.5 c.u., for a total range of one c.u. The range between a water and oil zone is 24.4–15.5 = 8.9 c.u. This means a resolution of saturation of about 1 part in 10. Clearly, this is a useful measurement in such an environment.

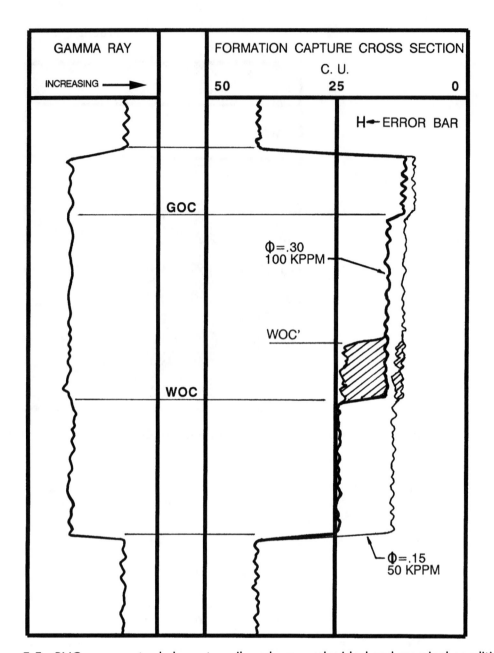

Figure 5.5 PNC response to shale, water, oil, and gas, under ideal and marginal conditions

What about the marginal conditions of 15 percent porosity and 50,000 ppm NaCl formation water salinity? Recomputing the water, oil, and gas zones using the same numbers as above except that $\phi_e = .15$ and $\Sigma_w = 38.0$ c.u. (50,000 ppm NaCl), yields

$$\text{water zone: } \Sigma_{LOG} = 14.2 \text{ c.u.}$$

$$\text{oil zone: } \Sigma_{LOG} = 12.2 \text{ c.u.}$$

$$\text{gas zone: } \Sigma_{LOG} = 10.6 \text{ c.u.}$$

Clearly, an error bar of +/- .5 c.u. cannot yield good quantitative results with a resolution of one part in two, but is useful only in a qualitative sense. The error bar may be reduced somewhat with the newer tools. It may further be reduced with multiple passes and/or reduced logging speed.

Before PNCs should be discounted as ineffective for marginal conditions, they are useful under certain conditions. If a proper base log is run, changes in saturation as would occur when a zone waters out may be detected by time lapse logging, i.e., comparing a current PNC log with one run earlier. For such comparisons it is important to compare only the same model tools from the same service companies. The movement of the WOC to the new position, WOC', is shown for both computed conditions on Figure 5.5.

MEASUREMENT TECHNIQUE

Exponential Decay

After the burst of neutrons from the source, the neutrons move into the formation and become thermalized after a few 10s of microseconds. This population of thermal neutrons decays exponentially according to the following equation:[11-13]

$$N = N_o e^{-(t-t_o)/\tau} + B \qquad \text{(Equation 5.4)}$$

where

N = Neutron population at time t

N_o = Neutron population at time t_o

t_o = Initial reference time

t = Time since t_o

τ = Thermal decay time—63% decay of neutron population

B = Background counts

Initially following the burst of neutrons, many of the counts are associated with borehole neutrons. After a few hundred microseconds, the borehole effects diminish and the formation neutrons assume the decay described by equation 5.4. When plotted on a semi-log plot, the count rate associated with the neutron population decline is shown on Figure 5.6. This figure shows that the greater the capture cross section of the formation, the more rapid the decline of thermal neutrons. Salt water in the pore space results in a higher formation cross section than if filled with oil or gas.

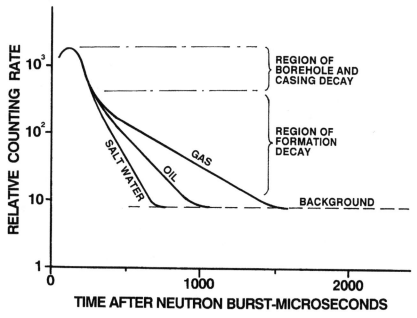

Figure 5.6 Typical count rate decay for clean fluid bearing formations

Computation of Capture Cross Section

The objective of the PNC hardware is to measure the slope of the linear (exponential) portion of the decay curve shown on Figure 5.6. The service companies set electronic gates over the range of this linear decay, make counts of gamma rays in these gates, and reconstruct the slope of the decay curve. From the slope they can compute the thermal decay time, T, or neutron half life, L, in microseconds. This is simply related to the bulk formation capture cross section through the equation

$$\Sigma_F = 4550/\tau = 3150/L$$

where

Σ_F = Formation bulk capture cross section, c.u.

τ = Thermal decay time, microseconds

L = Neutron half life, microseconds

This computation is done automatically and capture cross section is presented on the log. τ may be presented upon request.

The gating and analytical techniques used by the service companies are highly marketed and acclaimed by their makers. These techniques have evolved over the years and are the main means of differentiating the service company's products.

CHRONOLOGICAL DEVELOPMENT OF TOOLS AND LOG PRESENTATIONS

Neutron Lifetime Log (NLL, DNLL)

The Dresser Atlas Neutron Lifetime Log (NLL) made its debut about 1963 and was the first of this type of tool commercially available. It was a single detector version later superseded by the dual detector tool initially called the DNLL, and later called the NLL.[1,12] The gating for the NLL was fixed during a run in the hole, typically between 400-600 and 700-900 microseconds, although this could be changed. The source pulse frequency was every 1,000 microseconds. The NLL gating is shown on Figure 5.7. This tool suffered somewhat due to the fact that the gates are fixed. While the schematic shows the gates well positioned over the linear decay curve, they could in fact be affected by borehole effects adding counts to the first gate, or by background counts which were not measured. Furthermore, since the gate position could be changed, the selection of optimal positions may well have become an art with many local variations.

Figure 5.8 shows the NLL presentation for the dual detector tool.[7] This example dates from about 1975 and is influenced by the thermal decay time (TDT) of Schlumberger (see next section). On the left track is shown the gamma ray, uncorrected collar locator, and the monitor curve for proper tool functioning. The corrected collars are shown in the depth track. The business curves in the right two tracks include the counts in gates 1 and 2 (G1, G2), the ratio of counts in the SS detector divided by counts in the LS detector, which is essentially an uncalibrated porosity curve, the formation capture cross section, sigma, and the counts at the short and long spaced detectors (SS, LS). This log shows shale at the very top, in the middle, and at the bottom of the logged interval, with sigma shale reaching 36 c.u. This shale is clearly visible on the GR. This log also shows water, oil, and gas zones. The water zone has a response of about 27 c.u., the oil about 21 c.u., and the gas about 12 c.u. The gas zone is further indicated by a drop in ratio and a separation between SS and LS, with the LS reading to the right of the SS reading.

Thermal Decay Time (TDT-K)

The TDT tool came out in about 1968 as a single detector tool and in 1972 as the dual detector TDT-K.[13] This tool was set up with two gates, I and II, to measure the linear part of the

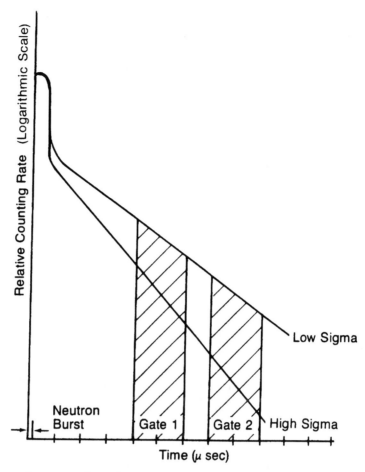

Figure 5.7 Gate placement for Atlas Neutron Lifetime Log (NLL) (Courtesy Western Atlas, Ref. 7)

decay, followed by a third gate, III, to measure background count rates (see Figure 5.9).[8] The unique feature of this tool is that after each measurement of thermal decay time, T, it would move the gate positions, which were preset in terms of τ, to be optimally located over the linear part of the decay curve. This feature was called the tau loop and worked very well. The pulse rate therefore varied depending on the formation cross section, but generally was between 1,000 to 2,000 microseconds between bursts.

A typical TDT-K presentation is shown on Figure 5.10.[14] On the left-hand track is typically shown a gamma ray log, although this example shows a traced open hole self potential (SP) log. F3 is the count rate at the far detector in gate III. This is similar to an insensitive GR log, except that it will be used for oxygen activation water movement detection. On the right, the formation capture cross section covers both tracks, and clearly shows a water, oil, and gas zone topped off by a shale across the interval logged. In the middle track is shown the ratio of near to far counts in gate I, N/F, compensated for background. This ratio is essentially an uncalibrated porosity. The overlay of F1 and N1, as shown on the right-hand track, separates over a gas zone, with the F moving to the left of N. Notice that the scales are typically in a ratio of about six to one to achieve a good overlay in a water zone.

Thermal Decal Time (TDT-M)

Introduced in 1980, the TDT-M had 16 fixed gates which could be spread out in four scale factors.[15] The basic gate set-up is shown on Figure 5.11.[15] The background measurement is taken between bursts. The burst frequency could be varied from less than 1,100 to more that 5,000 microseconds between bursts, depending upon which scale factor was selected. The gate selection for the capture cross section measurement varied depending upon the most recently observed τ. The later model, the TDT-Mb, provided certain borehole

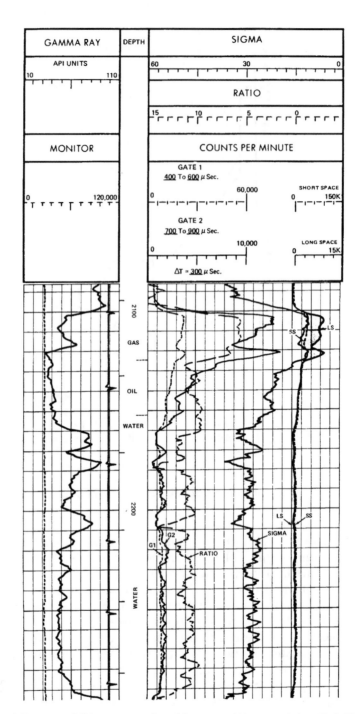

Figure 5.8 Dual Detector NLL presentation (Courtesy Western Atlas, Ref. 7)

information and in this regard was influenced by the Thermal Multigate Decay (TMD) of Welex (Halliburton). Processing was available to allow the TDT-M response to mimic the TDT-K for purposes of more effective monitoring. The neutron source of the TDT-M put out a significant increase of neutrons compared to the TDT-K version.

The log presentation was quite similar to the TDT-K and is shown on Figure 5.12.[15] As before, the GR is shown on the left track. Now both a near and far background are shown. The business curves are much the same (although they did not overlay the TDT-K). There is now a Σ_{NEAR} and Σ_{FAR} shown, with the far measurement considered more accurate. The N1 and F1 separation continues, except that the degree of separation is not as large as the earlier TDT-K. There are two gas zones shown in this log example, and the responses of the Ratio, capture cross section, and near far overlay are quite predictable.

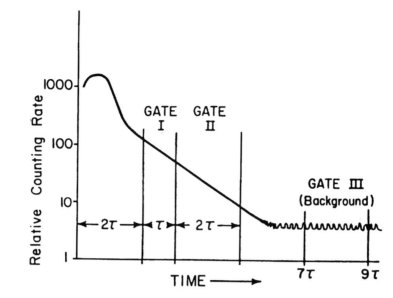

Figure 5.9 Gate positions for the Schlumberger TDT-K (Courtesy Schlumberger, Ref. 8)

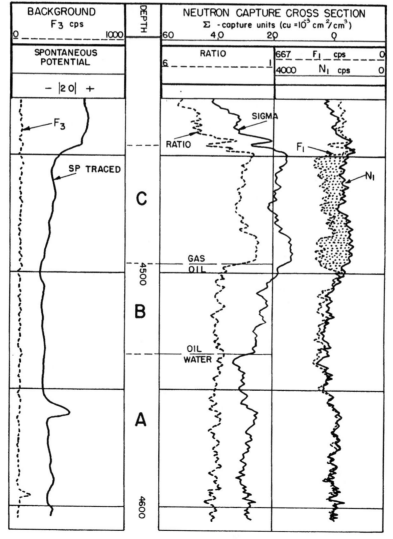

Figure 5.10 TDT-K log presentation (Courtesy Schlumberger, Ref. 14)

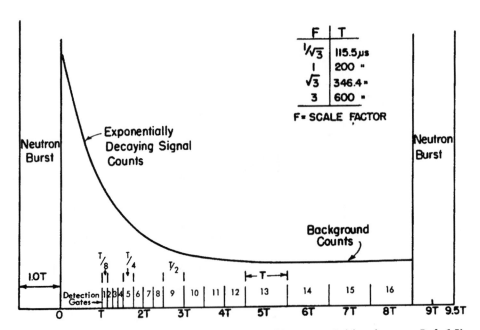

F	T
$1/\sqrt{3}$	115.5 μs
1	200 "
$\sqrt{3}$	346.4 "
3	600 "

F = SCALE FACTOR

Figure 5.11 Gate positions for the Schlumberger TDT-M (Courtesy Schlumberger, Ref. 15)

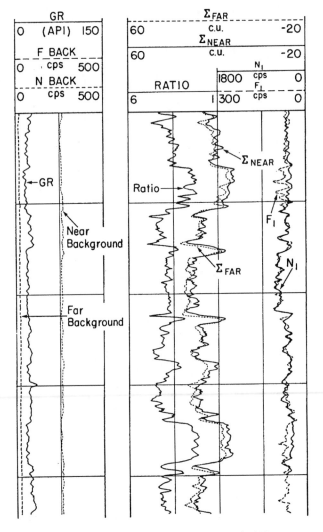

Figure 5.12 TDT-M log presentation (Courtesy Schlumberger, Ref. 15)

Thermal Multigate Decay (TMD)

The TMD was introduced by Halliburton (Welex) in about 1982 and essentially redefined the direction of future PNC use and development.[3,11,16,17] The TMD only has six gates. It bursts at a frequency of 800 microseconds, pausing every 1,250 bursts for a background measurement. This technique for measuring background after a number of bursts is called the "skip a beat" technique, and most tools today utilize such a system. The schematic of Figure 5.13 illustrates the Halliburton system.[11]

Of primary importance is the modeling concept introduced by Halliburton with the TMD. Up to the introduction of the TMD, measurements were made late in time after the burst to determine the slope of the formation portion of the decay curve. This avoided borehole counts spilling into the formation count gates. Furthermore, it was never clear how the borehole affected the log. Halliburton proposed that the actual decay of neutrons is comprised of two separate populations of neutrons, one in the formation and one in the borehole, and that each is decaying at its own exponential rate. This is shown on the leftmost decay sequence of Figure 5.13. When these two rates are added together, they equal the actual observed decay rate. This model is called the "dual exponential fit" model.

A Halliburton TMD log is shown on Figure 5.14.[11] In this case, the borehole is filled with standing water, oil and gas in the well bore, and the well is shut in. The presentation consists of two groups of parameters, the usual business logs labeled "Primary" and the quality control logs labeled "Quality." The primary presentation contains much the same as other logs, with the gamma ray in the left track. This track also contains Σ_{QUAL} which, if less than one may indicate a washed out hole and Σ_{BH-SS} which is the measured capture cross section of the borehole. Note that Σ_{BH-SS} reads about 100 c.u. when in water, 75 in oil, and becomes ratty in response in gas due to low count rates. The right-hand tracks are similar to previous log examples, with Σ^{Corr}_{FM} being the formation capture cross section, corrected for borehole effects, the uncalibrated porosity or ratio between the N and F detectors, $R_{N/F}$ and lastly the near and far overlay indicated by counts in gates 3-6 (G3-6) for the SS and LS detectors. Notice the gas zone from about X260 to X270 and the classic response of the curves of the right hand tracks.

Of the quality logs, the $G4_{ERROR}$ curve is the difference between the actual counts in gate 4 and the sum of the two exponentials. This is a goodness of fit measure for the dual exponential model and should remain near zero. Formation sigmas are presented for both the

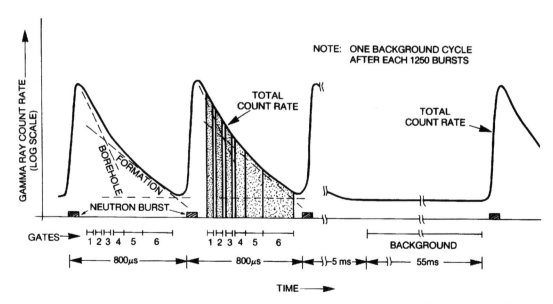

Figure 5.13 Gate positions and pulse sequence for the Halliburton Energy Services TMD (Courtesy SPWLA, Ref. 11)

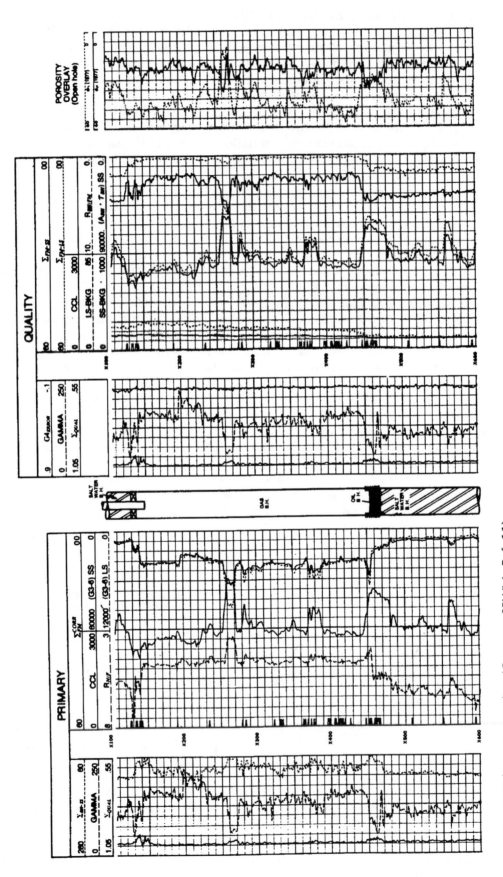

Figure 5.14 TMD log presentation (Courtesy SPWLA, Ref. 11)

near and far detectors, Σ_{FM-SS} and Σ_{FM-LS}, and the N and F background counts are presented as SS-BKG and LS-BKG, respectively. $R_{(BH/FM)}$ is the ratio of borehole to formation counts and reads low where gas is in the borehole. The last term, $(A_{BH}*T_{BH})SS$ indicates borehole count rate, again lower in gas. For more specific information on these variables or new computer mnemonics, consult Halliburton.

PDK-100

The PDK-100 of Western Atlas (Dresser Atlas) came out shortly after the TMD, about 1983.[18,19] This tool has a pulse rate of 1,000 microseconds between bursts, and measures background after 28 pulse cycles. The counts are taken in one hundred 10 microsecond gates from the beginning of the burst to the beginning of the next burst. A Schematic of the pulse sequence is shown on Figure 5.15.[4] This tool provides inelastic count rates, i.e., those of very high energy taken during and very shortly after the neutron burst. These inelastic count rates are helpful to analyze the in and near borehole area.

The log of Figure 5.16 is an Atlas PDK-100.[4] The basic presentation is similar to other logs with the addition of some new curves. Gamma ray remains in the left track. MSD is the mean standard deviation of sigma from a fixed number of prior measurements and indicates stability of sigma. MON monitors the tool functions. SGMA is the measured formation capture cross section. RATO is the ratio of near to far count rates and is an uncalibrated porosity. G1 and G2 correspond to the old NLL gate positions and BKS and BKL are background counts from the short and long spaced detectors, respectively. RICS is the ratio of inelastic to capture count rates for the near detector. RICS will track RATO except in a gas zone where RATO separates to the right of RICS. Note that at XX550 such separation exists, possibly indicating a gas zone. Due to the high shale content, this is not confirmed on other logs such as the expected separation in the SS-LS overlay on the supplemental presentation to the right. The RIN is the ratio of inelastic counts from the near divided by the far detectors. RIN is highly responsive to borehole fluid changes, especially gas. RBOR is the indicator of borehole capture cross section.

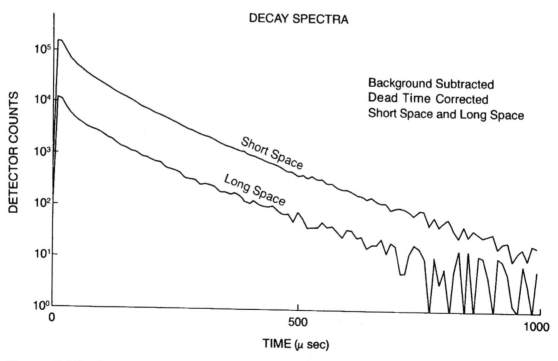

Figure 5.15 Gate positions and pulse sequence for the Western Atlas PDK-100 (Courtesy SPWLA, Ref. 4)

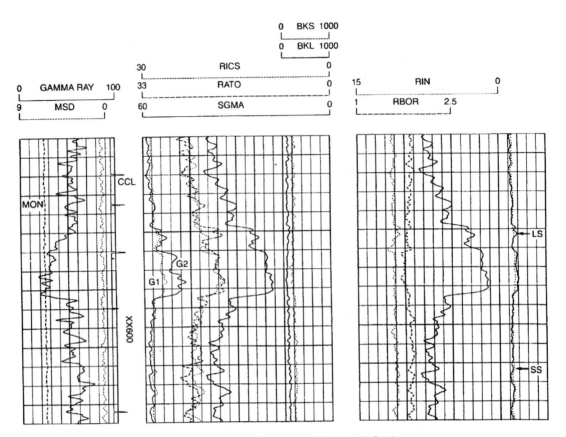

Figure 5.16 PDK-100 log presentation (Courtesy SPWLA, Ref. 4)

Dual Burst Thermal Decay Time Tool

The Dual Burst TDT, sometimes referred to as the TDT-P, was introduced about 1985. It utilizes two separate bursts of neutrons from an improved neutron source.[20] The first burst is short and is used to evaluate the borehole while the second burst is long and used to evaluate the formation. This tool utilizes a skip-a-beat system with 128 bursts between background measurements. Counts are made from 16 gates which are spread out across the dual burst cycle as shown on Figure 5.17.[20] The formation capture cross section presented on field logs is based on readings from the short and log bursts coupled with an extensive empirical data base comprised of over 4,000 cased hole core measurements. Alternatively, a complex two-component diffusion processing may be performed on the data. This latter processing is not done at the wellsite, is highly computer intensive, and likely to produce only marginal if any improvements in the sigma measurement in most cases.

The standard presentation of the Dual Burst TDT is shown on Figure 5.18.[5,21] The gamma ray, GR, and borehole sigma, SIBH are shown in the left track. Porosity, TPHI, is now calibrated and shown in the middle track. This is much like a compenstated neutron log (CNL) porosity and may be recorded on a sandstone or limestone scale. Formation capture cross section, SIGM, in c.u. is shown across the two right-hand tracks. The total selected counts for the near and far detectors, TSCN and TSCF, are similar to the N-F overlay and the separation below about 4,900 appears to indicate gas. The INFD is the inelastic count rate from the far detector and should track TSCN in a gas zone. In this case, the zone of interest is not a gas, but a low porosity limestone bounded above and below by a more porous sandstone.

The Dual Burst TDT offers a quality control log (not shown), much like the Halliburton TMD. Such a log would contain the gamma ray, GR, background counts for the far detector, FBAC, and MMOF to monitor neutron source performance, in the left-hand track. The middle tracks indicate the capture cross sections computed from the near and far detectors,

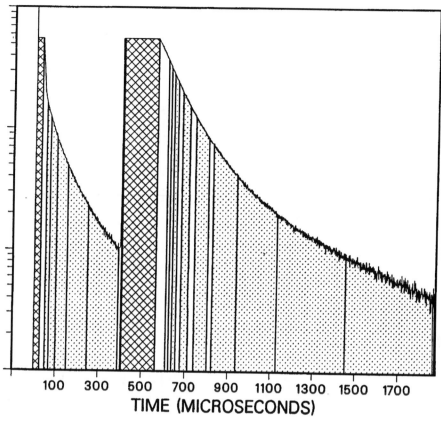

Figure 5.17 Gate timing for the Schlumberger Dual-Burst TDT-P tool (Courtesy Schlumberger, Ref. 20)

SFND and SFFD, and TCAF which indicates whether an adequate number of counts have been analyzed for the computations. The far right track contains the standard deviation of sigma, SDSI, and SIGC, the correction to SFFD for borehole effects.

Capture Tools with Spectral Capability

The Dual Burst TDT was the last of the classic capture cross section tools. From the NLL to the Dual Burst TDT, each new tool strived for an improved formation sigma measurement by improved counting gate schemes, processing, or hardware design. With each improvement often came new measurements such as the borehole sigma or inelastic counts. Another generation of tools is on the horizon. These include Computalog's PND-S,[22] Halliburton's TMD-L[23], and Schlumberger's RST series.[24] These tools do not offer improvements in capture measurements, but instead offer supplemental spectral information useful for carbon/oxygen logging. These tools offer an opportunity to approach areas where waters are too fresh or of unknown salinity with a single through tubing tool capable of both the sigma and C/O measurement. Further discussion of what these tools bring is deferred to the section on carbon-oxygen logging.

FACTORS INFLUENCING THE SIGMA MEASUREMENT
Borehole and Diffusion Effects

The service companies try to position the detectors of their tools to minimize borehole and diffusion effects. However, such effects remain. Diffusion occurs due to expansion of the neutron cloud with time. It also occurs when the capture cross section of the borehole

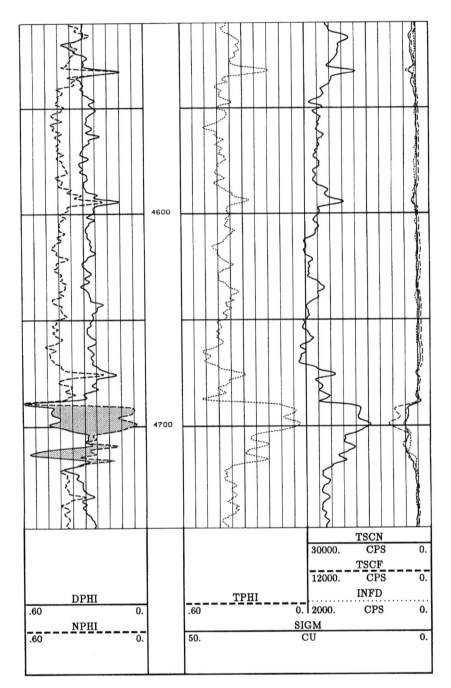

Figure 5.18 Dual Burst TDT log presentation (Courtesy Schlumberger, Ref. 21)

significantly differs from the formation. When the thermal neutrons are captured in one region, say the borehole, they then bleed or diffuse back into it from the formation, thereby introducing spurious counts not related to formation sigma, and hence the result is in error.

The ideal borehole fluid for a PNC log is a fairly high salinity water of 50,000 ppm NaCl or larger, thereby assuring that the thermal neutrons of the borehole are captured quickly. Lower capture cross section borehole fluids allow the borehole signal to continue for a long time and spill counts into the formation gates. Far and away, the worst condition for PNC logs is a gas filled borehole. One cannot always get optimal borehole fluids when logging. If the zone of interest is perforated or in communication with the borehole, the well cannot be shut in and/or loaded with salt water. To do this would cause the formation to take water and show an increased water saturation. Such contamination of the zone should be avoided.

The size of the borehole can be a strong factor, with the larger boreholes showing a significant degradation of signal. This is simply a case of getting less signal from the formation with larger boreholes. A similar effect occurs with thick cement sheaths around the casing which increase the distance between the tool and the formation to be measured. The depth of investigation of these tools depends on the formation sigma, but may be as high as 10 to 20 in. (25.4 to 50.8 cm) with larger depths of investigation associated with lower sigma formations.[25] Most service companies have environmental correction charts for the measured sigma to compensate for borehole and diffusion effects.[26-31] These corrected sigmas from the various vendors should all equal the intrinsic sigma of the formation. This may not actually be the case.

Logging Considerations

PNC logs are run at logging speeds of about 20-30 ft/min (6.1-9.1 M/min). Since discrete counts are measured, some statistical variation is inherent in the measurement which affects repeatability. Weighted average multiple logging passes with three to five logging runs is a technique commonly used to reduce such statistical variations. Reduced logging speeds also accomplish the same objective.

The vertical resolution of PNC tools is about two feet. The resolution is limited by the source to detector spacing and the algorithm used to smooth the data. The tool is not centralized (except the Dual Burst TDT when run as the oxygen activation Water Flow Log, WFL), and therefore is eccentralized.

Formation Water Salinity, Porosity, and Invasion

Formation water salinity and porosity are the most important factors affecting the suitability of PNC logs for an area. As a rule of thumb, the minimum porosity and salinity are .15 and 50,000 ppm NaCl, respectively. The effect, as discussed earlier in the discussion of the basic model, is that the tool error bar becomes large relative to the limits between water and hydrocarbon zones. Higher salinity formation water may allow meaningful data at lower porosities, and vice versa. Multiple runs, slower runs, or newer tools may reduce the error bar and thereby increase the useful range of PNC measurements. The presence of shale causes the situation to become worse in all cases.

Another factor is the water in the formation. With a depth of investigation of 10 to 20 in. (25.4-50.8 cm), the PNC is very sensitive to invasion. Invasion can occur when the interval of interest is in communication with wellbore fluids, either through perforations, channels, holes in casing, and the like. If the well is shut in, the fluid phases in the wellbore separate with the heavy phase, most likely water, settling to the lower part of the well. Perhaps the well is loaded with water to provide the best logging environment. This water may invade the formation, causing the logged formation to look wet. Similarly, if a base PNC log is to be run on a new well, it is important to wait some time before logging to allow for mud filtrate invasion and cement filtrate invasion to dissipate. A rule of thumb waiting period is about one month, although this may vary from one area to the next.

Which Sigma Is Presented?

Ideally, the formation sigma of interest is the true intrinsic sigma. This is not, unfortunately, what is usually displayed. Depending on the service company and what is requested, the sigma presented may be that measured from the near detector, may be diffusion but not borehole corrected, or may be the best try at intrinsic formation sigma. Only the Dual Burst TDT presents its measurement of intrinsic sigma at all times on the primary log. Furthermore, studies have shown that the measured sigma, even though processed by a service company to be intrinsic sigma, may be significantly in error, especially with low cross section borehole fluids and larger hole sizes.[32] It is recommended that an intrinsic measurement of sigma be requested with every run, even if it is necessary for off-well-site processing to obtain this log. Attempts have been made, with some success, to normalize logs of different service companies for better agreement.[33]

INTERPRETATION TECHNIQUES

Sigma-Porosity Cross Plot—Clean Formation

The sigma porosity cross plot for a clean formation is defined by three points: the capture cross sections of the matrix, water, and hydrocarbon.[34-37] These points define a triangular area as shown in Figure 5.19. The basic equation on which this analysis is based is equation 5.2, a linear equation for the model of Figure 5.4. The uppermost line of Figure 5.19 connects the matrix point at zero porosity with the water point at 100% porosity or ϕ_e =1.0. If these were the only two formation constituents present, at any porosity a PNC log would indicate a sigma defined by this line. Hence, this is the S_w =1.0 line. If only matrix and hydrocarbon were present, with no connate water, the lowermost line could similarly be constructed. At zero porosity the logging tool would indicate sigma matrix while at 100% porosity the tool would read Σ_h, i.e., about 21.0 c.u. for oil and less than 10 for a gas. For any intermediate porosity, a PNC log would read a sigma defined by this line if only hydrocarbon and matrix were present. This would be the S_w =0.0 line.

Once the triangular area has been defined, any combination of matrix, water, and hydrocarbon would define a point within this area. For a given porosity, the sigma read by a PNC logging tool would fall between the water (S_w=1) and hydrocarbon (S_w=0) lines. Since the system is linear, the distance between the water and hydrocarbon lines could be divided into four equal segments, separated by three lines corresponding to S_w=.25, .50, and .75 as shown on Figure 5.19. So, if at 30 percent porosity a PNC log read Σ A', this would be a point 100% water saturated. If at 25 percent porosity it read $\Sigma_{B'}$, this point would be 25% water saturated, i.e., a hydrocarbon point.

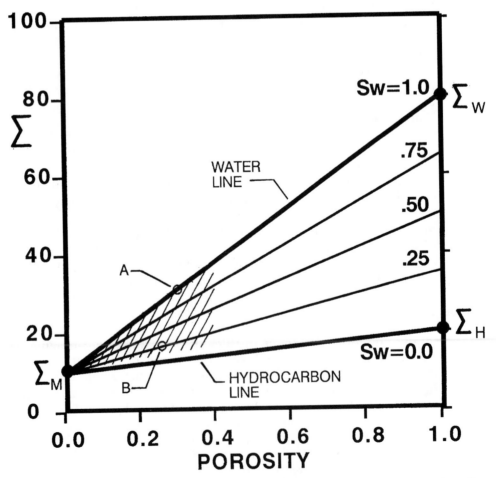

Figure 5.19 Layout for sigma-porosity cross plot showing water and hydrocarbon lines

The shaded area of Figure 5.19 shows the real region of interest. Realistically, porosities greater than about 40% are quite uncommon (except for some diatomites) and rocks are generally water wet and hence there is always some water saturation, even in a hydrocarbon zone.

Suppose that we know nothing about the reservoir, but have porosity data, either from the PNC log or from open hole data. We can develop the triangular shaped chart by cross plotting techniques. Note that porosity can be developed from the ratio curve of a PNC log. A number of porosity charts exist for each tool type and are available from the service companies. A calibrated porosity may be presented on the log instead of ratio with some tools. The coverage of each of these charts is beyond the scope of this text. If open hole porosities are available, they will generally be better and more accurate that porosity from a PNC.

To obtain the triangular chart, first plot sigma vs. porosity from the log over the clean formation interval, as shown in Figure 5.20. If both a water and a hydrocarbon zone are present, and there is a range of porosities over the interval of the data, two clusters will become apparent. The upper cluster is assumed to correspond to zones of 100% water-filled porosity, and a best fit straight line through that cluster defines the $S_w=1$ line. Extrapolation of that line to $\phi =1.0$ indicates the value of Σ_w. Extrapolation of that line to zero porosity indicates the value of Σ_M. The lower cluster is assumed to be a hydrocarbon zone. Since the matrix is likely water wet, some water saturation is present and it would be incorrect to draw a line through this cluster. Instead, the $S_w=0$ line is drawn from the value of Σ_M at zero porosity to Σ_H at $\phi =1.0$, here assumed to be oil with $\Sigma_H =21$. The $S_w= .25$, .50, and .75 lines are interpolated between the water and hydrocarbon lines. The lower cluster here has a saturation of about $S_w= .25$.

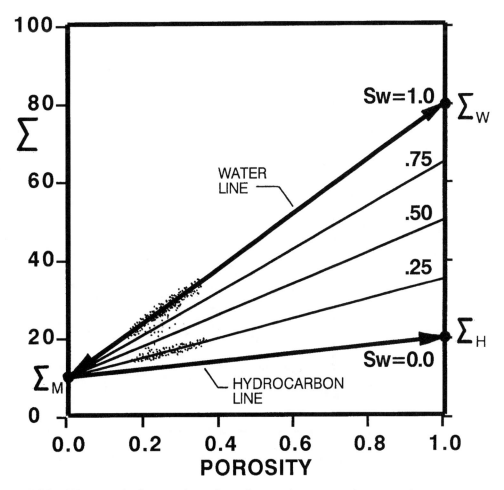

Figure 5.20 Water and oil zone data plotted on a sigma-porosity cross plot

Assumptions and Limitations of Sigma Porosity

Interpretations using this technique make some assumptions. It is assumed that the matrix, water salinity, and hydrocarbon type do not change over the interval being evaluated. If the area is on waterflood, the salinity will vary across the water zones when breakthrough occurs. If the salinity of the injected water is fresher than that of the formation, the point of breakthrough may look like an oil zone even though water saturation is high. If a gas cap is to be analyzed, it will require a new cross plot and new triangular area (although the water line will remain the same). It has further been assumed that no shale is apparent over the interval being logged. For shale zones, see the discussion which follows.

Cross Plot in Shaley Sands

When shale is encountered, an evaluation must be made of the shale porosity and capture cross section. The shale point plots on the sigma porosity cross plot as shown on Figure 5.21. The shale point is assumed to be 100% shale. When a line is drawn connecting the shale and the matrix point, that line is essentially a mixture of shale and matrix with no porosity other than that locked into the shale. To correct a point for shale, the shale volume must first be determined, typically from the open hole gamma ray log. If V_{SH} = .33 at a point, that point may be corrected by subtracting the shale. This is done by moving the point parallel to the shale matrix line and one third (V_{SH} = .33) of its length in such manner as to reduce both the sigma and porosity. The correction is expressed in equation form below.

$$\Sigma_{CORR} = \Sigma_{LOG} - V_{SH} (\Sigma_{SH} - \Sigma_M) \qquad \text{(Equation 5.5)}$$

and

$$\phi_{CORR} = \phi_{LOG} - V_{SH} \phi_{SH} \qquad \text{(Equation 5.6)}$$

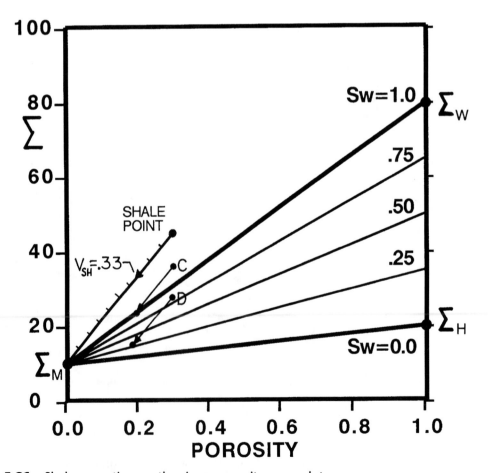

Figure 5.21 Shale correction on the sigma-porosity cross plot

The shale capture cross section is evaluated from the sigma log in a zone assumed to be 100% shale. The points C and D are both corrected for V_{SH} = .33. C is clearly a water point, falling on the S_w = 1.0 water line. Point D initially looks to be wet, but with shale correction of .33 shows a saturation of S_w = .30, i.e., a hydrocarbon bearing zone.

These shale corrected values of porosity and sigma are then used for the sigma porosity cross plot. They in effect are examining the porosity remaining after the shale has been removed to determine the saturation of the fluids in the non-shale pore space.

Interpretation Where Parameters Are Known

If the parameters are known, as would be the case in development wells of known reservoirs, or when running a monitor log with a known base log, the triangular area could be constructed and saturation calculated while logging. Of course, it would not be done graphically, but would use the equation for water saturation. If equation 5.2 is solved for water saturation, S_w, it becomes

$$S_w = \frac{(\Sigma_{LOG} - \Sigma_M) - \phi_e(\Sigma_H - \Sigma_M)}{\phi_e(\Sigma_w - \Sigma_H)} - V_{SH}\frac{(\Sigma_{SH} - \Sigma_M)}{\phi_e(\Sigma_w - \Sigma_H)} \qquad \text{(Equation 5.7)}$$

If the interval is clean and V_{SH}=0, this equation may be simplified to

$$S_w = \frac{(\Sigma_{LOG} - \Sigma_M) - \phi_e(\Sigma_H - \Sigma_M)}{\phi_e(\Sigma_w - \Sigma_H)} \qquad \text{(Equation 5.8)}$$

As long as the parameters do not change, the above equation may be used effectively. Notice in the above that the effective porosity, ϕ_e, is used, and it is shale corrected while the capture cross section, Σ_{LOG}, is for the bulk formation and is not shale corrected. If this seems strange, please go back to the model on which this equation is based.

Calculating Σ_w from a Water Zone

If the capture cross sections of the formation water is high, then one can assume a reasonable value for sigma matrix of, say eight c.u., without introducing too much error in saturation calculations. The capture cross section of the water may be estimated from a clean zone of 100% water saturation by rearranging equation 5.8 to become

$$\Sigma_w = \frac{1}{\phi_e S_w}\left\{\Sigma_{LOG} - (1 - V_{SH} - \phi_e)\Sigma_M - \phi_e(1 - S_w)\Sigma_H - V_{SH}\Sigma_{SH}\right\} \qquad \text{(Equation 5.9)}$$

This value may then be used in equation 5.6 or 5.7 for saturation calculations.

Using the Sigma Porosity Cross Plot

On Figure 5.22 is a TMD example from the Gulf of Mexico. In this well we have an open hole gamma ray, various resistivity logs, and a compensated neutron and density. The cased hole log is a TMD, with gamma ray, borehole sigma, ratio N/F, formation sigma, and N-F overlay. In this particular well, zones lower in the well below the logged interval shown have played out and the company is looking at possible zones uphole which might be productive. In this regard, shaley zones are considered. This analysis will look at six points, labeled A-F, at depths of 697, 705, 796, 816, 856, and 921, respectively. It is known that Σ_m = 8.0 c.u. from other wells in this field. The steps for this analysis will be outlined below and will be used generally to fill out the logged and computed data on the chart of Figure 5.23. Notice in this example that gas is indicated above about 692. A separate cross plot would be required for analysis of the gas points.

Step 1. Determine GR$_{shale}$, GR$_{clean}$, Σ_{shale}, and ϕ_{shale}. From the *open hole gamma ray log*, the values for the shale and clean sand line are GR$_{shale}$ = 117, GR$_{clean}$ = 48. The open hole GR should be used since the cased hole GR may be contaminated with radioactive salts from production. The large zone assumed to be 100% shale is taken at 770-775. Here, Σ_{SH} = 33.0 c.u.,

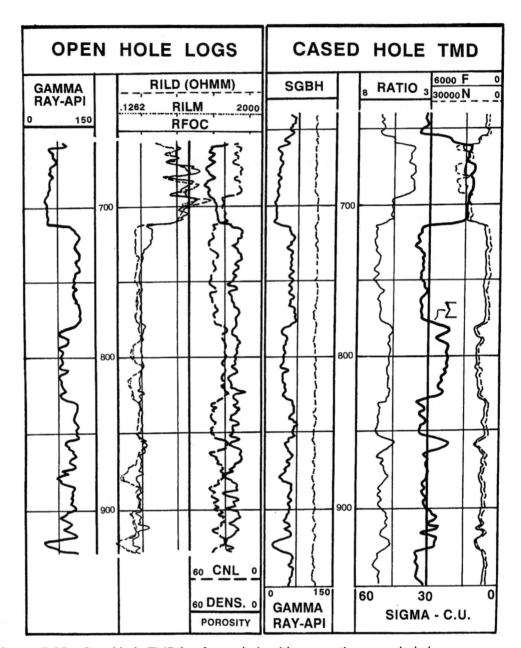

Figure 5.22 Cased hole TMD log for analysis with supporting open hole logs

and ϕ_{SH} = 29.0 porosity units (p.u.). The sigma value comes from the TMD sigma curve in the shale interval. The porosity is the average of the neutron density porosity in porosity units, i.e., (18.0 + 40.0)/2 = 29.0 p.u.

Step 2. Read the Σ_{log}, ϕ_{log}, GR_{log} at Each Depth of Interest. This step is straightforward enough. These data are tabulated on Figure 5.23 under the LOG DATA heading. These are the numbers which we use for further calculations.

In this case, open hole CNL-density logs are used for porosity by taking the average of the two readings. If a modern PNC log was used, it might have porosity displayed directly. Using older logs, we might use the PNC ratio curve to obtain porosity. The ratio and sigma are typically needed and porosity can be determined from service company charts for their respective tools and the borehole environment.

Step 3. Evaluate Shale Volume. Before we can perform this step, we need to consider two things. First, we must compute the relative gamma ray deflection, X, using equation

LOG DATA				COMPUTED DATA				RESULTS	
#	DEPTH	Σ_{LOG}	ϕ	GR	X	Vsh	$\phi_{COR.}$	$\Sigma_{COR.}$	Sw	
A	X697	14.0	34.0	32.0	.00	.00	34.0	14.0	.10	OIL
B	X705	14.5	32.5	33.5	.03	.01	32.1	14.2	.14	OIL
C	X796	22.5	25.0	55.0	.50	.31	16.1	14.8	.57	OIL?
D	X816	25.0	28.0	55.0	.50	.31	19.1	17.3	.75	WATER
E	X856	21.0	27.0	58.0	.57	.37	16.2	11.7	.25	OIL
F	X921	27.0	30.0	32.0	.00	.00	30.0	27.0	1.00	WATER
	X772	33.0	29.0	78.0	1.00	1.00				SHALE

$GR_{cl}=32.0, \quad GR_{sh}=78.0$

$SIGMA_{sh}=33.0, \quad POROSITY_{sh}=29.0, \quad SIGMA_{ma}=8.0$

Porosity and GR from opeh hole logs in this example

Figure 5.23 Values picked from log or computed for depths A–F on the log of Figure 5.22

4.1 for each point. This is the basis for computing the shale volume. To get V_{SH}, we must select an appropriate correlation curve. In this case, we know that the Clavier correlation provides reasonable estimates of shale volume (see Figure 4.3). Using the Clavier curve, we then get the shale volumes indicated on Figure 5.23.

Step 4. Correct Sigma and Porosity for Shale. The sigma and porosity values which come from the log each include a contribution from shale. Hence, to eliminate that contribution, we must apply the corrections expressed by equations 5.5 and 5.6. These corrections allow us to focus our measurement on the remaining porosity which may contain producible hydrocarbons. The results of this computation are tabulated on Figure 5.23.

Step 5. Cross Plot Sigma$_{CORR}$ and ϕ_{CORR}. This cross plot is shown on Figure 5.24. In this case, the matrix sigma is known to be eight c.u. The non-corrected raw points are plotted as circles, while the shale corrected points are plotted as dots. The saturations may be evaluated graphically. Interval F is clearly water, as is indicated by the open hole resistivity log and the cross plot. Points A, B, and E, indicate hydrocarbon, with water saturations of .25 or less. Point C is questionable, with S_w about .50, while point D looks wet with a .75 water saturation. Notice that the points C, D, and E have had rather substantial shale corrections. Extrapolation of the water line to 100 p.u. indicates that the $\Sigma_w = 74$ c.u. Once this value is known, equation (5.7) or (5.8) may be used to compute the saturation of each point.

A dual water model is sometimes used for this computation. A short discussion of this model and a solution of this log is presented later in this chapter.

APPLICATIONS ARISING FROM TIME-LAPSE PNC LOG RUNS

Why Multiple Runs Over Time?

To this point, the discussion showed numerous log examples with gas, oil, water, and shale zones. The GOC and WOC were identified. Computation of saturations was also performed

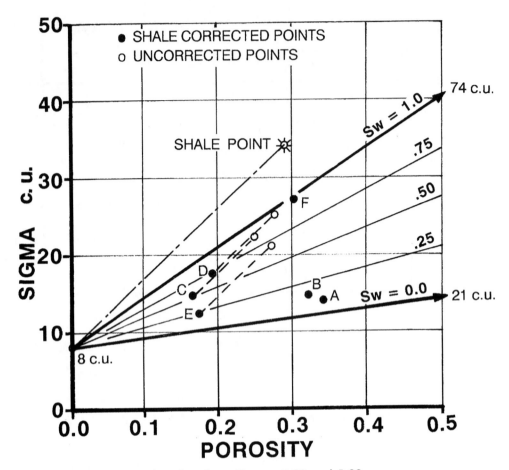

Figure 5.24 Cross plot for log data from Figures 5.22 and 5.23

for a number of zones. Multiple logging runs over time offers the operator a very powerful tool to monitor the changes in the reservoir as the well or nearby wells are produced. Changes in the locations of the GOC or WOC allow a means of predicting when such gas or water breakthrough is likely to occur. Zones which have been bypassed can be singled out for reperforation. If certain operations are to be performed between logging runs, the effects of those operations may be measured by successive runs. The examples which follow highlight some of the more important of such applications.

Monitor GOC and WOC Contact Movement

The example in Figure 5.25 is a well which is in an originally strong water drive reservoir.[38] The operator of this well began injecting gas into the gas cap to drive the oil down and improve recovery due to a lower residual oil saturation, (ROS), following displacement by gas. The water salinity is about 96 kppm NaCl and porosity is about 26-30 p.u. This well was drilled with a 7-7/8 in. (20 cm) bit and cased with 5 1/2 in. (14 cm) casing.

A base PNC and CNL was run in 1988, and monitor logs were run at approximately yearly intervals thereafter. In the middle track is shown the CNL porosity, NPHI, and the formation capture cross section is shown as SIGM in the right-hand track. The gamma ray is presented as usual in the left hand track. Recall that the ratio curve of a PNC log or the TPHI curve of the Dual Burst TDT closely follow the neutron porosity, and therefore, the comments which follow would apply if ratio or a PNC porosity was presented instead of NPHI. NPHI was used due to accuracy requirements for quantifying the saturation changes with a high degree of confidence.

From the NPHI curves, it is clear that the GOC is moving down, with the GOC being at about X690 on 4/89, at X708 on 4/90, at X728 at 10/90, and so on. However, notice that the sigma curve is dropping, indicating displacement of the WOC downward over time.

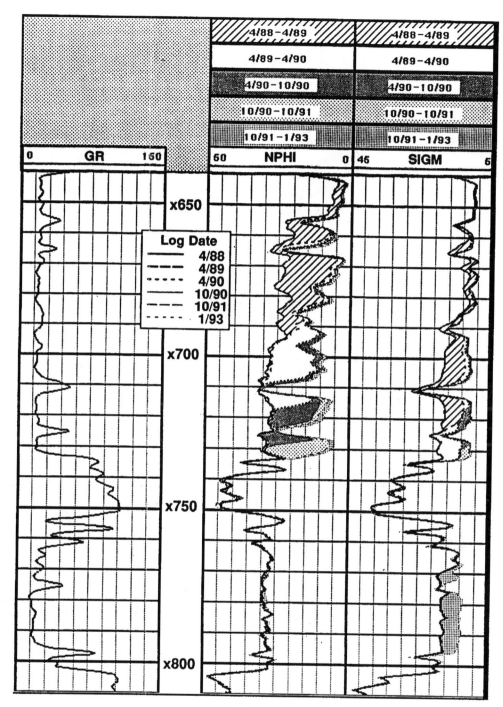

Figure 5.25 Example of PNC used to monitor the GOC and WOC movement over time (Courtesy SPWLA, Ref. 38)

For, example, the WOC is at about X722 on 4/89, X733 on 4/90, and finally appears in a zone from X770-X795 on the 1/93 logging run. The NPHI shows the gas cap growth while the changes in SIGM show the oil movement downward.

Steam Flood Monitoring

The example in Figure 5.26 shows a Dual Burst TDT log run in the Alberta Tar Sands.[39] The porosity varies over a narrow range of 35 to 38 p.u. The wells are monitor wells with a 6.875 in. (17.5 cm) borehole and 4.5 in. (11.4 cm) cemented casing. The well was not perforated and there was no liquid in the casing. The tool was contained in a Dewar (vacuum) container

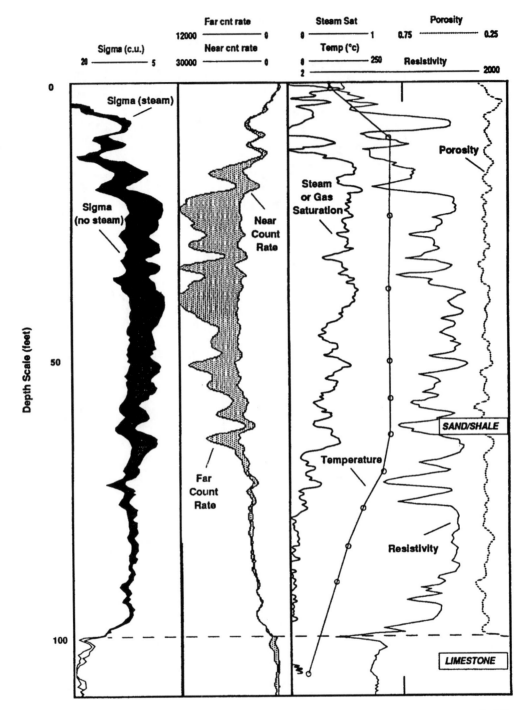

Figure 5.26 Example of steam flood monitoring with a PNC (Courtesy SPWLA, Ref. 39)

and was surrounded by a 3.5 in. (8.9 cm) boron excluder sleeve. The purpose of the sleeve is to provide a high capture cross section environment around the tool and avoid problems with gas in the borehole.

The logs show a base sigma curve, indicated as "no steam." A later run shows the effects of steam injection by a drop in the sigma readings and the near-far separation shown in the second track. Steam saturation was computed and is shown adjacent to a confirming temperature profile. Such monitoring is usually done at frequent intervals and the steam effect is often seen to increase with time. Steam saturations were calculated using computer modeling of the near and far count rates.

Locating Zones of Water Injection

The TMD runs in Figure 5.27 are used to locate zones of water injection.[40] Injection is accomplished through tubing, exiting at injection mandrels located at 8,570 and 8,825 as shown on the sketch adjacent to the logs. A radioactive tracer survey yielded ambiguous results and the TMD was brought in to confirm zones of injection. Notice that the log is run in tubing and making measurements out in the formation. This well was drilled with a 9 7/8 in. (25.1 cm) bit, with 4.5 in. (11.4 cm) cemented casing and 2 3/8 in. (6.0 cm) tubing.

Normal injection into this well was 1,700 barrels per day (BPD) (270 m³/day) of 4 kppm NaCl brine. A base log was run, shown as FM, PASS 1, the solid black line sigma curve. Then 180 bbls (28.6 m3) of 132 kppm NaCl brine was injected and the well relogged. This is PASS 2. Following this second run, one percent by weight borax was added to this brine

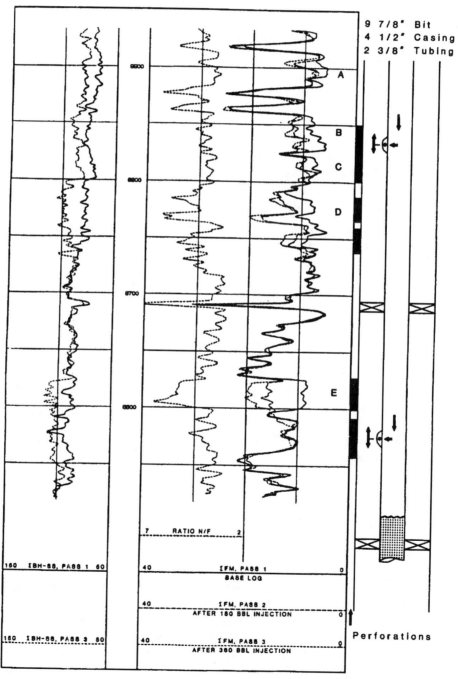

Figure 5.27 Example of a PNC used to locate zones of water injection (Copyright SPE, Ref. 40)

and again 180 bbls (28.6 m3) was injected. The log following this injection is listed as PASS 3. Zones B, C, D, and E adjacent to the perforations are clearly showing an increase in sigma and therefore are taking injected fluid. It would appear that E is taking the largest amount. Notice that Zone A is also indicating that it is being injected into. In the left-hand track is the borehole sigma from PASS 1 and PASS 3. Clearly, no flow exists between the upper and lower sets of perforations since the borehole sigma is not affected by the salty or borax containing brine. The packer shown is sealing. However, the borehole sigmas do not overlay above the top set of perforations, indicating that flow is moving up to and past zone A. Such movement is most likely in the annulus since the separation is nearly the same as that adjacent to the upper set of perforations. This indicates some leakage problem uphole, possibly at another packer not shown.

Residual Oil Saturation (ROS) from a Log-Inject-Log Technique

The log-inject-log technique is a means of evaluating the residual oil saturation, ROS, with a pulsed neutron log.[21,41] The basic technique is done using the following steps.

1. Run the PNC with the well producing (This is to get a base log and is not used in the ROS computation).
2. Inject a relatively fresh water of a known salinity and capture cross section in order to displace native waters out to about 36 in. (1 m) or at least 2 depths of investigation of the tool. Injection rates should be slow enough to avoid stripping residual oils away from the borehole.
3. Run the PNC and determine sigma formation for the zones of fresh water injection.
4. Repeat step 2, except inject a very salty water of known and high capture cross section.
5. Run the PNC and determine sigma formation for the zones of salty water injection.
6. Compute the residual oil saturation using the equations which follow.

The capture cross section of the formation after fresh and salt water injection are given by the equations below.

$$\Sigma_{LOG\text{-}FR} = (1-\phi)\Sigma_M + \phi\Sigma_{WFRESH} \, S_W + \phi(1-S_W)\Sigma_H \qquad \text{(Equation 5.10)}$$

$$\Sigma_{LOG\text{-}S} = (1-\phi)\Sigma_M + \phi\Sigma_{WSALT} \, S_W + \phi(1-S_W)\Sigma_H \qquad \text{(Equation 5.11)}$$

Where

$$\Sigma_{WFRESH} = \text{Sigma of the fresh water}$$

$$\Sigma_{WSALT} = \text{Sigma of the salty water}$$

$$\Sigma_{LOG\text{-}FR} = \text{Logged sigma after the fresh water injection}$$

$$\Sigma_{LOG\text{-}S} = \text{Logged sigma after the salty water injection}$$

Subtracting equation 5.11 from equation 5.10 and solving for S_w yields

$$S_w = \frac{\Sigma_{LOG-S} - \Sigma_{LOG-FR}}{\phi(\Sigma_{WSALT} - \Sigma_{WFRESH})} \qquad \text{(Equation 5.12)}$$

Noting that $1-ROS = S_w$, the expression for ROS becomes

$$ROS = 1 - \frac{\Sigma_{LOG-S} - \Sigma_{LOG-FR}}{\phi(\Sigma_{WSALT} - \Sigma_{WFRESH})} \qquad \text{(Equation 5.13)}$$

Figure 5.28 shows a Dual Burst TDT run in a log-inject-log mode for ROS determination. In this example, 7 in. (17.8 cm) casing is cemented in a 8.5 in. (21.6 cm) borehole. The first log is a composite of three logging passes following 4 hours of injection of 31 kppm NaCl water. The well was again logged after a second 10 hour injection of 166 kppm NaCl water. ROS analysis was performed only across the perforated intervals.

The well sketch is shown on the left-hand track of Figure 5.28. The second track shows the Σ_{LOG} fresh and Σ_{LOG} salt runs. The calculated ROS is shown on the third track as the

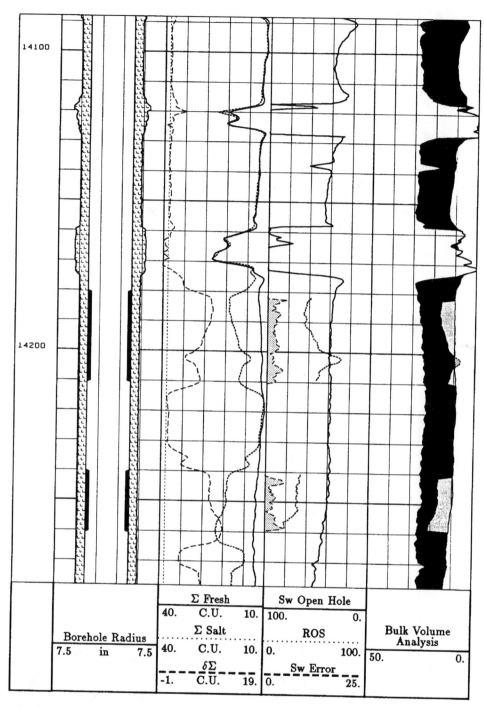

		Σ Fresh		Sw Open Hole			
		40.	C.U.	10.	100.		0.
			Σ Salt			ROS	
	Borehole Radius						Bulk Volume Analysis
	7.5	in	7.5	40.	C.U.	10.	0.	100.	50.		0.
			δΣ		Sw Error		
		-1.	C.U.	19.	0.		25.

Figure 5.28 Example of residual oil saturation (ROS) determination from a log-inject-log-technique (Courtesy Schlumberger, Ref. 5)

finely dotted line. A Bulk Volume Analysis presentation is presented on the right hand track. The Bulk Volume Analysis is presented in porosity units, p.u. The left boundary of the black area is the porosity. The black area is oil while the white area is water. So, at a point between the perforations, say at depth 230, the water fills about 40% of the porosity while the oil fills about 60% of the porosity, for $S_w = .40$. The black area is reduced across the perforated intervals. This is a result of the flushing of the injected waters. The remaining black area is the Residual Oil Saturation and the shaded is the moved oil.

The technique for Log-Inject-Log discussed above is the simplest approach. Other more complicated approaches to ROS using logs are also used.

Dual Water Model

The Dual Water model of a formation is shown on Figure 5.29. It is used primarily by Schlumberger, although not necessarily by those running PNC logs. In this model, the only difference from the model discussed throughout this chapter is that the shale is now broken into two components.[42,43] One component, the shale matrix, is thrown in with the rock matrix and assumed to be about the same capture cross section. Sometimes, the shale matrix is broken out as a fifth component. The other part of the shale is called "bound water," and is water tied into the shale either in shale porosity (which cannot flow) or chemically as a hydrate. The effective porosity, ϕ_e, is the porosity as we had used previously, and is made up of the movable and connate water, called the "free water," plus the hydrocarbon. The total porosity, ϕ_t, is the sum of the effective porosity plus the fractional volume of the bound water. The terms used on Figure 5.29 are defined on Table 5.3.

Like the earlier model, the bulk formation cross section, Σ_{log}, is the linear combination of the contributions from each formation constituent, and is shown below.

$$\Sigma_{LOG} = V_M \Sigma_M \; + \; V_{WF} \Sigma_{WF} \; + \; V_{WB} \Sigma_{WB} \; + \; V_H \Sigma_H \qquad \text{(Equation 5.14)}$$

$$\text{MATRIX} \qquad \text{FREE} \qquad \text{BOUND} \qquad \text{HYDRO-}$$
$$\text{WATER} \qquad \text{WATER} \qquad \text{CARBON}$$

If Σ_{Wa} is defined as $S_{WF}\Sigma_{WF} + S_{WB}\Sigma_{WB} + S_{HT}\Sigma_H$ then

$$\Sigma_{LOG} = (1 - \phi_t)\Sigma_M + \phi_t \Sigma_{WA} \qquad \text{(Equation 5.15)}$$

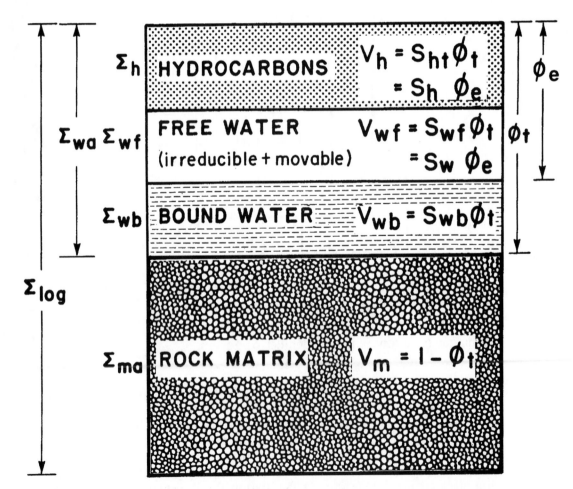

Figure 5.29 Dual water model for cased reservoir analysis (Courtesy Schlumberger, Ref. 43)

TABLE 5.3 Definition of Terms

Σ_{log}	Value of Sigma in Capture Units as read from the TDT log
Σ_h	Sigma of Hydrocarbons
Σ_{wf}	Sigma of Free Water
Σ_{wb}	Sigma of Bound Water
Σ_{ma}	Sigma of Rock Matrix
Σ_{wa}	Sigma of apparent of all the Pore Fluids
ϕ_t	Total Porosity (Pore Volume)
V_{ma}	Total Rock Matrix Volume
V_h	Hydrocarbon Volume
V_{wf}	Free Water Volume
V_{wb}	Bound Water Volume
S_{ht}	Hydrocarbon Saturation of the Total Propensity
S_{wf}	Free Water Saturation of the Total Propensity
S_{wb}	Bound Water Saturation of the Total Propensity
S_{wt}	Sum of Swf and Swb
ϕ_e	Effective Porosity—contains only Free Water and possible Hydrocarbons
S_h	Hydrocarbon Saturation of the Effective Porosity
S_w	Free Water Saturation of the Effective Porosity

Manipulation of the above equations can put various parameters in a form which can be easily used to compute the saturation from the log data.

Rather than discuss this model in detail, it may be more illustrative to solve the log of Figure 5.22 using this model. The table of data which we will use is shown on Figure 5.30. That group headed "Log Data" is identical to the values tabulated in the earlier computation of this example. The total porosity is that computed from the average of the neutron and density logs. The GR and Σ_{LOG} are read directly from the logs.

For the "Computed Data" information, Σ_{WA}, sigma water apparent, is computed from the following equation which is derived from equation 5.15 by solving it for Σ_{WA}

$$\Sigma_{WA} = \frac{\Sigma_{LOG} - \Sigma_M}{\phi_t} + \Sigma_M \qquad \text{(Equation 5.16)}$$

The term S_{wb}, the bound water saturation of the total porosity is simply the relative gamma ray deflection, X, that we had computed earlier, i.e.,

$$S_{wb} = \frac{GR_{LOG} - GR_{CLEAN}}{GR_{SHALE} - GR_{CLEAN}} \qquad \text{(Equation 5.17)}$$

At this point, we can use a cross plot technique. It is a plot of sigma water apparent, Σ_{WA}, vs the bound water saturation, S_{wb}. This plot is shown on Figure 5.31. Points A, B, C, D, E, and F are shown. If there was a spread of data points with shale in an otherwise 100% water zone, the upper water line would be defined by a cluster of such data points. As it is here, it is defined on one end by a clean 100% water sand, and so Σ_{wa} is equal to the capture cross section of the free water, i.e., 71.3 c.u. (Note what the formation water sigma was from the earlier calculated example). At the other end, the shale water is calculated to be 94.2 c.u. This upper line defines all combinations of water and shale with no hydrocarbon. Along this line, the water saturation based on total porosity is $S_{wt} = 1.0$.

The lower line is defined on the left by a clean sand with its pore space filled with hydrocarbon. If that hydrocarbon is oil, then the $S_{wt} = 0.0$ line, i.e., line of no water, begins at the oil sigma, about 21.0 c.u. A family of lines is then drawn parallel to the initial water line. In this figure the family defines the lines $S_{wt} = .25, .50.$ and $.75.$ The dashed line connecting the clean oil sand to the bound water point represents the maximum possible bound water. Due to connate water also being present, a point should theoretically not get too close to that diagonal line.

#	DEPTH	Σ_{LOG}	ϕ_T	GR	Σ_{WA}	S_{wb}	S_{wt}	ϕ_e	S_w	
		LOG DATA			COMPUTED DATA				RESULTS	
A	X697	14.0	34.0	32.0	25.6	.00	.08	34.0	.08	OIL
B	X705	14.5	32.5	33.5	29.0	.03	.11	31.5	.08	OIL
C	X796	22.5	25.0	55.0	66.0	.50	.66	12.5	.32	OIL
D	X816	25.0	28.0	55.0	68.7	.50	.72	12.5	.44	OIL?
E	X856	21.0	27.0	58.0	56.1	.57	.45	11.6	.00	OIL
F	X921	27.0	30.0	32.0	71.3	.00	1.00	30.0	1.00	WATER
	X772	33.0	29.0	78.0	94.2	1.00				SHALE

$GR_{cl}=32.0$ $GR_{sh}=78.0$ $SIGMA_{ma}=8.0$ c.u.

Figure 5.30 Values picked from log or computed for depths A–F on the log of Figure 5.22 for the dual-water model

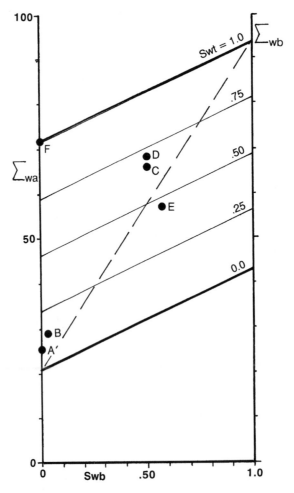

Figure 5.31 Dual water model cross plot for data from Figures 5.22 and 5.30

Based on the cross plot, we can then determine the S_{wt} for each point. Alternatively, S_{wt} can be computed. The water saturation based on total porosity is $S_{wt} = S_{WF} + S_{WB}$. Combining this identity with equation (5.14), it can be shown that

$$S_{wt} = \frac{\Sigma_{WA} - \Sigma_H - S_{WB}(\Sigma_{WB} - \Sigma_{WF})}{\Sigma_{WF} - \Sigma_H} \qquad \text{(Equation 5.18)}$$

Once we have determined S_{wt}, we may then convert these back to the usual terms, S_w and O_e, by the equations

$$S_W = \frac{S_{Wt} - S_{WB}}{1 - S_{WB}} \qquad \text{(Equation 5.19)}$$

and

$$\phi_e = \phi_t(1 - S_{WB}) \qquad \text{(Equation 5.20)}$$

Computing these saturations and effective values for porosity finishes the data in Table 5.3. The saturations using the Dual Water model appear, in this case, to be somewhat more optimistic than the conventional sigma-porosity cross plot technique.

POINTS TO REMEMBER

- PNC logs are through tubing devices.
- Wells do not have to be shut in (subject to safety considerations) prior to logging.
- PNC logs are used to measure porosity, water saturation, and detect gas.
- PNC logs can locate gas/oil and water/oil contacts.
- Time lapse PNC logging is useful to monitor changes in oil or gas saturation, movement of GOC or WOC.
- PNC logs can determine Residual Oil Saturation, ROS, using a log-inject-log approach.
- PNC logs are frequently run to locate bypassed hydrocarbons in uncompleted zones prior to plugging and abandoning.
- PNC logs are useful for fluid movement monitoring.
- PNC tools use a pulse neutron source which emits 14 mev neutrons.
- Gamma rays from inelastic collision occur during the first tens of microseconds and are not used for capture measurements.
- Capture gamma rays arise primarily a few 10 of microseconds after the burst until about 1000 µs or more.
- The Formation Sigma curve is used for saturation measurements.
- The N/F ratio curve is similar to an uncalibrated porosity.
- The Near-Far count rate overlay is used for gas detection.
- The borehole sigma curve monitors the capture cross section of the wellbore fluids.
- The inelastic count rates are sensitive to wellbore fluids, especially gas.
- The capture cross section of fresh water and oil are nearly identical, and hence the PNC tools cannot discriminate oil from fresh water in a formation.
- The rule of thumb minimum conditions for satisfactory quantitative PNC saturation computations is 15 p.u. and 50 kppm NaCl.
- The PNC tools are not good lithology tools.
- The depth of investigation of the Sigma measurement is about 8 to 18 in. (20 - 45 cm).
- The best fluid in the wellbore is a high (greater than 50 kppm NaCl) salinity water.
- The worst fluid in the wellbore is a gas.
- Weighted average of multiple logging passes may be run to reduce statistics inherent in a nuclear measurement.
- Typical logging speeds are in the neighborhood of 20-30 ft/min (6-9 m/min).
- Cross-plot techniques are available to compute saturations if conditions are not well known—if parameters are known, saturation may be computed while logging.
- The two main cross-plot techniques include the sigma-porosity cross plot and the "Dual Water" cross plot.

References

1. Youmans, A.H., Hopkinson, E.C., Bergan, R.A.,and Oshry, H.I., "Neutron Lifetime, A New Nuclear Log," 1963 SPE of AIME Fall Meeting, March, 1964.
2. Jameson, J.B., McGee, B.F., Blackburn, J.S., and Leach, B.C., "Dual-Spacing TDT Applications in Marginal Conditions," *JPT*, pp. 1067-1077, September, 1977.
3. Smith, H.D.Jr., Arnold, D.M., and Peelman, H.E., "Applications of a New Borehole Corrected Pulsed Neutron Capture Logging System (TMD)," 24th Annual CWLS-SPWLA Symposium, Calgary, June, 1983.
4. Randall, R.R., Lawrence, T.D., Frost, E., and Fertl, W.H., "PDK-100 Log Examples in the Gulf Coast," 26th Annual SPWLA Symposium, June, 1985.
5. Rogers, L.T., Schimkowitsch, E.B., and Watson, J.T., "Field Log Examples of the Dual Burst TDT Tool," 11th Annual CWLS Formation Evaluation Symposium, Calgary, Canada, Sept., 1987.
6. Schlumberger, "Log Interpretation Charts," SMP-7006, Schlumberger Wireline & Testing, Houston, Texas, 1994.
7. Dresser Atlas, "Neutron Lifetime Log," Document 3319, Dresser Industries, Houston, Texas, 1983.
8. Sclumberger, "The Essentials of Thermal Decay Time Logging," Houston, Texas, March, 1976.
9. Schlumberger, "Interpretation of TDT Logs," Houston, Texas, 1975.
10. Hart, P.E., and Pohler, M.A., "Pulsed Neutron Log Analysis Techniques and Results for Gulf Coast and East Texas Sandstones," 30th SPWLA Annual Symposium, June 11-14, 1989.
11. Schultz, W.E., Smith, H.D.Jr., Verbout, J.L., Bridges, J.R., Garcia, G.H., "Expiremental Basis for a New Borehole Corrected Pulsed Neutron Capture Logging System (TMD)," 24th Annual CWLS-SPWLA Symposium, Calgary, Canada, June, 1983.
12. Serpas, C.J., Wichmann, P.A., Fertl, W.H., DeVries, M.R., and Randall, R.R., "The Dual Detector Neutron Lifetime, Log-Theory and Practical Applications," 18th Annual SPWLA Symposium, June 5-8, 1977.
13. Dewan, J.T., Johnstone, C.W., Jacobson, L.A., Wall, W.B., and Alger, R.P., "Thermal Neutron Decay Time Logging Using Dual Detection," 14th Annual SPWLA Symposium, May 6-9, 1973.
14. McGhee, B.F., McGuire, J.A., and Vacca, H.L., "Examples of Dual Spacing Thermal Neutron Decay Time Logs in Texas Coast Oil & Gas Reservoirs," 15th Annual SPWLA Symposium, June 2-5, 1974.
15. Hall, J.E., Johnstone, C.W., Baldwin, J.L., and Jacobson, L.A., "A New Thermal Neutron Dacay Logging System—TDT-M," SPE Paper 9462, 55th Annual Fall Conference of the SPE of AIME, Dallas, Texas, Sept. 21-24, 1980.
16. Welex, A Halliburton Company, *Thermal Multigate Decay Logging,* Houston, Texas, 1983.
17. Buchanan, J.C., Clearman, D.K., Heidbrink, L.J., and Smith, H.D.Jr., "Applications of TMD Pulsed Neutron Logs in Unusual Downhole Logging Environments," 25th Annual SPWLA Symposium, New Orleans, Louisiana, June, 1984.
18. Randall, R.R., Gray, T., Craik, G., and Hopkinson, E.C., "A New Digital Multiscale Pulsed Neutron Logging System," SPE Paper 14461, 60th Annual SPE Technical Conference, Las Vegas, Nevada, Sept. 22-25, 1985.
19. Vasilev, M.L., Randall, R.R., Oliver, D.W., and Fertl, W.H., "Applications of a New Pulsed Neutron Capture Log to Middle East Reservoirs," SPE Paper 15717, SPE Middle East Oil Technical Conference, Manama, Bahrain, March 1987.
20. Steinman, D.K., Adolph, R.A., Mahdavi, M., Marienbach, E., Preeg, W.E., and Wraight, P.D., "Dual-Burst Thermal Decay Time Logging Principles," SPE Paper 15437, SPE Annual Technical Conference, 1986.
21. Schlumberger, "Dual-Burst TDT Logging," Document SMP-9130, July, 1988.
22. Streeter, R.W., Hogan, G.P. II, and Olson, J.D., "Oil or Fresh Water? A New Through-Tubing Measurement to Determine Water Saturation," Paper SPE 27646, SPE Permean Basin Oil and Gas Recovery Conference, Midland, Texas, March, 1994.
23. Jacobson, L.A., Ethridge, R., and Wyatt, D.F.Jr., "A New Thermal Multigate Decay-Lithology Tool," 35th Annual SPWLA Symposium, June, 1994.
24. Scott, H.D., Stoller, C., Roscoe, B.A., Plasek, R.E., and Adolph, R.A., "A New Compensated Through-Tubing Carbon/ Oxygen Tool for Use in Flowing Wells," 32nd SPWLA Annual Symposium, Midland, Texas, June, 1991.
25. Antkiw, S., "Depth of Investigation of the Dual-Spacing Thermal Neutron Decay Time Logging Tool," 17th SPWLA Annual Symposium, June, 1976.
26. Locke, S., and Smith, R., "Computed Departure Curves for the Thermal Decay Time Log," 16th SPWLA Annual Symposium, June, 1975.
27. Randall, R., Hopkinson, E., and Youmans, A.H., "A Study of the Effects of Diffusion on Pulsed Neutron Capture Logs," SPE Paper 6786, 52nd Annual SPE Conference, Denver, Colorado, October, 1977.
28. Preeg, W.E. and Scott, H.D., "Computing Thermal Neutron Decay Time (TDT) Environmental Effects Using Monte Carlo Techniques," Paper SPE 10293, 56th Annual SPE Conference, San Antonio, Texas, October, 1981.
29. Smith, H.D.Jr., Wyatt, D.F. Jr., and Arnold, D.M., "Obtaining Intrinsic Formation Capture Cross Sections with Pulsed Neutron Capture Logging Tools," 29th SPWLA Annual Symposium, June, 1988.
30. Murdoch, B.T., Hunter, C.J., Randall, R.R., and Towsley,C.W., "Diffusion Corrections to Pulsed Neutron Capture Logs: Methodology," 31st Annual SPWLA Symposium, June, 1990.
31. Welex, A Halliburton Company, "Thermal Multigate Decay Log Interpretation Charts," Houston, Texas, 1983.
32. Bonnie, R.J.M., "Evaluation of Various Pulsed Neutron Capture Logging Tools Under Well-Defined Laboratory Conditions," 32nd Annual SPWLA Symposium, June, 1991.
33. Neuman, C.H., "Programs to Process and Display Raw Data from Pulsed Neutron Capture Logs," 34th Annual SPWLA Sumposium, June, 1993.

34. Clavier, C., Hoyle, W.R., and Meunnier, D., "Quantitative Interpretation of TDT Logs," Paper SPE 2658, 44th Annual SPE Meeting, Denver, Sept. 28-Oct. 1, 1969.

35. Clavier, C., Hoyle, W., and Meunier, D., "Quantitative Interpretation of Thermal Decay Time Logs: Part I. Fundamentals and Techniques," *JPT*, June, 1971.

36. Dresser Atlas, "Neutron Lifetime Log Interpretation," Document No. 3319, 1979.

37. Jones, R.L., "Bootstrapping Your Way into TDT Analysis," *The Log Analyst*, July-August, 1982.

38. Fitz, D.E. and Ganapathy, N., "Quantitative Monitoring of Fluid Saturation Changes Using Cased-Hole Logs," 34th Annual SPWLA Symposium, June 13-16, 1993.

39. Dunn, K.J., Guo, D.S., and Zalan, T.A., "Gas/Steam Saturation Effects on Pulsed Neutron Capture Count Rates," 32nd Annual SPWLA Symposium, June 16-19, 1991.

40. Tesarek, P.B., and Heysse, D.R., "Water Injection Profiles in Eola Field Determined by Changes in Thermal Neutron Capture Cross Section," Paper SPE 16816, 62nd Annual SPE Technical Conference, Dallas, Texas, September 27-30, 1987.

41. Dresser Atlas, "LIL—Log-Inject-Log Measurements of Residual Oil Saturations," Dresser Petroleum Engineering Services, Paper PES-0104, Houston, Texas, December, 1982.

42. Clavier, C., Coates, G., and Dumanoir, J., "The Theoretical and Expiremental Bases for the 'Dual Water' Model for the Interpretation of Shaly Sands," Paper SPE 6859, 52nd Annual SPE Conference, Denver, Colorado, October 9-12, 1977.

43. Schlumberger, "Dual Water Model Cased Reservoir Analysis," Houston, Texas, March, 1980.

Other Suggested Reading

Mills, A.A., "Reservoir Monitoring in Low-Salinity Environments with Pulsed-Neutron-Capture and Gamma Ray Logs," SPE Formation Evaluation, pp. 177-183, September, 1993.

Noblett, B.R., and Gallagher, M.G., "Utilizing Pumpdown Pulsed Neutron Logs in Horizontal Wellbores to Evaluate Fractured Carbonate Reservoirs," 34th Annual SPWLA Symposium, June, 1993.

Thompson, and Klimo, C.R., "Pulsed Neutron Log Interpretation Case Histories for Inland and Offshore South Louisiana," Paper SPE 20587, 65th Annaul SPE Conference, New Orleans, Louisiana, September, 1990.

Butsch, R.J., and Vacca, H.L., "Experimental Pulsed Neutron Porosity for Gas-Filled Boreholes in the Gulf Coast," Paper SPE 20588, 65th Annual SPE Conference, New Orleans, Louisiana, September, 1990.

Kimminau, S.J., and Plasek, R.E., "The Design of Pulsed Neutron Reservoir Monitoring Programs," Paper SPE 20589, 65th Annual SPE Conference, New Orleans, Louisiana, September, 1990.

Jeckovich, G.T., and Olesen, J.R., "Enhancing Through-Tubing Formation Evaluation Capabilities with the Dual Burst Thermal Decay Tool," Paper SPE 19580, 64th Annaul SPE Conference, San Antonio, Texas, October, 1989.

Johnson, A., and Scanlon, M.E., "Pulsed Neutron Monitoring in the Bunter Sands of the Esmond Complex," Paper SPE 19294, Offshore Europe 89 Conference, Aberdeen, Scotland, September, 1989.

Smith, M.P., and Wyatt, D.F. Jr., "Quantitative Flood Monitoring Utilizing a New Pulsed Neutron Tool Modeling Concept," Paper SPE 16815, 62nd Annual SPE Conference, Dallas, Texas, September,1987.

Taneja. P.K., and Carroll, J.F., "Abnormal Pressure Detection Using a Pulsed Neutron Log," 26th Annual SPWLA Symposium, June, 1985.

Wiese, H.C., "TDT Log Applications in California," *JPT*, pp 429-444, February, 1983.

Libson, T.E., Vacca, H.L., and Meehan, D.N., "Stratton Field, Texas Gulf Coast: A Successful Cased-Hole Re-Evaluation of an Old Field to Determine Remaining reserves and Increase Production Levels," Paper SPE 12184, 58th Annual SPE Conference, San Francisco, California, October, 1983.

Jones, R.L., "Pulsed Neutron Logs Through Drill Pipe," *The Log Analyst*, pp. 5-9, Sept.-Oct., 1983.

DeVries, M.R., and Fertl, W., "Time Lapse Cased Hole Reservoir Evaluation Based on the Dual-Detector Neutron Lifetime Log—The CHES II Approach," Document 3330, Dresser Atlas, Houston, Texas, 1980.

Youngblood, W.E., "The Application of Pulsed Neutron Decay Time Logs to Monitor Waterfloods with Changing Salinity," *JPT*, pp. 957-963, June, 1980.

Smith, R.L., and Menzel, R.M., "Flood Evaluation by Log Analysis for Improved Recovery in Producing Wells," Paper SPE 6371, SPE Permian Basin Oil and Gas Recovery Conference, Midland, Texas, March, 1977.

Beaver, M.H., Lindley, B.W., McGhee, B.F., and Vacca, H.L., "Fractional Flow Curves Derived by TDT Logs and Production Tests," Paper SPE 6193, 51st Annaul SPE Technical Conference, New Orleans, October, 1976.

Smith, R.L., and Patterson, W.F., "Production Management of Reservoirs through Log Evaluation," Paper SPE 5507, 50th Annual SPE Conference, Dallas, Texas, September, 1975.

CHAPTER 6

FORMATION EVALUATION: NEUTRON LOGGING

NEUTRON LOGGING OVERVIEW

Applications and Mode of Operation

Neutron logs are used to evaluate formation porosity, detect gas, and as correlation logs between open and cased hole when the gamma ray lacks character. The neutron logs are primarily responsive to two parameters, the porosity and the hydrogen density of the pore fluid downhole. Hydrogen is present in water, oil, and gas. Both water and oil have about the same hydrogen density, called hydrogen index, while a gas has a substantially lower hydrogen index. When the pore space is filled with a liquid, either water or oil, the neutron logs measure porosity.

Neutron Interactions

Neutron logging tools use a chemical neutron source such as Americium-Beryllium, AmBe. This provides a continuous source of neutrons with initial energies of about 4.5 million electron volts (mev). Neutrons with such energies are called "fast neutrons." These fast neutrons collide with various nuclei in the formation and are slowed down. When a collision with a neutron occurs, energy is lost to the other nucleus. It turns out that the energy loss tends to be small unless the mass of the nucleus with which the neutron collides is about the same as the neutron. Hydrogen is the only atom fitting this description and therefore the slowing down of such fast neutrons is primarily a function of the number of hydrogen atoms per unit formation volume.

Within about a microsecond after the emission of a neutron, it is reduced significantly in energy to the so called "epithermal" state. Such epithermal neutrons have energies ranging from .2 to 10 ev. A few microseconds later, these neutrons are reduced in energy to the "thermal" state, with energies of about .025 ev. These neutrons essentially "rattle around" in the formation until which time they are captured. From the section on pulsed neutron logging, is was observed that this period of time could be up to about 1,000 microseconds or longer.

Ideally, the neutron tools would like to measure how far, on the average, these energetic neutrons propagate into the formation. This is a function of the formation matrix and pore fluids and is called the "slowing-down" length. Rather than measure this length, the tools measure a neutron density a fixed distance from the source. This neutron density can be related to the slowing down length.[1-4]

SINGLE DETECTOR NEUTRON TOOLS

Tool Configuration

The basic single detector neutron tool is shown on Figure 6.1.[4] This figure is split to show the difference in tool response to low and high hydrogen content formations. After fast

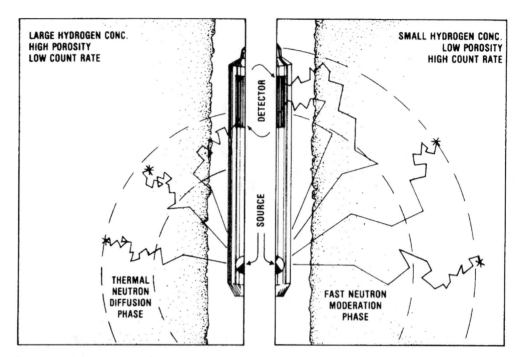

Figure 6.1 Single detector neutron tool in high (left) and low (right) hydrogen content formation (Courtesy SPWLA, Ref. 4)

neutron moderation phase, the neutrons are epithermal for a very short time, after which they are thermal for a long time and experience diffusion. The propagation of neutrons is shown on the left for large concentration of hydrogen, and on the right for a low concentration of hydrogen. Where the neutrons have moved farther out, there are more neutrons in the vicinity of the detector and hence more counts. While the fast moderation phase is insensitive to the thermal capture cross section of formation atoms, the thermal phase is sensitive to such high capture cross section atoms as hydrogen, chlorine, boron, and others. Excessively high salinities will cause the number of neutrons and hence count rates to be reduced at the detector due to the capture of the thermal neutrons.

The detector may be designed to count neutrons or capture gamma rays. Most tools focus on detecting thermal neutrons using a high pressure He^3 thermal neutron detector. If the tool was to measure epithermal neutrons, the He^3 thermal neutron detector would be wrapped in a cadmium sleeve. This sleeve would block the low energy thermal neutrons from the detector while the epithermal neutrons would pass through and be detected. Due to the low count rates with an epithermal measurements, most neutron tools are thermal neutron detectors.

The Log Response

The greater the neutron propagation or slowing down length, the more counts detected at the detector. This is the case since more neutrons are adjacent to the detector when they move farther from the source. The greater the hydrogen concentration, the smaller the neutron propagation. Figure 6.2 schematically shows the response of such a single detector neutron log to a variety of formation and pore space conditions.[5] Notice that the shales record the fewest counts. This is due to the presence of hydrogen in the form of OH hydroxyl radicals associated with the shale matrix and a more limited shale porosity. The gas response indicates a large neutron count rate, since gas has a low hydrogen index and the neutrons propagate deeper into the formation. Notice also that the lowest zone has essentially the same response to both water and oil at the same porosity. The hydrogen indices of water and oil are approximately equal.[5]

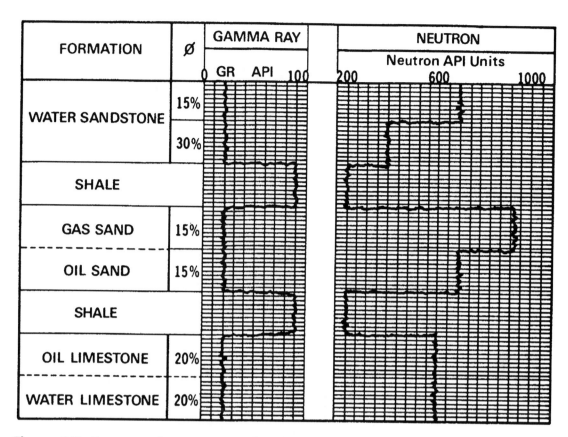

Figure 6.2 Response of gamma ray and single detector neutron log to various formations (Reproduced with permission of Halliburton Co., Ref. 5)

Applications of a Single Detector Tool

When the gamma ray lacks character, it cannot be used to tie in the cased hole collar locator to the open hole logs. The neutron log serves as an excellent substitute for this application. When run with a collar locator in cased hole, it can easily be correlated with the open hole neutron log. (See Chapter 4.) Other applications of the neutron log are the location of clean gas zones. However, the presence of the highly moderating shales could and often does mask gas zones.

The single detector tool can be used to estimate the porosity of the formation. The following technique, shown on Figure 6.3, assumes that the formation is free of both shales and gas, and that the borehole conditions do not change over the interval of interest.[5] First, draw a shale base line through the lowest neutron count values (Shale Line). Second, draw a line through the highest readings, corresponding to the densest zones (Dense Zone Line). Third, the relationship between American Petroleum Institute (API) Units on the neutron log and porosity is a logarithmic one, and therefore a logarithmic scale can be set up to span the dense zone line and the shale line. The limits are chosen to be 2% porosity at the dense zone line and 42% porosity at the shale line. Fourth, this logarithmic scale can then be drawn as shown and porosity values read off the log. Experience in an area may suggest different end values for the log scale for use on old logs.[5]

COMPENSATED DUAL DETECTOR NEUTRON LOGS

Configuration—Dual Detector Compensated Neutron Tools

The dual detector compensated neutron log utilizes an AmBe chemical source yielding about 4×10^7 neutrons/sec. The detectors are high pressure He[3] detectors. A schematic of the

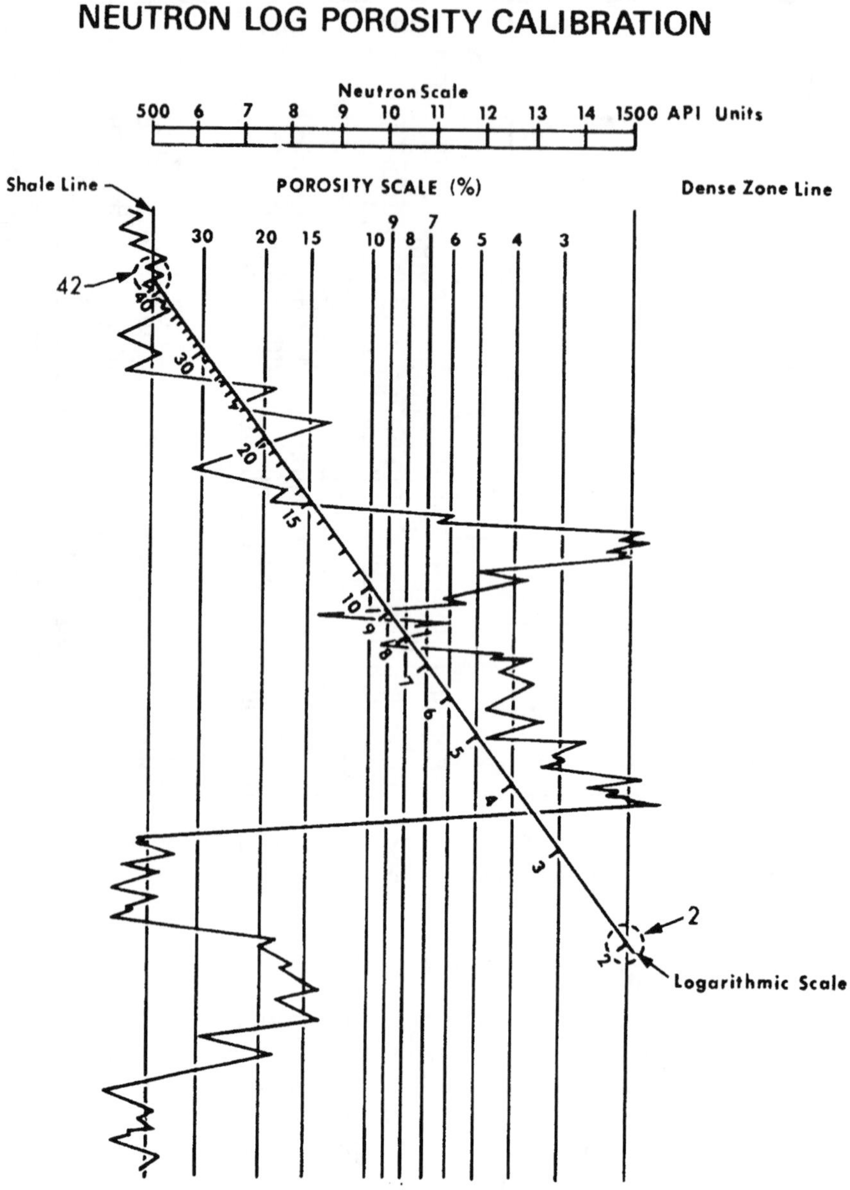

NEUTRON LOG POROSITY CALIBRATION

Figure 6.3 Estimation of porosity using a single detector neutron log (Reproduced with permission of Halliburton Co., Ref. 4)

dual spaced tool is shown on Figure 6.4.[4] The closer detector is referred to as the near or short spaced detector and the other as the far or long spaced detector. Certain tools are available which have both thermal and epithermal detectors. However, most tools detect thermal neutrons, especially for cased hole applications. The count rate of epithermal neutrons is simply too low to be of use in cased hole. Tool diameters are generally larger diameter (greater than 2.75 in. or 7.0 cm). These tools are available from most companies under a variety of trade names.

Dual Detector Tool Response

The response of a compensated neutron log (CNL*) (*CNL is a mark of Schlumberger), is shown from lab tests in Figure 6.5.[6] These early tests show that the ratio of the near/far detector count rates can be related directly to porosity units if the lithology is known. Note that separate response curves are shown for sandstone, limestone, and dolomite. As a result of this observation, logs are typically recorded in porosity units on the limestone scale and

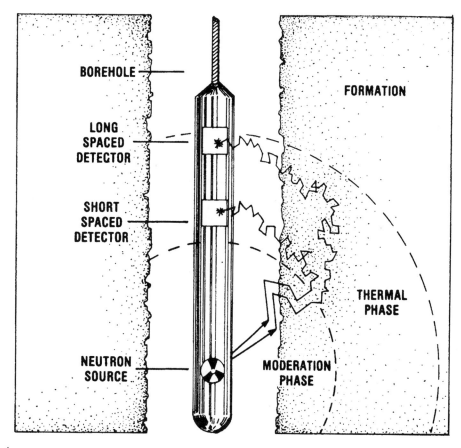

Figure 6.4 Dual detector compensated neutron tool (Courtesy SPWLA, Ref. 4)

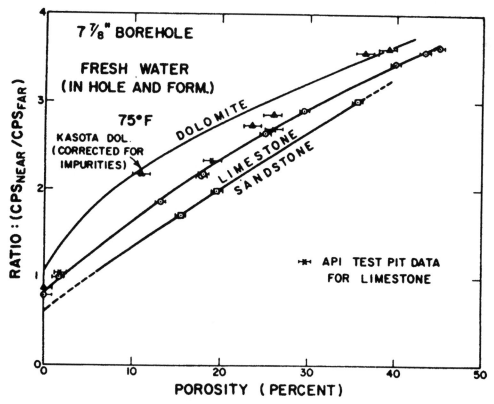

Figure 6.5 CNL response from laboratory tests (Courtesy Schlumberger, Ref. 6)

corrected for lithology changes. Figure 6.6 shows a chart typical of those in chart books for this correction.[2] For example, if the limestone response is 21 porosity units (p.u.), and the matrix is, in fact, sandstone, then the sandstone porosity is 26 p.u. Compare this to the chart of Figure 6.5. These charts will not match up exactly since they represent different tools from different service companies.[6,7]

ENVIRONMENTAL EFFECTS

Environmental Correction (Departure) Curves

The environmental effects on the dual detector tool response can be fairly large. The chart of Figure 6.7 shows an early environmental correction chart for the CNL.[6] Such charts are tool specific and the appropriate chart should be used for a particular tool model or design. This chart is useful since it shows the various factors and their relative magnitude on the porosity measurement. From this chart, borehole size and cement thickness have quite large effects, as does temperature. Other effects are less significant. Notice on each chart that there appears a heavy line noted by an asterisk. These lines refer to standard calibration conditions.

Consider the following corrections shown on Figure 6.7. Suppose that a porosity of 29 p.u. is noted on the log. This porosity must be corrected to standard calibration conditions for an accurate assessment of porosity downhole. If the open hole diameter is 10 3/4 in. (27.3 cm) the data must be corrected to the standard calibration condition of 8 3/4 in. (22.2 cm). Chart A of Figure 6.7 indicates a correction of about -2.7 p.u., while the casing thickness correction of chart B is about -1.3 p.u., and so on. The standard conditions for cased hole are:

1. 8 3/4 in. (22.2 cm) borehole.
2. 5 1/2 in. (14 cm), 17 lb/ft (25.5 kg/m) casing having a wall thickness of .304 in. (.77 cm).

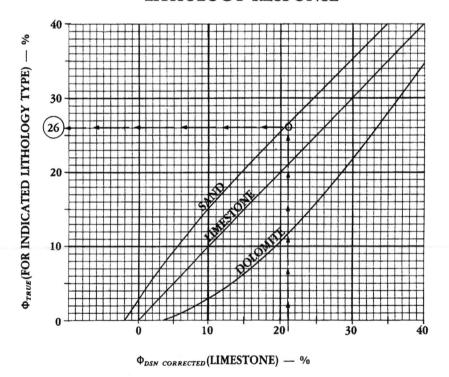

DUAL SPACED NEUTRON LITHOLOGY RESPONSE

Φ_{TRUE} (FOR INDICATED LITHOLOGY TYPE) — %

$\Phi_{DSN\ CORRECTED}$ (LIMESTONE) — %

Figure 6.6 Chart to correct porosity measurement for lithology effect (Reproduced with permission of Halliburton Co., Ref. 2)

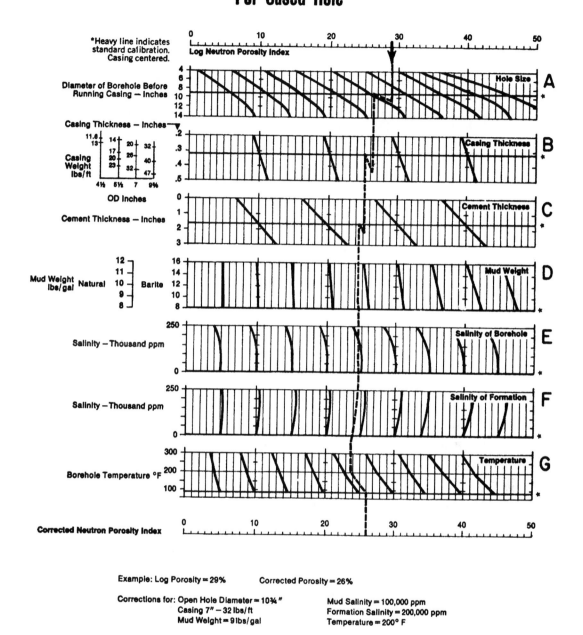

Figure 6.7 Early CNL correction chart for environmental effects (Courtesy Schlumberger, Ref. 6)

3. Cement thickness of 1.62 in. (4.1 cm.)—Assumes casing is centralized in borehole.
4. Fresh (0 ppm salinity) water in borehole and formation.
5. 75°F (24°C) and 1 atmosphere pressure.
6. Tool eccentralized in casing.

Notice that the new chart books do not contain all of the cased hole parameters shown here on one chart. Furthermore, the new charts are used somewhat differently and also have an indicated pressure dependence which is not indicated on the chart of Figure 6.7.[6,7]

Hydrogen Index of Formation Fluids

The hydrogen index, H, of a fluid is a measure of its hydrogen content per unit volume, i.e., hydrogen density. As indicated earlier, neutron tools respond to hydrogen in the pore space and oil and water are nearly equal in their hydrogen index. Some expressions for hydrogen index are shown below.[8,9]

$$\text{Water, Salt Water: } H_w = \rho_\omega \times (1 - P) \qquad \text{(Equation 6.1)}$$

where ρ_ω is the specific gravity of the salt water (or density, gm/cc), and P is the NaCl concentration in ppm (for example, if salt water has a NaCl concentration of 100,000 parts per million, use P = .1). The hydrogen index has no units and is equal to one for fresh water. The above equation has no temperature dependence.

$$\text{Hydrocarbons (Oil): } H_o = 1.28 \times \rho_o \qquad \text{(Equation 6.2)}$$

$$\text{Heavy Hydrocarbons } (\rho_h > 0.25): H_h = \rho_h + 0.3 \qquad \text{(Equation 6.3)}$$

Another proposed equation

$$H_h = 9 \times \left(\frac{4 - 2.5\rho_h}{16 - 2.5\rho_h} \right) \times \rho_h \qquad \text{(Equation 6.4)}$$

where ρ is the density of the hydrocarbon, and o or h refers to the oil or hydrocarbon respectively.

For gases or light hydrocarbons ($\rho_h < .25$):

$$H_h = 2.2 \times \rho_h \qquad \text{(Equation 6.5)}$$

Gas Effect

When gas is present in a zone, the effect is to show an erroneously low porosity. At first glance, this is due to the reduced hydrogen content or hydrogen index of the gas. While this is true, the observed porosities are too low, even when the lower hydrogen index is taken into account. The reason is that when gas is present, the neutron cloud is significantly larger, causing more neutrons than expected to appear at the far detector. This results in a lower than expected ratio, and lower than expected porosity indication. [10,11]

The incremental increase in porosity due to the greater slowing down length and hence larger cloud of neutrons has been called the "excavation effect." A correction based on this

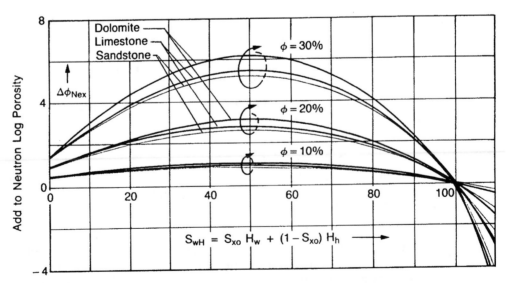

Figure 6.8 CNL excavation effect correction (Courtesy Schlumberger, Ref. 8)

excavation effect proposes to add back rock and reduce the porosity. This correction adjusts the porosity to be what it would be if only the hydrogen index of the pore fluid was the controlling parameter of the porosity measurement. Figure 6.8 shows the excavation effect correction.[8] The horizontal axis is plotted in terms of S_{xo} and the hydrogen indices of the water and hydrocarbon phase. When dealing with cased hole logs, unless the cement was only recently circulated, there is usually no flushed or invaded zone. So, for cased hole environments with no invasion, this equation may be written as

$$S_{wH} = S_w \times H_w + (1 - S_w) \times H_h \qquad \text{(Equation 6.6)}$$

where S_w is the water saturation of the zone and S_{wH} is the apparent bulk hydrogen saturation of the pore fluid. This chart may be expressed in equation form as:

$$\Delta\phi_{Nex} = K \times (2\phi^2 S_{WH} + 0.04\,\phi) \times (1 - S_{WH}) \qquad \text{(Equation 6.7)}$$

where $\Delta\phi_{nex}$ is the neutron log correction, ϕ is the formation porosity, and K is a constant whose value is 1.0 for sandstone, 1.046 for limestone, and 1.173 for dolomite.

The actual porosity, ϕ, and the measured neutron porosity, ϕ_N, can be related by the following equation when the hydrocarbon phase is gas and water is the heavy phase.[12,13]

$$\phi_N = \phi\left[\frac{H_h}{H_w}S_g + 1 - S_g\right] - \Delta\phi_{Nex} \qquad \text{(Equation 6.8)}$$

Response to Shales

As discussed earlier in this chapter, shales have hydrogen both as water in the pore space and as hydroxyl (OH) radicals attached to the matrix (bound water). This excessive amount of hydrogen results in an erroneously high indication of porosity in shales. In a nutshell, shales are rocks with a lot of hydrogen content not related to porosity.

PULSED SOURCE NEUTRON TOOLS

Tool Configuration

Schlumberger has recently introduced the Accelerator Porosity Sonde* (APS*). The * indicates that these names are marks of Schlumberger. This tool is presently being characterized in its response for the open hole environment. The characterization of this tool for the cased hole environment has not been done at this time.[14-18]

The APS sonde uses a pulsed neutron source similar to the Thermal Decay Time (TDT)/Reservoir Saturation (RST) tool series of Schlumberger. This source is pulsed 30 times in 10 microsecond bursts at 40 usec intervals for neutron porosity and slowing down time measurements. This is followed by a 100 microsecond burst and 700 microsecond period for measurement of formation capture cross section, sigma. This tool has the advantage of no active sources and therefore improved safety and possible risk to the environment. All detectors are He^3 detectors and measure epithermal neutrons, except for the capture detector which measures thermal neutrons.

A schematic of the APS sonde is shown on Figure 6.9.[18] Four sets of neutron detectors are indicated, the near epithermal detector, the array epithermal detector, the array thermal detector, and the far epithermal detector. Notice that the array detectors are eccentered within the sonde and shielded from borehole signals. The principal porosity measurement of the sonde is based on the ratio of near/array epithermal count rates. The array thermal count rate is used for the capture cross section measurement. The remaining detectors are used to measure the neutron slowing down time.

This tool has certain advantages in that it measures epithermal neutrons and is not affected by formation capture cross section. The spacings of the detectors have been optimized to provide excellent hydrogen index measurements and to substantially reduce any effects of formation atom density, especially in shaley formations. Furthermore, certain

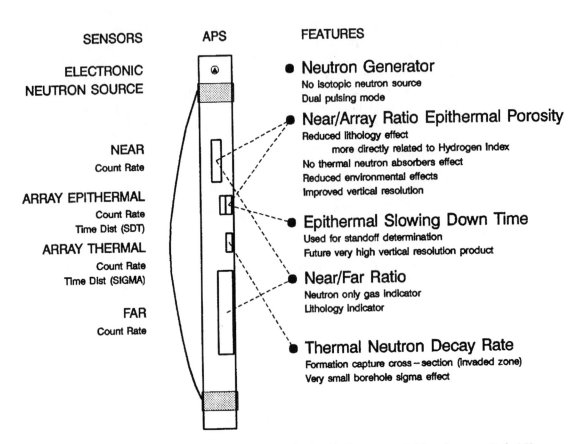

Figure 6.9 Schlumberger Array Porosity Sonde (APS) (Courtesy Schlumberger, Ref. 18)

detectors are shielded from the borehole to minimize its effects. While the measurement of epithermal neutrons has not been useful in cased holes, this sonde uses a pulsed source. This source puts out about 10 times more neutrons than a AmBe chemical source. As a result, it may be possible that sufficient counts are available for cased hole use. This sonde is 3 5/8 in. (9.21 cm) in diameter.

APPLICATION EXAMPLES OF DUAL DETECTOR NEUTRON LOGS

Effect of Invasion

Figure 6.10 shows porosity in the right hand track. Here is an open hole density porosity, ϕ_D, and open hole CNL porosity, ϕ_{OH}.[19] The interval from 6,438 to 6,442 is a low pressure gas zone of about 20–25% porosity. This well was drilled with a heavy mud and significant invasion was expected and indicated from the resistivity logs. The neutron density overlay shows the characteristic crossover indicating a gas zone, apparently a result of residual gas saturation.

A cased hole CNL was run 30 days later and is shown overlaying the open hole CNL over most of the interval. The cased hole log reads a significantly lower porosity, ϕ_{CH}, over the gas interval, indicating that the invaded fluid had largely dissipated.[19]

Monitoring of Gas Movement

The logs of Figure 6.11 show a series of CNL logs run in Alaska for the purpose of monitoring gas cap movement and gas encroachment under shale layers.[20] The gamma ray in the left track shows shale breaks at about 9,395, 9,430, and 9,580. Production perforations are shown below about 9,610. On the right are neutron logs to monitor gas movement. The open hole CNL was run in July of 1979, and based on the open hole data of that time,

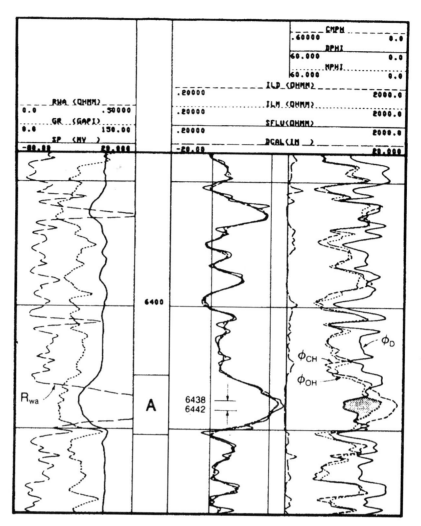

Figure 6.10 CNL showing reduced effects of mud filtrate invasion after well is cased (Courtesy Schlumberger, Ref. 19)

the initial gas-oil contact (GOC) is at about 9,310. In November, 1983, a monitor log was run. It shows, by the darkly crosshatched reduction in porosity seen by the log due to the gas effect, that the gas cap has moved to about 9,410. However, two other gas indications are detected under the shales at 9,430 and 9,580. Gas is indicated under these shales down to local GOCs at 9,475 and 9,595 respectively. When a monitor was run again in June, 1986, the various gas oil contacts had all moved down somewhat with a most significant change in the interval below the 9,430 shale. Its local GOC had moved from 9,475 to 9,550 in the time between these logs.

The shales were situated with a significant dip relative to horizontal, and the gas was creeping under these shale barriers. The operator was concerned that lenses of oil above these shale barriers would be left behind. Such a potentially bypassed lens exists from 9,420 to 9,430. There is also the possibility of a bypassed lens above the 9,580 shale. Such bypassed zones would be perforated and produced based on these and later logs.

Locating Sand Production Sources

Figure 6.12 shows a series of CNL logs run over a perforated interval known to be producing sand.[20] An open hole base CNL was run in March, 1979. This was followed by monitor logs run in 1984, 1986, and 1987. The more recent logs show an increase in apparent porosity. This is interpreted as an indication of loss of matrix in the vicinity of the wellbore and therefore sand production. The sand productive intervals are indicated.

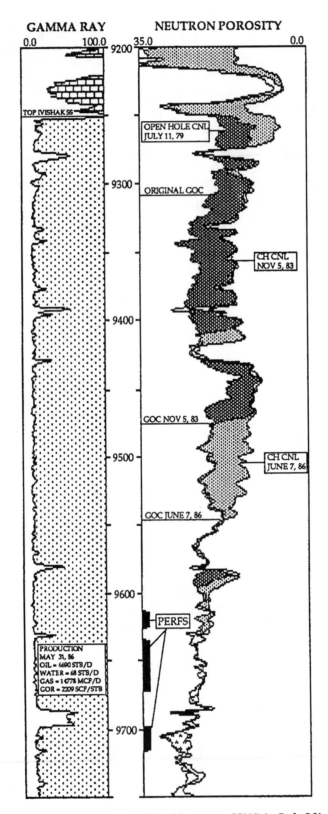

GAMMA RAY
0.0   100.0

NEUTRON POROSITY
35.0   0.0

9200

TOP IVISHAK SS

OPEN HOLE CNL
JULY 11, 79

9300

ORIGINAL GOC

CH CNL
NOV 5, 83

9400

GOC NOV 5, 83

9500

CH CNL
JUNE 7, 86

GOC JUNE 7, 86

9600

PERFS

PRODUCTION
MAY 31, 86
OIL = 6690 STB/D
WATER = 68 STB/D
GAS = 14778 MCF/D
GOR = 2209 SCF/STB

9700

Figure 6.11   Example of gas monitoring with a CNL (Courtesy SPWLA, Ref. 20)

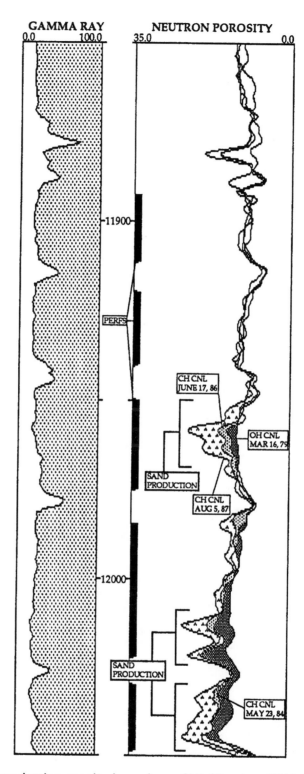

Figure 6.12 Sand production monitoring using a CNL (Courtesy SPWLA, Ref. 20)

POINTS TO REMEMBER

- Neutron logs are used as correlation logs between open and cased hole.
- Neutron logs are used to measure porosity.
- Neutron logs can detect gas.
- Neutron logs are used to monitor gas cap movement or gas encroachment.
- Neutron logs may be helpful at locating sources of sand production.

- Neutron logging tools use chemical AmBe neutron sources.
- Neutrons primarily are affected by hydrogen in the formation.
- The main hydrogen in the formation is in pore fluids.
- Most neutron tools detect thermal neutrons.
- Epithermal neutrons are not affected by capture cross section.
- Single detector neutron tools may be used for porosity evaluation.
- Dual detector neutron tools measure liquid filled porosity.
- Shales appear to have erroneously high porosities.
- Gas zones appear to have erroneously low porosities.
- Porosity measurements must be corrected to the appropriate matrix—limestone, sandstone, or dolomite.
- The borehole environment corrections are available.
- Hydrogen index is an indication of the bulk hydrogen density of a fluid.
- Water and oil have hydrogen indices of about 1.0.
- Gas has a low hydrogen index.
- The excavation effect causes a greater reduction in measured porosity than expected on the basis of the hydrogen index of a gas.
- Pulsed neutron sources may replace chemical sources in the future and such tools are available in open hole today.

References

1. Ellis, D.V., *Well Logging for Earth Scientists,* Elsevier Science Publishing Company, New York, New York, 1987.
2. Welex (Now Halliburton Energy Services), "Neutron Logging," Document EL-1016, Houston, Texas, 1983.
3. Edmundson, H., and Raymer, L.L., "Radioactive Logging Parameters for Common Minerals," 20th Annual SPWLA Symposium, June, 1979.
4. Arnold, D.M., and Smith, H.D. Jr., "Experimental Determination of Environmental Corrections for a Dual-Spaced Neutron Porosity Log," 22nd Annual SPWLA Symposium, Mexico City, Mexico, June, 1981.
5. Gearhart (Now Halliburton Energy Services), "Basic Cased Hole Seminar," Document WS-530, Fort Worth, Texas, 1982.
6. Alger, R.P., Locke, S., Nagel, W.A., and Sherman, H., "The Dual Spacing Neutron Log—CNL," Paper SPE 3565, 46th Annual SPE Meeting, New Orleans, Louisiana, 1971.
7. Schlumberger, "Log Interpretation Charts," Document SMP-7006, Houston, Texas, 1994.
8. Schlumberger, "Log Interpretation Principles/Applications," Document SMP-7017, Houston, Texas, 1987.
9. Schlumberger, "Cased Hole Log Interpretation Principles/ Applications," Document SMP-7025, Houston, Texas 1989.
10. Segesman, F., and Liu, O., "The Excavation Effect," 12th Annual SPWLA Symposium, Dallas, Texas, May, 1971.
11. Ullo, J.J., "Response of the Dual Spacing Neitron Log (CNL) To Gas," Paper SPE 10295, 56th Annual SPE Conference, San Antonio, Texas, October, 1981.
12. Truman, R.B., Alger, R.P., Connell, J.G., and Smith, R.L., "Progress Report on Interpretation of the Dual-Spacing Neutron Log (CNL) in the U.S.," 13th Annaul SPWLA Symposium, May, 1972.
13. Flolo, L.H., "Calculating Injected Gas Saturation from Cased Hole Compensated Neutron Logs in an Oil Reservoir," 31st Annual SPWLA Symposium, June, 1990.
14. Mills, W.R., Stromswold, D.C., and Allen, L.S., "Pulsed Neutron Porosity Logging," 29th Annual SPWLA Symposium, June, 1988.
15. Scott, H.D., Wraight, P.D., Thornton, J.L., Olesen, J-R, Hertzog, R.C., McKeon, D.C., DasGupta, T., and Albertin, I.J., "Response of a Multidetector Pulsed Neutron Porosity Tool," 35th Annual SPWLA Symposium, June, 1994.
16. Olesen, J-R, Flaum, C., and Jacobsen, S., "Wellsite Detection of Gas Reservoirs with Advanced Wireline Logging Technology," 35th Annual SPWLA Symposium, June, 1994.
17. Schlumberger, "IPL Integrated Porosity Lithology—A New Look at Neutron Porosity Logging," Document SMP-9282, Houston, Texas, 1992.
18. Schlumberger, *Integrated Porosity Lithology System—An Overview,* Houston Product Center, Texas, August, 1993.
19. Vacca, H.L., and Walker, C.M., "The Cased Hole Compensated Neutron Log Enhances Shaly Gas Identification," 26th Annual SPWLA Symposium, June, 1985.
20. Dupree, J.H., "Cased-Hole Nuclear Logging Interpretation, Prudhoe Bay, Alaska," *The Log Analyst,* pp. 162-177, May-June, 1989.

CHAPTER 7

FORMATION EVALUATION: CARBON/OXYGEN LOGGING

TYPICAL APPLICATIONS OF CARBON/OXYGEN LOGGING

Overview

Carbon/oxygen logging has application in determining the presence of water and oil and their saturations behind casing in formations whose waters are fresh or of unknown salinity. Where salinities are high and known, the pulsed neutron capture logs are superior in saturation determinations. Except for the new generation of small diameter spectral tools, the carbon/oxygen (C/O) tools have traditionally been large diameter tools and require the shut-in of the well and removal of tubing prior to logging. Smaller diameter tools such as the Reservoir Saturation Tool (RST) of Schlumberger and other tools which are designed for both capture and C/O measurements are currently being introduced.

The applications of these induced spectral gamma ray (carbon-oxygen, C/O) tools includes the following:

- Discriminate water/oil contact when salinity of formation waters is low or unknown.
- Evaluate hydrocarbon zones and saturations in fresh, mixed, or unknown water salinity environments.
- Locate water and oil zones in waterfloods where mixed salinities exist between formation and flood waters.
- Evaluate saturations in formations behind casings when open hole logs are not available.
- Monitoring of steam and CO_2 flood fronts/breakthrough.
- Numerous other applications arising from the yields of elements other than carbon and oxygen.

Carbon and Oxygen Are Not Enough

While at first glance the carbon/oxygen ratio seems like a useful measurement to evaluate formation fluid saturations, it is of little use by itself. Carbon and oxygen arise from a number of sources downhole such as borehole fluids, formation matrix and, of course, formation fluids. While the C/O ratio of formation fluids is desired, the actual C/O ratio measured is the ratio of counts coming from a number of sources.[1]

$$\frac{C}{O} = \frac{C_{\text{formation fluid}} + C_{\text{matrix}} + C_{\text{borehole}}}{O_{\text{formation fluid}} + O_{\text{matrix}} + O_{\text{borehole}}}$$ (Equation 7.1)

From the above equation it appears that water as the borehole fluid is probably the best choice since it eliminates extra carbon counts from the borehole. In this regard, an exclusion sleeve may also be placed around the tool to minimize the borehole effect. While water may be best, a single and known fluid phase in the wellbore may be accounted for and is sufficient to get good results.[2] The larger diameter RST of Schlumberger measures the borehole C/O ratio and is capable of compensating for oil and water in the wellbore. Knowledge of the formation type is also critical since Limestone ($CaCO_3$) and Dolomite ($CaMg(CO_3)2$) are carbonates with both C and O contributions while sandstone (SiO2) has no C contribution.

Figure 7.1 shows how the atomic C/O ratio, when the borehole effects are excluded, varies with porosity, oil saturation, and matrix material. The lines of the fan charts are stated in units of oil saturation. Clearly, if sandstone alone is the matrix, C/O is equal to zero at all porosities if $S_o=0$, i.e., 100% water in the pore space. If oil is present, the C/O ratio increases with porosity and oil saturation.

When in a pure limestone or dolomite, the atomic C/O ratio is .333 (one carbon to three oxygen atoms) at zero porosity due to the presence of CO_3 in a carbonate. However, notice that the $S_o=0.0$ line slopes downward with increasing porosity. This is due to the fact that carbonate, CO_3, is being replaced with water, H_2O, causing a decrease in the C/O ratio. When at about 27% porosity, a C/O ratio of .3 could indicate a water zone in a carbonate or an oil zone in sandstone.

As a result of the foregoing relationship, tools which measure the C/O ratio also must address the formation matrix. Therefore, as a minimum, such a tool must also discriminate between limestone/dolomite, i.e. a carbonate, and sandstone. Tools available today not only measure silicon (Si) and Calcium (Ca) to discriminate sandstone from carbonate, but also measure elemental concentrations or yields of such elements as hydrogen (H), chlorine (Cl), sulfur (S), and iron (Fe).

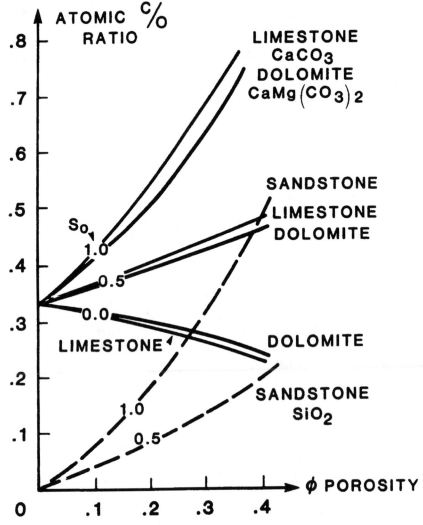

Figure 7.1 Atomic C/O ratio for sandstone and carbonates (Courtesy Schlumberger)

Carbon/Oxygen Equipment

The service companies which run carbon/oxygen equipment and the name of their respective equipment is indicated below. The following Table 7.1 shows a list of equipment 3 3/8 in. (8.57 cm) or larger. Unless tubing is quite large, this group of tools is not designed for through tubing operation. The * indicates that the name and letter designations are marks of the company with which they are associated.

A new generation of equipment is being introduced which is to be capable of making both the C/O and a quality capture measurements in thru-tubing sizes. The Schlumberger RST tool, originally designed for C/O measurements, is now being programmed for an alternate neutron burst pulse sequence for capture logging. The other companies have upgraded their capture tools to make a spectral measurement. These thru-tubing tools are listed below in Table 7.2.

These tools are rated to at least 300 °F (150 °C) and 13,500 psi (93.1 mPa). If operating anywhere near to these conditions, contact your service company for specific ratings and limitations.

METHODS OF OPERATION TO GET CARBON/OXYGEN RATIO

Pulsing Sequence

All carbon/oxygen tools are pulsed neutron tools using a minitron or similar neutron source like those used in the capture tools. The tools emit periodic bursts of high energy 14 million electron volts (mev) neutrons at a much higher frequency than capture tools. The period between bursts is 50 microseconds for the MSI C/O and 100 microseconds for the PSG, GST, and RST. Measurements are made in both the inelastic and capture modes.[3]

The so called inelastic gamma rays arise from interactions between these high energy neutrons and atoms of the wellbore and formation in the vicinity of the source. The energy spectrum of these inelastic gamma rays is measured within a timed gate which is open during the neutron burst period.

Within a few microseconds after the source is turned off, inelastic collisions cease and the neutrons slow down to their thermal state. Gamma rays now arise primarily in the thermal or capture mode.[4] The gamma ray energies in this mode are similar to those detected by pulsed neutron capture logging tools. A timed gate(s) is opened during this

TABLE 7.1 Large Diameter C/O Tools

Schlumberger
  • Gamma Spectrometry Tool*, GST*, 3 5/8 in. (9.21 cm)
Western Atlas
  • Multiparameter Spectroscopy Instrument-Continuous
      Carbon/Oxygen Log*, MSI C/O*, 3 1/2 in. (8.89 cm)
Halliburton Energy Services
  • Pulsed Spectral Gamma Log*, PSG*, 3 3/8 in. (8.57 cm)

TABLE 7.2 Through Tubing Pulsed Neutron Tools with Spectral Capability

Schlumberger
  • Reservoir Saturation Tool*, RST*
      RST-A*: 1 11/16 in. (4.29 cm)
      RST-B*: 2 1/2 in. (6.35 cm)
Halliburton
  • Thermal Multigate Decay-Lithology tool*, TMD-L*
      1 11/16 in. (4.29 cm)
Computalog
  • Pulsed Neutron Decay-Spectrum*, PND-S*
      1 11/16 in. (4.29 cm) and 1 5/8 in. (4.13 cm)

capture period between bursts to measure the energy spectrum of gamma rays arising from the capture mode. Since this capture mode continues for nearly 1,000 microseconds, its gamma rays also appear in the inelastic spectrum measured during the burst period. This capture spectrum must be subtracted from the inelastic spectrum to correct it for these capture counts in the background. In this manner, the gamma ray energies in the corrected inelastic spectrum are resulting only from inelastic collisions and not capture events.

Figure 7.2 shows the source and gating sequence for the Schlumberger GST tool.[5] The inelastic gate is open during the neutron source burst period, after which a capture gate opens to measure background counts. The spectrum of the gamma ray energies is measured during these gates and the capture spectrum is subtracted from the inelastic spectrum as shown. There are two such capture gates between neutron bursts. The second is used to measure the formation capture cross section. This is not a good quality measurement and cannot substitute for the superior formation capture measurement available on the new thru-tubing C/O tools.

Windows Method of Obtaining C/O Ratio

The measurements taken during the gating period are somewhat like those shown in Figure 7.3.[9,10] These are the measured inelastic and capture gamma ray energy spectrums. A carbon/oxygen ratio is derived from the inelastic spectrum by setting windows to collect counts associated with carbon and oxygen energy levels.[7,8] The inelastic spectrum shown on Figure 7.3 shows the Halliburton positions of the C, O, Ca, and Si windows.[9,10] Notice that one spectrum is 5x the other. From this inelastic spectrum, both the C/O and Ca/Si ratios are obtained. Due to the intrinsic resolution of the detectors, the competing reaction processes, and to Compton scattering of the gamma rays by atoms in the vicinity of the detector, the peaks of the spectrum are not sharp, but are spread out. As a result, any gate selected will have counts from spurious sources. Therefore, a C/O tool's response will have to be calibrated against known formation types with known saturations. This technique is discussed later in this section.

Figure 7.3 shows the difficulty inherent in the C/O measurement. Two solid curves are shown, one for a 36 porosity unit (p.u.) oil sand and the other for a 35 p.u. water sand. The

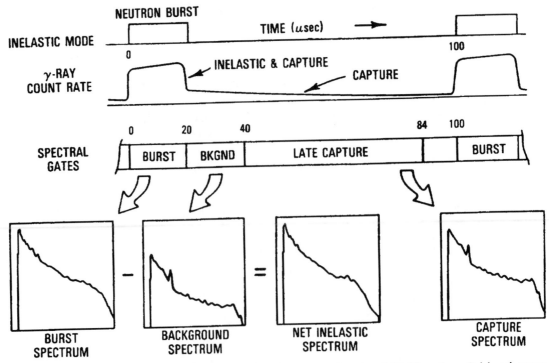

Figure 7.2 Source and gating sequence for the Schlumberger GST (Courtesy Schlumberger, Ref. 5)

Inelastic Spectrum

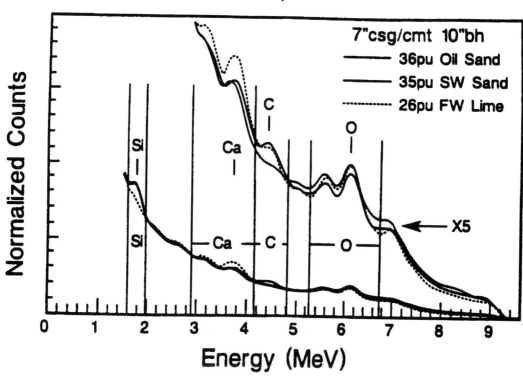

Capture Spectrum

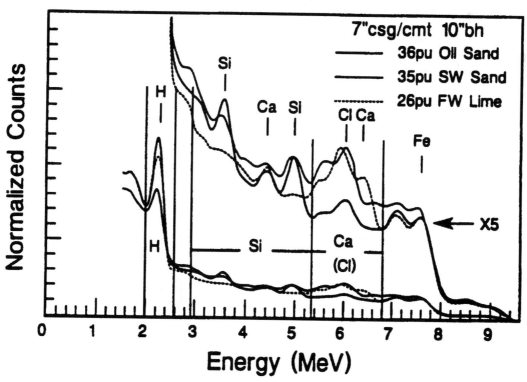

Figure 7.3 Inelastic and capture spectra, with Halliburton Energy Services PSG gate positions (Reproduced with permission of Halliburton Co. and SPWLA, Ref. 9)

oil sand shows increased counts in the carbon gate and fewer counts in the oxygen gate. However, the differences are quite small, and this spectrum shows the total dynamic range for the conditions shown. The C/O ratio is based on the areas within the windows.

The energy spectrum in the capture mode is also shown on Figure 7.3. It shows the Halliburton gating scheme for the capture spectrum. Calcium and silicon gates are also shown, but these are more sensitive to salinity in the capture mode. Hydrogen, H; chlorine, Cl; and iron, Fe, are also shown on this spectrum. Sulfur, S, is also commonly detected. The well defined hydrogen and iron peaks are used for calibration points of the energy levels in the spectrum.

When gates like those above are used, the actual C/O and Ca/Si ratios observed must be correlated back to empirical data. Such correlation charts for the Halliburton PSG are shown on Figure 7.4. and are unique to the type of equipment used.[9,10] Families of such charts exist for various borehole environments. It is clear that if porosity is known, the fraction of carbonate may be determined from the Ca/Si ratio, and then the saturation from the appropriate positioning of the fan on the C/O chart.

For example, suppose the PSG indicates a Ca/Si ratio of 1.43 and a C/O ratio of .45 over a 30 p.u. interval with 7 in. (17.8 c.m) casing and fresh water cement in a 10 in. (25.4 cm) borehole. From the chart of Figure 7.4, part b, Vca is determined to equal zero, hence the interval is 100% sandstone. From the chart in Figure 7.4, part a, the C/O ratio clearly shows this to be a water zone. Higher in the interval, the Ca/Si remains about the same, but now the C/O reads .53. From the chart in Figure 7.4, part a, this is now determined to be an oil interval having a saturation of about So=.80.

Note that the carbonate/sandstone proportions may be determined from either the inelastic or capture spectrum. Each has disadvantages. The inelastic spectrum has few counts and a very shallow depth of investigation. The capture has more counts and a greater depth of investigation, but may be contaminated by chlorides. Most saturation computations are done using the inelastic Si/Ca information because its depth of investigation matches that of the C/O measurement.

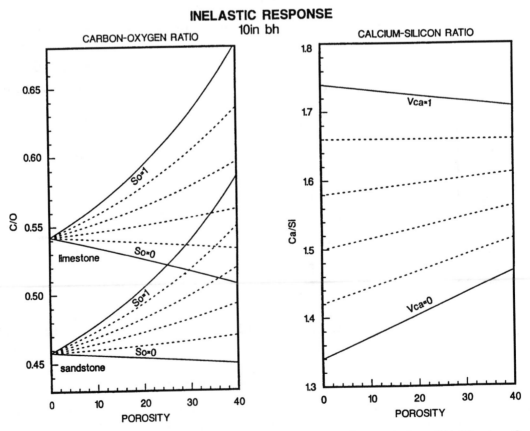

Figure 7.4 PSG inelastic response charts (Reproduced with permission of Halliburton Co. and SPWLA, Ref. 10)

   Ideally, good open hole information should be available for such things as porosity, shale volume, salinity, and the like. If not, then the capture data may be used for evaluation of porosity using hydrogen as an indicator, shale volume using iron as an indicator, and salinity using chlorine as an indicator. These, however, are not good quantitative indicators. As a result, C/O surveys are often run with a compensated neutron log (CNL) for a better porosity and a spectra natural gamma ray for better shale volume indications.

Elemental Yields Method

   Both Schlumberger and Halliburton utilize this technique. The spectra from both the net inelastic and capture count time gates are used to compute yields or relative concentrations of elements downhole. These tools also compute a thermal decay time and hence formation capture cross section during the period between bursts.

   Compared to the gamma ray energy windows technique for analysis, the elemental yield technique is more complex. In both the inelastic and capture modes, elemental fitting standards are developed for the tool.[5, 11] These responses correspond to the tool's response to a formation comprised of 100% of that element. For example, Figures 7.5a and 7.5b show the elemental standards for elements in both the capture and inelastic modes.[5,11] The carbon and oxygen elemental standards are shown in the inelastic standards of Figure 7.5a, along with silicon, iron, calcium, and sulfur. Figure 7.5b shows the elemental standards in the capture mode. The standards include iron, Fe; chlorine, Cl; calcium, Ca; silicon, Si; sulfur, S; and hydrogen, H.

   When the inelastic spectrum is evaluated, it is processed using an algorithm which least square best fits the inelastic elemental standards over the 256 energy gates of the spectrum measurement. This best fit then determines the elemental yields of each of the elements contributing to the spectrum. A similar measurement is made for the capture mode.

   Accurate information from open hole logs and regarding casing and cement is helpful in improving the accuracy of such evaluations. When other information is not available, the C/O tool is frequently run with a CNL for porosity and spectral gamma ray for shale identification.

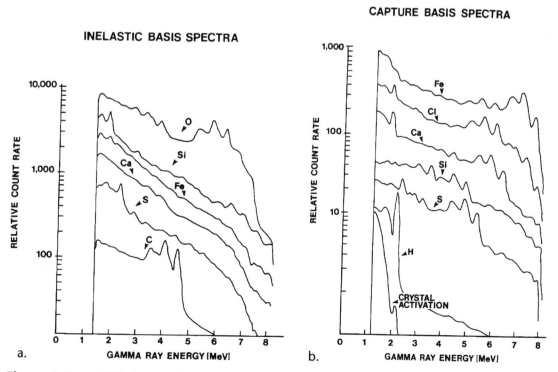

Figure 7.5 a. GST elemental standards for inelastic basis spectra and b. GST elemental standards for capture basis spectra (Courtesy Schlumberger, Ref. 5)

Depth of Investigation and Logging Considerations

Some C/O logging programs consist of separate logging passes (same tool in the hole) for the inelastic and capture modes. The capture passes are made first at logging speeds of about 10 ft/min. (3 m/min.), followed by the inelastic passes which are taken as stationary readings at discreet depths or at speeds of 1-3 ft/min. (.3-1.0 m/min.). The Halliburton PSG takes both the inelastic and capture spectrum during the same logging pass. This pass is typically done at speeds of about 5 ft/min. (1.5 m/min.). As with other nuclear devices, multiple passes are made in each mode to reduce statistics inherent in the nuclear measurement. Carbon/Oxygen logs are run very slowly and are quite time consuming.

The depth of investigation for carbon oxygen tools is quite shallow, typically 4 to 7 in. (12.7 to 17.8 cm). This is much less than the depth of investigation for a capture tool. As a result, good hole conditions are required. You should have good cement coverage and centralized casing over the interval to be logged. If the well is to be shut in, beware of wellbore fluids invading the zones to be logged. Windows or elemental yields, this is an inherently difficult measurement, and the more certain knowledge one has about the borehole region, the better.

With larger diameter tools, the tubing must be removed before logging. This is an expensive operation. Furthermore, production is lost during this time and the time of logging, with the risk that the well will not come back onto production. Shutting in also means allowing borehole fluid to invade those zones in communication with the wellbore, thereby rendering logs run across such intervals to be questionable.

INTERPRETATION AND PRESENTATION OF CARBON/OXYGEN LOGS

C/O and Ca/Si Charts

This is a basic and accurate technique to get saturations, and was discussed earlier in the windows method section. All of the service companies perform an interpretation similar to it. If porosity is known, the matrix is determined from the Ca/Si ratios and the fan chart is positioned accordingly on the C/O porosity plot. The C/O ratio yields saturation directly from this appropriately positioned fan chart.

Clean Uniform Zones

When zones are clean with constant lithology and porosity, the C/O ratio curve can be used to estimate saturation.[12] This quick look technique assumes a linear relationship between C/O ratio and saturation. The oil saturation can be estimated from the following equation:

$$S_o = \frac{(C/O)_{LOG} - (C/O)_w}{(C/O)_o - (C/O)_w}$$
(Equation 7.2)

where $(C/O)_o$ is the ratio in an oil saturated zone and $(C/O)_w$ is the ratio in a water saturated zone. An estimate of $(C/O)_o$ may be made by estimating the oil saturation in an oil zone and solving the above equation for $(C/O)_o$. This technique cannot work when changing lithologies and porosities are present.

Overlay Method

Test pit data developed by Atlas shows that there is an approximately linear relationship between the C/O and Si/Ca ratios at constant saturations.[13] This linear relationship is the basis of the overlay technique. This relationship is shown on Figure 7.6.[13] In this figure, note that the water line is nearly the same for a variety of porosities. Due to this linear relation, the C/O will overlay a reversed and appropriately scaled Si/Ca. Sometimes Ca/Si is presented to overlay the C/O curve. When this is done, the reversed Si/Ca or Ca/Si is scaled to overlay the C/O in a water zone. When the C/O moves to the right of this overlay, this is an indication of excessive carbon counts and hence the presence of a hydrocarbon. In

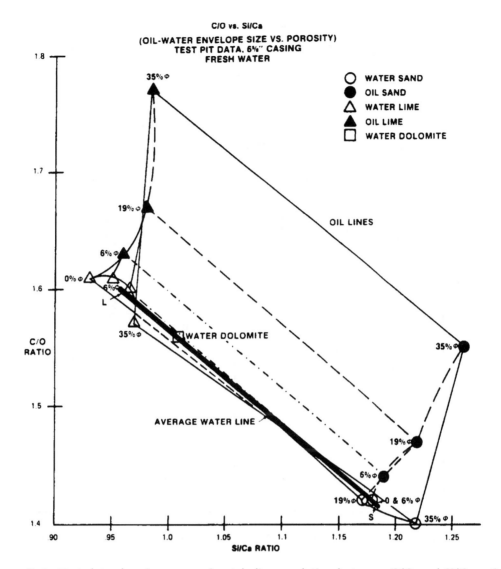

Figure 7.6 Test data showing approximately linear relation between C/O and Si/Ca ratios (Courtesy Western Atlas, Copyright SPE, Ref. 13)

some presentations, Si/Ca is presented and the C/O is reversed. Excessive carbon counts increase to the left in this case.

Figure 7.7 shows an Atlas C/O log with the C/O curve and the reversed Si/Ca overlay.[13] The reversed Si/Ca is scaled to overlay in the water zones. In this example, there are clearly separations of the two curves. The Iwa curve, which is related to the water saturation, shows a shift to the right indicating decreased water saturation. The overlay technique can be used quantitatively, but such applications are beyond the scope of this manual. Such interpretations depend on the individual company's tool design and response characteristics.

Yield Ratio Presentation

Rather than present logs for eight elemental yields, a series of ratios may be presented. These ratios, used by the GST, are shown on Table 7.3 below.[14,15] The Halliburton ratios are slightly different, especially the lithology indicator ratio. The C/O comes from the inelastic run while the others come from the capture run. These ratios make it simpler to relate the yields to formation parameters of interest.

Figure 7.8 shows a GST run in both the inelastic and capture modes.[14] The various indicator ratios are shown along with the capture cross section, Σ, and the sulfur yield. The capture mode was run at 10 ft/min. (3 m/min.) and the inelastic measurements was taken

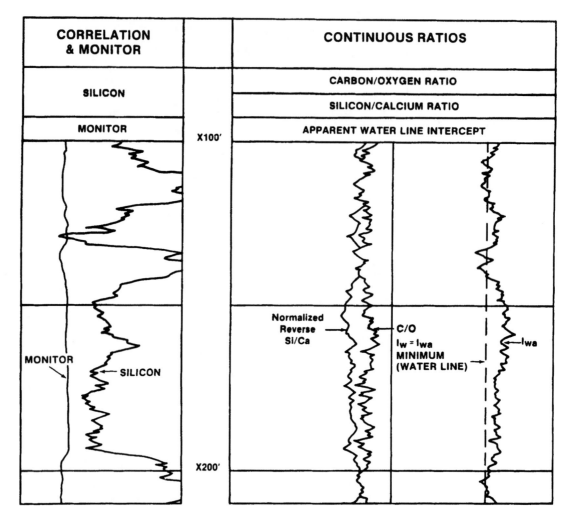

Figure 7.7 Example of C/O and reversed Si/Ca overlay method (Courtesy Western Atlas, Copyright SPE, Ref. 13)

TABLE 7.3

Yield Ratio	Interaction	Name	Label
C/O	Inelastic	Carbon-Oxygen Ratio	COR
CI/H	Capture	Salinity-Indicator Ratio	SIR
H/(Si+Ca)	Capture	Porosity-Indicator Ratio	PIR
Fe/(Si+Ca)	Capture	Iron-Indicator Ratio	IIR
Si/(Si+Ca)	Capture and Inelastic	Lithology-Indicator Ratio	LIR

by a series of closely spaced five minute stationary readings. Comparison of this log with a borehole fluid identification device (gradiomanometer) run before the GST indicates that the borehole oil/water contact is moving during the logging runs. As a result, its position is different between the inelastic and capture runs.

This log shows:

- The LIR, reading near zero over the interval logged, shows this to be a carbonate. It is, in fact, a dolomitic limestone.
- The higher sulfur yield and higher sigma above 4,942 ft. are indicative of the anhydrite caprock.
- The COR increases from .20 to .27 at 4,986 ft, indicating the water/oil contact in the formation. This is confirmed by a simultaneous decrease in sigma and decrease in SIR.

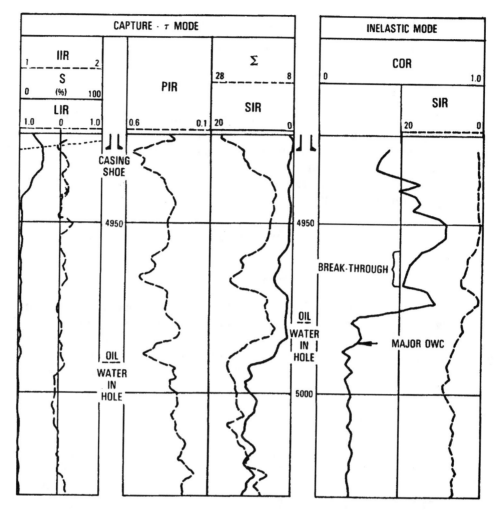

Figure 7.8 Yield ratio presentation of GST spectral data (Courtesy Schlumberger, Ref. 14)

- Above the wellbore fluid interface, the high COR, along with low sigma and SIR, shows undepleted hydrocarbons.
- Premature breakthrough is indicated from 4,970 to 4,959 ft, as indicated by the low COR and increases in the sigma SIR. Notice that this breakthrough occurs in a high porosity stringer as indicated by the PIR.
- The IIR is visible only at the top of the logged section. It readily shows the change from tubing to casing.

Carbon/Oxygen Envelope Presentation

The C/O ratio can be related to formation parameters with the following equation (see the first page of this section):[1,16]

$$C/O = A \frac{\alpha(1 - \phi) + \beta\phi S_o + B_c}{\alpha(1 - \phi) + \delta\phi S_w + B_o} \qquad \text{(Equation 7.3)}$$

where

A = Ratio of average C and O fast neutron (gamma ray producing) cross sections

Bc, Bo = C and O contributions from borehole

Sw, So = Water and oil saturation

α = Atomic concentration of C in matrix

β = Atomic concentration of C in formation fluid

γ = Atomic concentration of O in matrix

δ = Atomic concentration of O in formation fluid

ϕ = Porosity

To compute the C/O envelope, the minimum and maximum values of the C/O ratio must be calculated given the lithology and porosity. These are preferably known from external sources, such as open hole logs, but may be determined from the capture elemental yields. C/Omin occurs at 100% water saturation, i.e., Sw=1.0 and So=0. C/Omax occurs at 100% oil saturation, i.e., Sw=0 and So=1.0. C/O$_{min.}$ and C/O$_{max}$ are expressed in the following equations:

$$C/O_{MIN} = A \frac{\alpha(1 - \phi) + B_c}{\gamma(1 - \phi) + \delta\phi + B_o} \qquad \text{(Equation 7.4)}$$

$$C/O_{MAX} = A \frac{\alpha(1 - \phi) + \beta\phi + B_c}{\gamma(1 - \phi) + B_o} \qquad \text{(Equation 7.5)}$$

Figure 7.9 shows the formation of the C/O envelope. For this example, assume that the borehole fluid is water. The lower section is a pure sandstone, SiO2. Over the sandstone interval, the C/Omin line is straight and reads zero since the pore fluids are water and no formation carbon is present. The C/Omax line in the sandstone interval falls to the right of

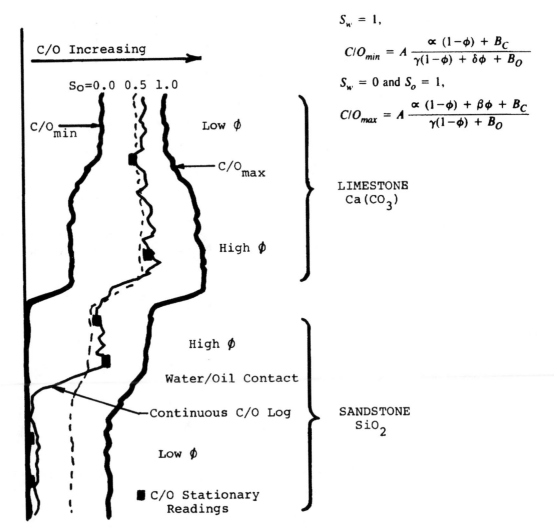

Figure 7.9 Effects of matrix and porosity on C/O envelope

C/Omin since carbon is present in the pore fluid. The carbon content increases with porosity and therefore the C/Omax line also increases with porosity. When moving into the limestone, the C/Omin line shifts to the right due to the carbon content of the carbonate, even though the pore fluid is water. Notice here that higher porosity causes a reduction of carbonate matrix and hence C/Omin is slightly lower in higher porosity. When hydrocarbon is the pore fluid, the C/Omax is again displaced to the right, more so with greater porosity. Now, carbons fill the pore space and it becomes the dominant factor. The thin line in the middle of the envelope corresponds to the 50 percent saturation point, So=Sw=.5.

The actual C/O log may be run continuously or by stationary measurements. Such a log is plotted with the envelope in Figure 7.9. When run stationary, as is common with the GST in certain areas, the measurements are shown as small black squares. Clearly the lower two points show water, falling on the C/Omin line, while the others show hydrocarbon, at oil saturations of about 50% to 65%. If the porosity is high enough and the interval relatively clean, the C/O log may be run continuously. The solid line depicts such a log. It shows a water oil contact about midway up the sandstone interval, just below the point of increased porosity.

Figure 7.10 shows the C/O envelope with stationary readings taken between the depth of 700 and 800 ft. With the exception of the two lowest readings and perhaps two more above at 758, the stationary readings indicate hydrocarbons with saturations of So>.5. Note that the interval tested is primarily sandstone with less than 20% carbonate by volume. Little carbonate is present in the interval above 700 ft., which is quite shaley.

THROUGH TUBING CARBON/OXYGEN LOGGING

Reservoir Saturation Tool (RST)

In 1991 Schlumberger introduced a small diameter through-tubing carbon/oxygen tool. It is available in 2 1/2 in. and 1 11/16 in. diameter sizes and is called the Reservoir Saturation Tool, RST-B and RST-A, respectively. This tool differs from other tools in that it has two detectors, one which evaluates the borehole C/O and hence borehole fluid holdup (water/oil fraction by volume), and the other for evaluation of the formation C/O ratio, as well as other parameters available from other C/O instruments. The small diameter is made possible through the use of high-density gadolinium oxyorthosilicate (GSO) scintillation gamma ray detectors and other hardware improvements.[17,18]

The tool configuration is shown on Figure 7.11 for both the 1 11/16 in. and 2 1/2 in. sizes. The extra shielding in the 2 1/2 in. size allows this tool to be run in producing wells where water and oil are both being produced and flowing in the well-bore. Such two-phase flow must be well mixed for the RST to be effective. Measurements may be made in the inelastic mode and in the capture-sigma modes, just as in the GST. The tool can also be run in the sigma mode alone, thereby imitating a TDT.

RST Interpretation Technique

C/O ratio interpretation charts for the RST using both the near and far detectors for a limestone formation having a porosity of 43 p.u. are shown on Figure 7.12.[19] These charts are for 7 in. (17.8 c.m) cemented casing in a borehole size of 8 1/2 in. (21.6 cm). A family of such charts would be required for a proper interpretation. The family of charts would be a function of porosity, matrix, and borehole/casing sizes. The upper and lower charts of Figure 7.12 are for the RST-A and RST-B, respectively, chart. It is clear that the near detector is strongly responsive to borehole fluids while the far detector is affected significantly by both borehole and formation fluids.[17, 18, 19]

To use the charts of Figure 7.12, the C/O response from the near and far detectors are cross-plotted on the appropriate chart. Oil holdup, y_o, increases from 0 to 1 from left to right across the quadrangle and oil saturation increases from 0 to 1 from the bottom to top across the quadrangle. For example, if an RST-B near C/O reads .2 and far C/O reads .3, this would indicate an oil holdup in the borehole of about .25 and a formation oil saturation of about .60.

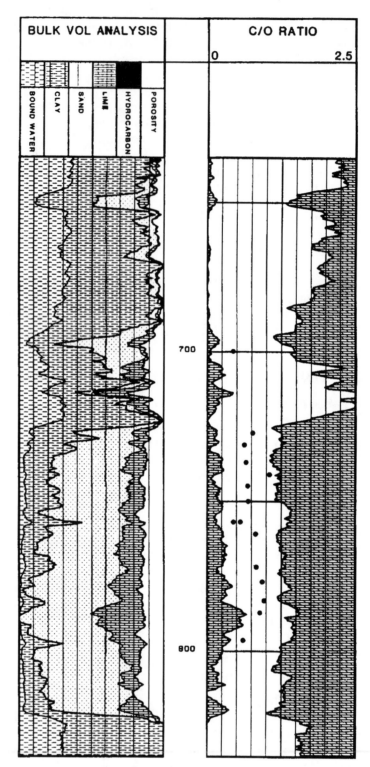

Figure 7.10 Example of Schlumberger GST stationary tests and C/O envelope (Courtesy Schlumberger)

RST Sondes

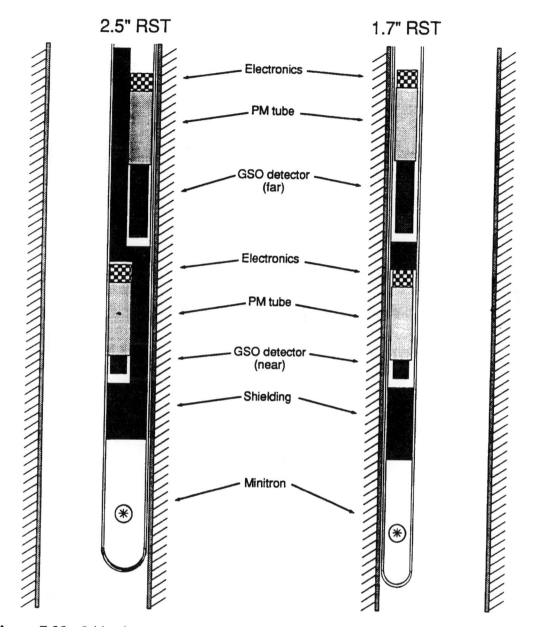

Figure 7.11   Schlumberger Reservoir Saturation Tool (RST) (Courtesy Schlumberger)

RST Data Presentation

The log and cross plot example of Figures 7.13 and 7.14 show results from a smaller RST-A.[20] This example is from a Middle East well, shut in and filled with formation water over the interval logged. The borehole was 6.5 in. (16.5 cm) with 4.5 in. (11.4 cm) 9.5 lbm/ft (14.14 kg/m) casing. The model chart used for the logged interval assumes a limestone of 28 p.u. overlaying a sandstone of 33 p.u. The cross-plotted data is shown on Figure 13, with the upper quadrilateral being suitable for the limestone and the lower one suitable for the sandstone. Across the sandstone, the cluster of points is concentrated at an area indicating water in the borehole and in the formation. The upper cluster, which is plotted on the limestone quadrilateral, shows data points clustered to indicate water in the borehole, and significant oil in the formation.

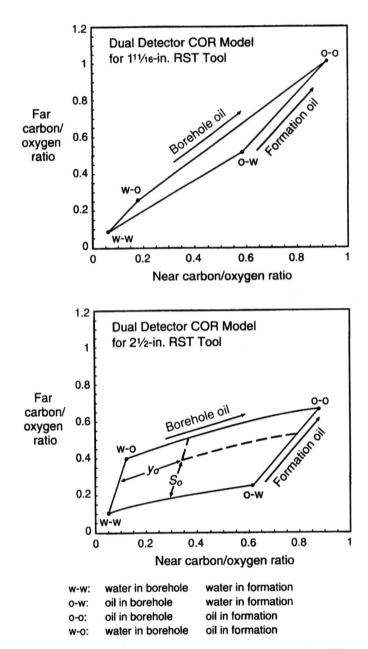

Figure 7.12 Effects of formation oil saturation and borehole oil holdup on RST response (Courtesy Schlumberger, Ref. 19)

The log of Figure 7.14 shows the interpretation based on the crossplot. The C/O ratios (COR) for the near and far detectors are shown in the left-hand track with the lithology from open hole date on the right-hand track. The middle track show the original pore volume analysis based on open hole logs. The white area is water while the shaded area is the original oil in place. The heavy black area shows the oil in place detected by the RST. This log clearly shows significant depletion all across the original oil bearing limestone, except for the very top 10 feet (3 m) and another interval about 20 feet (6 m) below that.

Computalog PND-S and CATO Measurement

Computalog introduced the Pulsed Neutron Decay-Spectrum (PND-S) tool to measure both capture and inelastic gamma ray parameters. This tool uses a NaI crystal and would generally not offer the resolution necessary to make conventional C/O ratio measurements.

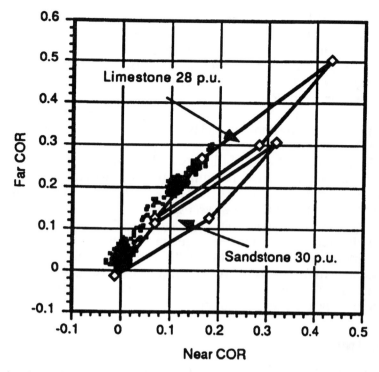

Figure 7.13 Cross plot evaluation of RST-A data in sandstone and limestone layers (Courtesy Schlumberger, Ref. 20)

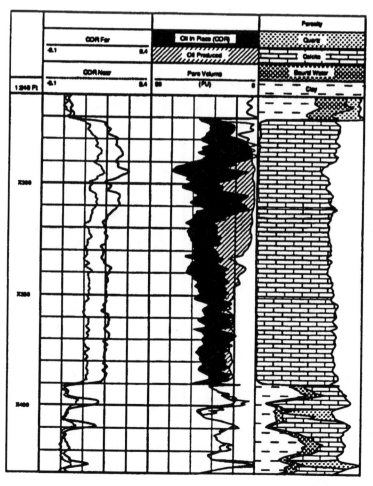

Figure 7.14 Example of an RST-A log presentation (Courtesy Schlumberger, Ref. 20)

Instead, the PND-S measures a parameter called CATO. CATO is a ratio of inelastic gamma ray energies less than 4.5 mev, those associated with carbon, silica, calcium, etc., divided by those inelastic gamma ray energies greater than 4.5 mev which are associated with oxygen.[21,22] So, CATO is the ratio of all inelastic gamma rays (excluding oxygen) to those of oxygen. Note that counts used for the CATO ratio are normalized for background using a proprietary technique.

The response of the CATO measurement is illustrated on Figures 7.15 and 7.16.[21] Figure 7.15 shows the response of inelastic oxygen, total inelastic, and CATO in a Texaco test well. CATO shows a clear and strong response difference between oil and water sands. Figure 7.16 shows the effect of lithology and porosity on the response. The effect of sandstone (SS), and limestone (LS), is minimized with the CATO measurement. On the basis of these data, it is claimed that the CATO measurement provides a much improved dynamic range over conventional C/O measurements.

Figure 7.17 shows a PND-S log example.[21] The left-hand track shows the gamma ray, the middle track CATO, and the right track the bulk volume analysis. For this example, the Clearfork Limestone formation of Texas, average porosity is 15 p.u. Open hole porosity was available from the client. Water salinity was less than 20,000 ppm NaCl. The casing size is 5 1/2 in. (14.0 cm) in a 7 7/8 in. (20.0 cm) borehole. The zone from 5,146 to 5,157 shows 15 to 20 percent water saturation.

Halliburton TMD-L

The Halliburton TMD-L is clearly a transition tool primarily designed for capture, but adapted for rudimentary spectral evaluation.[23] At present, a weighted least squares fitting

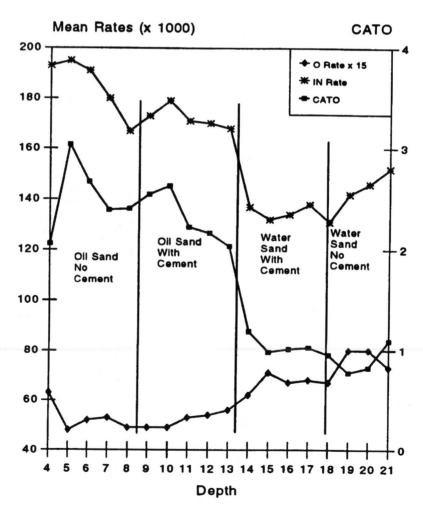

Figure 7.15 Response of Computalog CATO measurement in test well (Copyright SPE, Ref. 21)

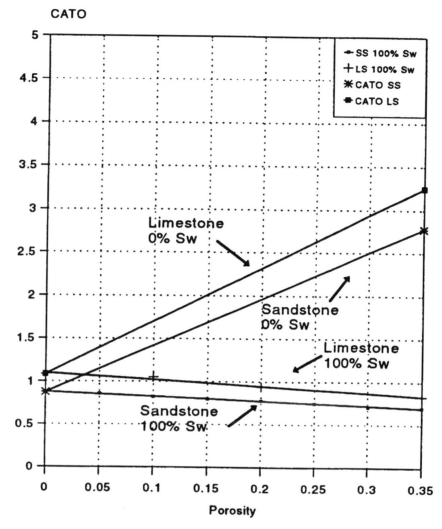

Figure 7.16 CATO response in sandstone and limestone (Copyright SPE, Ref. 21)

to obtain relative elemental yields from the capture spectrum for Fe, Si, and Ca+Cl can be done. At this time, only examples of the Si yield have been published. An example of the Halliburton TMD-L is shown on Figure 7.18.[23] Of the five tracks shown, not counting the depth track on the far left, the left track contains the gamma ray, GR, along with the silicon index, SI. The higher SI readings shows zones C and D to be limes and zone F to be a limey-sand while zones A, B, and E are sandstone. Track 2 shows the gas indicator, which is the ratio of inelastic count rate at the far detector to the near count rate (capture). It shows little or no gas over this interval. RNFC is similar to the N/F ratio, but corrected for sigma. Tracks 3 and 4 show the borehole and formation capture cross sections, respectively, which Σ unc shows the statistical uncertainty of Σ_{FM}. Track 5 shows the near, NTMD, and far, FTMD, count rates along with the inelastic count rate from the far detector.

EXAMPLES OF CARBON/OXYGEN LOG APPLICATIONS

C/O Envelope or Alley in a Sandstone Reservoir

Figure 7.19 shows a C/O envelope in the second small track. On the right, the bulk volume analysis indicates that this interval is essentially sandstone, with a few lime streaks and only very small amounts of lime and shale throughout. Water saturated sandstone theoretically has no carbon and hence the tool should read C/Omin and be an essentially

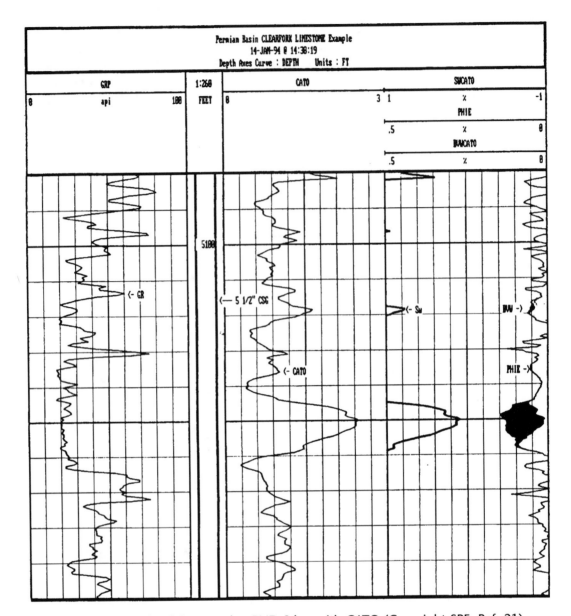

Figure 7.17 Example of Computalog PND-S log with CATO (Copyright SPE, Ref. 21)

straight line. Except for deviations due to the small amount of limestone and the two large lime streaks, this is nearly the case. The C/Omax line is a function of the amount of limestone and pore volume since that porosity would now be filled with oil. Notice the dramatic effect that the limestone has on the c/o envelope.

Stationary measurements were taken with a Schlumberger GST over this interval and the black squares show the measure depths. In this example, the GST indicates water breakthrough at about 3,010, 3,030, and 3,180, with some nearby points looking quite wet.

Detection of Water Influx

The log of Figure 7.20 is from the Ivishak formation in Prudhoe Bay.[24] The water saturation from open hole logs is shown on the right-hand track. This well was not immediately perforated, although intervals were proposed. When time came to perforate, stationary GST measurements were made and showed that the well had an influx of water over the interval from about 9,220 to 9,260 due to significantly increased water saturations over this interval. As a result, only the upper two sets of perforation were done.

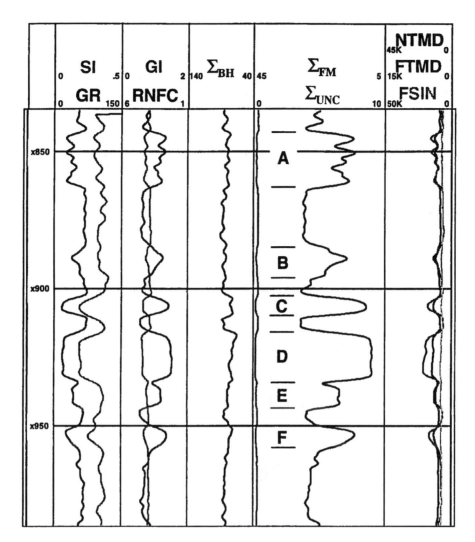

Figure 7.18 Example presentation of Halliburton Energy Services TMD-L log (Reproduced with permission of Halliburton Co. and SPWLA, Ref. 23)

Detection of Injected Water Breakthrough

The example of Figure 7.21 shows data from a 2 1/2 in. (6.4 cm) RST- B run in an observation well in Saudi Arabia.[25] The borehole fluid is water and the RST shows the saturation indicated in the pore volume track under the "Carbon Oxygen" heading. The C/O measurement here has provided a measure of the oil volume. If injected water breakthrough had occurred, the remaining volume after the oil is subtracted from the porosity would be a mixture of formation water and injected water. Furthermore, the RST measures a formation sigma. If these waters are of differing salinities, their volumes may be estimated by solving the following two equations simultaneously for V_w and V_{Winj}:

$$\Sigma_{LOG} = (1 - \phi)\Sigma_M + V_O\Sigma_O + V_W\Sigma_W + V_{Winj}\Sigma_{Winj} \qquad \text{(Equation 7.5)}$$

$$\phi = V_O + V_W + V_{Winj} \qquad \text{(Equation 7.6)}$$

Such a computation has yielded the zones of injection water breakthrough on Figure 7.21 as shown on the right under the heading "carbon/Oxygen and Σ." The upper zone, with a breakthrough of fresher water, would have been interpreted as hydrocarbon with only capture measurements. This combined C/O and capture capability not only shows correct saturations, but now shows where injected waters are breaking through.

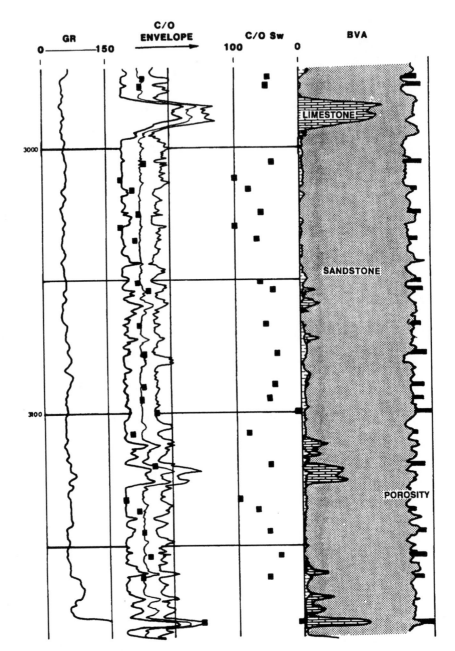

Figure 7.19 GST measurements plotted on C/O envelope, with saturation and bulk volume analysis (Courtesy Schlumberger)

C/O Response to Coal Beds

The Atlas continuous C/O log was run over the McAlester formation of Oklahoma. This formation contains a number of coal beds. The log of Figure 7.22 clearly shows these coal beds on the C/O log.[26] Apparently, the atomic ratio of carbon to oxygen can be correlated with the rank and Btu content of coals.[26]

C/O Used for Steam Flood Monitoring

The example of Figure 7.23 is an Atlas MSI-C/O log used to monitor steam flood performance in heavy oil reservoirs.[27] This figure shows the hydrocarbon analysis for three observation wells spaced at 40, 160, and 325 ft. (14.6, 58.3, 118.5 m) from the steam injection

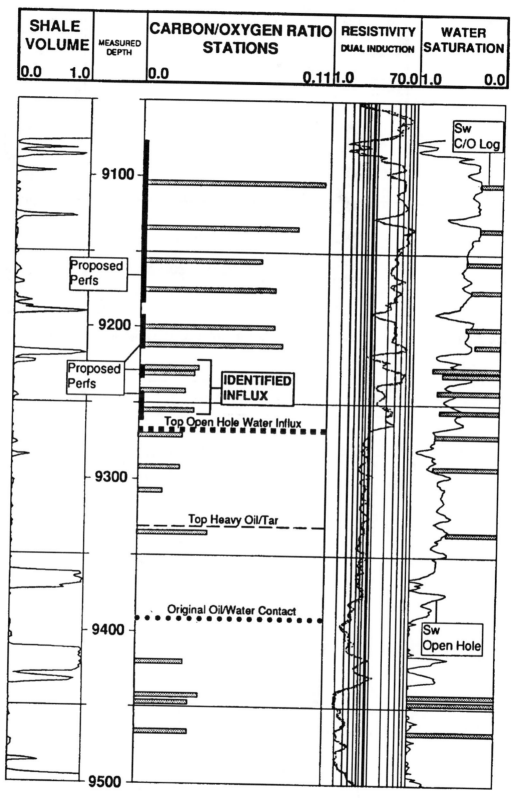

Figure 7.20 Detection of water influx with a GST (Copyright SPE, Ref. 24)

well. The nearest well has clearly been effectively swept except for the top 10 feet of the interval. The intermediate well shows significant steam development over the lower half of the hydrocarbon interval, and the farthest well shows steam just breaking into this zone. The flood front movement both laterally and vertically is being monitored using the C/O log in these observation wells.

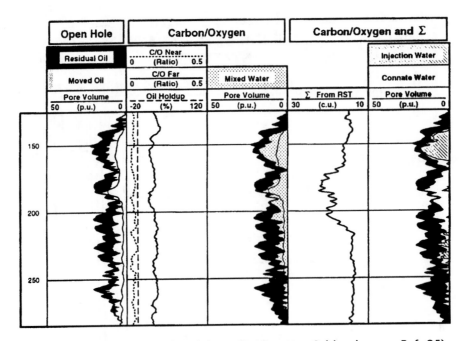

Figure 7.21 Detection of injected water breakthrough (Courtesy Schlumberger, Ref. 25)

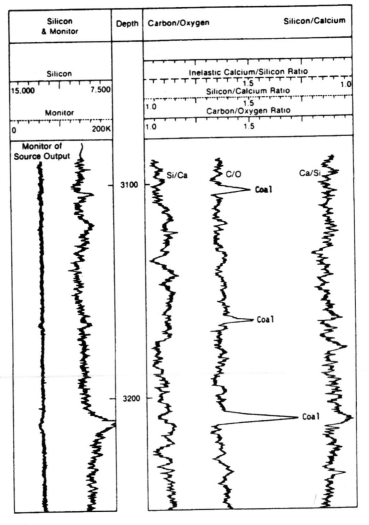

Figure 7.22 C/O response to coal beds (Copyright SPE, Ref. 26)

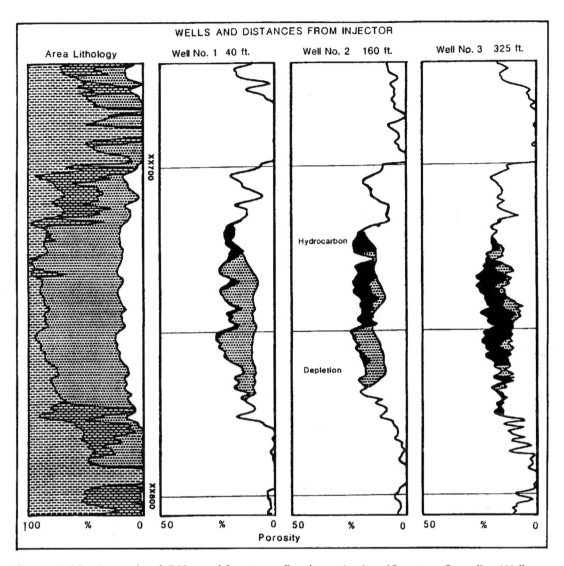

Figure 7.23 Example of C/O used for steam flood monitoring (Courtesy Canadian Well Logging Society, Ref. 27)

POINTS TO REMEMBER

- C/O logging is useful to determine saturations when formation waters are of low, unknown, or mixed salinity.
- C/O is useful to monitor water floods, steam floods, CO_2 floods and the like.
- C/O is useful for locating coal beds.
- Carbonates contribute to the carbon/oxygen ratio, and therefore sandstones must be differentiated from carbonates.
- Tools are typically capable of evaluating relative elemental yields of C, O, Ca, Si, Fe, S, H, and Cl.
- The C and O counts come from the inelastic spectrum.
- The inelastic spectrum must be corrected for background.
- While basic equipment is larger diameter, newer tools are available in thru-tubing sizes.
- Two techniques are used to obtain a C/O ratio, the Windows technique and the Elemental Yields technique.
- The inelastic data must be gathered at very slow logging speeds, typically from 0 to 5 ft/min. (0 to 1.5 m/min.).
- Multiple passes are typically required, typically 3 to 5 logging passes.

- Depth of investigation is very shallow compared to capture measurements. Typical depth of investigation is 5 to 7 in. (12.7 to 17.8 cm)
- Small diameter tools are extremely desirable since they do not require pulling tubing and shutting in the well.
- While numerous interpretation techniques are available, the C/O ratio must fall between $C/O_{min.}$ and C/O_{max} in the C/O envelope presentation.
- The C/O alley or envelope is affected by lithology and porosity, as well as the fluid in the pore space.
- The RST, especially the RST- B, is capable of performing a formation C/O evaluation even when two phase flow is in the wellbore.
- The CATO measurement is new and appears to have a larger dynamic range than C/O.

References

1. Gilchrist, W. A. Jr., Rogers, L. T., and Watson, J. T., "Carbon/Oxygen Interpretation—A Theoretical Model," SPWLA 24th Annual Logging Symposium, June, 1983.
2. Woodhouse, R., and Kerr, S. A., "The Evaluation of Oil Saturation through Casing Using Carbon/Oxygen Logs," *The Log Analyst,* Jan.–Feb., 1992.
3. Oliver, D. W., Frost, E., and Fertl, W. H., "Continuous Carbon/Oxygen (C/O) Logging—Instrumentation, Interpretive Concepts and Field Applications," SPWLA 22nd Annual Logging Symposium, June, 1981.
4. Wichmann, P. A., Hopkinson, E. C., and McWhirter, V. C., "The Carbon/Oxygen Log Measurement," Publication 3324, Dresser Atlas, Houston, Texas, Sept., 1976.
5. Hertzog, R.C., "Laboratory and Field Evaluation of an Inelastic-Neutron-Scattering and Capture Gamma Ray Spectroscopy Tool," Paper SPE 7430, 53rd Annual SPE Conference, Houston, Texas, October, 1978.
6. Schultz, W. E., and Smith, H. D. Jr., " Laboratory and Field Evaluation of a Carbon/Oxygen (C/O) Well Logging System," *J. Pet. Tech.,* Oct., 1974.
7. O'Brien, W. J., et al, "Comprehensive Analysis of the Carbon Oxygen Log," Paper DE83 006014, U. S. Department of Energy, Washington, DC, Jan., 1983.
8. Hearst, J.R., Conaway, J.G., Troka, D.E., and Grau, J.A., "A Comparison of Energy-Window and Spectral-Fitting Methods for the Estimation of Carbonate Content in Rocks Using Neutron-Induced Gamma Rays," *The Log Analyst,* pp. 11–19, November–December, 1993.
9. Jacobson, L. A., Beals, D. F., Wyatt, D. F. Jr., and Hrametz, A., "Response Characterization of an Induced Gamma Spectrometry Tool Using a Bismuth Germanate Scintillator," SPWLA 32nd Annual Logging Symposium, Midland, Texas, June, 1991.
10. Wyatt, D., Jacobson, L.A., Durbin, D., and Lasseter, E., "Logging Experience with a New Induced Gamma Spectrometry Tool," 33rd Annual SPWLA Symposium, June, 1992.
11. Westaway, P., Hertzog, R. and Plasek, R. E., "The Gamma Spectrometer Tool Inelastic and Capture Gamma-Ray Spectroscopy for Reservoir Analysis," SPE Paper 9461, 55th SPE Annual Fall Technical Conference, Dallas, Texas, Sept., 1980.
12. Heflin, J.D., Lawrence, T., Oliver, D., and Koenn, L., "California Applications of the Continuous Carbon/Oxygen Log," API Joint Chapter Meeting, Bakersfield, California, October, 1977.
13. Lawrence, T. D., "Continuous Carbon/Oxygen Log Interpretation Techniques," *J. Pet. Tech.,* pp 1394–1402, August, 1981.
14. North, R. J., "Through-Casing Reservoir Evaluation Using Gamma Ray Spectroscopy," SPE Paper 16356, SPE California Regional Meeting, Ventura, California, Apr., 1987.
15. Schlumberger, "Gamma Ray Spectrometry Tool," Document M-086302, Montrouge, France, September, 1983.
16. Roscoe, B.A., and Grau, J.A., "Response of the Carbon/Oxygen Measurement for an Inelastic Gamma Ray Spectroscopy Tool," Paper SPE 14460, 60th Annual SPE Conference, Las Vegas, Nevada, September, 1985.
17. Roscoe, B. A., Stoller, C., Adolph, R. A., Boutemy, Y., Cheeseborough, J. C. III, Hall, J. S., McKeon, D. C., Pittman, D., Seeman, B., and Thomas, S. R. Jr., "A New Through-Tubing Oil-Saturation Measurement System," SPE Paper 21413, SPE Middle East Oil Show, Bahrain, (canceled), Nov., 1991.
18. Scott, H. D., Stoller, C., Roscoe, B. A., Plasek, R. E., and Adolph, R. A., "A New Compensated Through-Tubing Carbon/Oxygen Tool for Use in Flowing Wells," SPWLA 32nd Annual Logging Symposium, Midland, Texas, June, 1991.
19. Schlumberger, "RST-Reservoir Saturation Tool," Document SMP-9250, June 1993.
20. Stoller, C., Scott, H.D., Plasek, R.E., Lucas, A.J., and Adolph, R.A., "Field Tests of a Slim Carbon/Oxygen Tool for Reservoir Saturation Monitoring," Paper SPE 25375, SPE Asia Pacific Oil & Gas Conference, Singapore, February, 1993.
21. Streeter, R.W., Hogan, G.P. II, and Olson, J.D., "Oil or Fresh Water? A New Through-Tubing Measurement to Determine Water Saturation," Paper SPE 27646, SPE Permian Basin Oil & Gas Recovery Conference, Midland, Texas, March, 1994.
22. Odom, R.C., Streeter, R.W., Hogan, G.P. II, and Tittle, C.W., "A New 1.625 in. Diameter Pulsed Neutron Capture and Inelastic/Capture Spectral Combination System Provides Answers in Complex Reservoirs," 35th Annual SPWLA Symposium, Tulsa, Oklahoma, June, 1994.

23. Jacobson, L.A., Ethridge, R., and Wyatt, D.F. Jr., "A New Thermal Multigate Decay-Lithology Tool," 35th Annual SPWLA Symposium, Tulsa, Oklahoma, June, 1994.

24. Dupree, J. H., and Cunningham, A. B., "The Application of Carbon/Oxygen Logging Technology to the Ivishak Sandstone, Prudhoe Bay, Alaska," SPE Paper 19615, 65th Annual SPE Technical Conference, San Antonio, Texas, Oct., 1989.

25. Audah, T.A., and Chardac, J.L., "Reservoir Fluid Monitoring Using Through-Tubing Carbon-Oxygen Tools," 34th Annual SPWLA Symposium, June, 1993.

26. Rieke, H. H., Oliver, D. W., Fertl, W. H., and McCord, J. P., "Successful Application of Carbon/Oxygen Logging to Coalbed Exploration," *J. Pet. Tech.*, Feb., 1983.

27. Chace, D. M., Schmidt, M. G., and Ducheck, M. P., "The Multi-parameter Spectroscopy Instrument Continuous Carbon/Oxygen Log—MSI C/O," 10th CWLS Formation Evaluation Symposium, Calgary, Alberta, Canada, Sept., 1985.

Other Suggested Reading

Barton, W., and Flynn, J.M., "Applications of the Carbon/Oxygen Log in Johnson (Gloriets) Field," Paper SPE 8453, 54th Annual SPE Conference, Las Vegas, Nevada, September, 1979.

Cannon, D.E., and LaVigne, J.A., "Through-Casing Reservoir Evaluation," SPE Formation Evaluation, pp. 201–208, June, 1987.

Cannon, D. E., and Rossmiller, J. W., "Oil Saturation Evaluation for EOR in a Carbonate," SPE Paper 13288, 59th Annual Technical Conference, Houston, Texas, Sept., 1984.

Fertl, W.H., and Frost, E., "Recompletion, Workover, and Cased Hole Exploration in Clastic Reservoirs Utilizing the Continuous Carbon/Oxygen (C/O) Log—The CHES III Approach," Paper SPE 9028, SPE Southwest Texas Regional Meeting, Corpus Christi, Texas, April, 1980.

Freeman, D. W., and Fenn, C. J., "An Evaluation of Various Logging Methods for the Determination of Remaining Oil Saturation in a Mixed Salinity Environment," SPE Paper 17976, SPE Middle East Oil Technical Conference, Manama, Bahrain, March, 1989.

Georgi, D.T., "Application of Time-Series Analysis to Induced Gamma Ray Spectroscopy Logs From Two Cold Lake Heavy-Oil Observation Wells," Paper SPE 19602, 64th Annual SPE Technical Conference, San Antonio, Texas, 1989.

Girrell, B.I. and Greenberg, M.L., "Application of the Gamma Spectrometry Service to Illinois and Michigan Basin Carbonates," 26th Annual SPWLA Symposium, June, 1985.

McGuire, J.A., Rogers, L.T., and Watson, J.T., "Improved Lithology and Hydrocarbon Saturation Determination Using the Gamma Spectrometry Log," Paper SPE 14465, 60th Annual SPE Conference, Las Vegas, Nevada, September, 1985.

Ostermeier, R.M., "Comparison of Core and Log Carbon/Oxygen Ratios in South Belridge Diatomite," 34th Annual SPWLA Symposium, June, 1993.

Schlumberger, "Cased Hole Log Interpretation Principles/Applications," Document SMP-7025, Houston, Texas, 1989.

Schweitzer, J.S., Manente, R.A., and Hertzog, R.C., "Gamma-Ray Spectroscopy Tool Environmental Effects," Paper SPE 11144, 57th Annual SPE Conference, New Orleans, Louisiana, September, 1982.

Thompson, J.B., Frost, E., and Fertl, W.H., "Evaluation and Monitoring of Secondary and EOR Projects in California Based on the Continuous Carbon/Oxygen (C/O) Log," SPE Paper 10739, SPE California Regional Meeting, San Francisco, California, March, 1982.

CHAPTER 8

FORMATION EVALUATION: SONIC AND ACOUSTIC TECHNIQUES

SONIC OVERVIEW

Background

In this section, the words sonic and acoustic are used interchangeably. Within the last few years acoustic logs have been run through cased hole with increasing frequency. This has occurred because of the development of new tools capable of determining reliable information regarding the propagation of both the compressive and shear waves in the formation and a better understanding of the information available from such data. Traditionally, the sonic logs provided a compressive wave velocity in an open hole environment, and this velocity is used to compute a formation porosity. Such measurements are now accurately available through casing, both for the compressive and shear wave, as well as information on pore fluid in high porosity unconsolidated clastic formations. These parameters also provide a mechanism for computing formation rock mechanical properties necessary for fracturing operations. Sonics are also used to tie in to seismic sections with density logs and to view deeper into the earth with downhole geophones arrays.

The study of sonic wave propagation in both and open and cased hole promises to tap into a wealth of new information heretofore not available. Much research is currently being done by both the service companies and oil companies to understand and apply sonic information.

BASIC PRINCIPLES

Wave Properties Review

Acoustic waves are usually depicted as a sinusoidal wave train with peaks and valleys. The time between peaks or the time between valleys is called the period, T, and has units of seconds. The frequency of a wave is related to this period as its reciprocal, and

$$f = 1/T \qquad \text{(Equation 8.1)}$$

where f is in cycles per seconds or hertz.

Whereas T is the time period between adjacent peaks in seconds, the wavelength, λ, is the distance between adjacent peaks in units of length. The velocity, V, of an acoustic wave is related to the wavelength and frequency by the equation

$$V = f\lambda \qquad \text{(Equation 8.2)}$$

Velocity is usually expressed in units of ft/sec or m/sec.

Types of Waves

There are three main types of waves of interest to acoustic logging at this time. These are the compressional, shear, and Stoneley waves.[1-4] Compressional waves, sometimes called primary or "P-waves," propagate along the wellbore parallel to the direction of particle displacement. Such direction of motion is shown in the schematic of Figure 8.1.[5] The velocity of the compressional wave, V_c, in an isotropic near elastic solid is expressed as

$$V_c = \left[\frac{K + (4/3)\mu}{\rho} \right]^{\frac{1}{2}}$$

(Equation 8.3)

where

K = Bulk modulus

μ = Shear modulus

ρ = Density

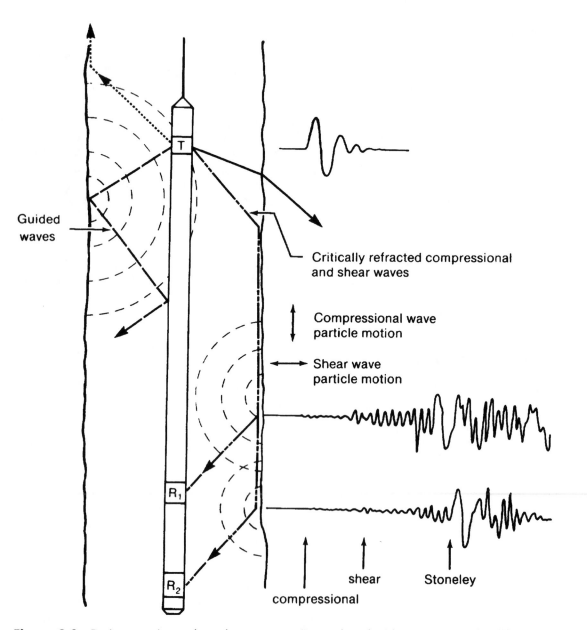

Figure 8.1 Basic acoustic sonde and wave types (Reproduced with permission of Halliburton Co. and SPWLA, Ref. 5)

Shear waves, also called secondary or "S-waves," propagate in a direction perpendicular to the particle motion. Such motion for a wave propagating along a wellbore is shown on Figure 8.1. For a near elastic isotropic solid, the velocity may be expressed as

$$V_s = \left[\frac{\mu}{\rho}\right]^{\frac{1}{2}}$$

(Equation 8.4)

Shear waves move more slowly than the compressional waves. Typically, shear waves propagate at about one half to two thirds the velocity of the compressional waves in formations downhole.

Stoneley waves propagate along the borehole wall and are a result of the interaction of the borehole wall and the borehole fluids. They have velocities typically less than the velocities of the borehole fluid. The wave train which arrives at the receiver as a function of time is shown on the lower right in Figure 8.1. The compressional, shear, and Stoneley waves are labeled.

Other waves exist, such as the Pseudo-Rayleigh Waves and the leaky mode waves, but these are not of significant interest at this time.

Wave Propagation and Refraction

If acoustic waves are plotted as rays propagating perpendicular to the wave front, the rays can be looked upon as and treated just like light waves. Figure 8.2 shows the incident acoustic ray and the refracted compressional and shear waves propagating into the formation.[1,4] Where the subscript i refers to the incident and r to the refracted waves, Snell's Law applies and may be written

$$\frac{\sin \phi_i}{\sin \phi_r} = \frac{V_f}{V_b}$$

(Equation 8.5)

where

ϕ = Angle relative to line perpendicular to wall

i = Subscript for incident angle

r = Subscript for refracted angle

V_b = Velocity of shear or compressional wave

V_f = Velocity of borehole fluid

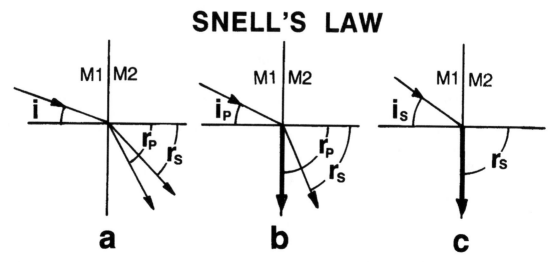

SNELL'S LAW

Figure 8.2 Snell's law for primary and shear waves (Reproduced with permission of Halliburton Co., Ref. 1)

Notice in Figure 8.2 that at some angle the compressional or shear wave is fully refracted and moves only along the borehole wall. This is the critical angle and only those waves refracted at the critical angle are detected at the receiver of logging tools, unless reflections or refractions occur from deeper in the formation. The critical angle at which this occurs is

$$\sin \phi_{icrit} = V_f/V_b \qquad \text{(Equation 8.6)}$$

This equation shows that if $V_f > V_b$, the value of the sin of the critical angle is greater than one. In this case, there is no refracted shear wave. This accounts for the occasional disappearance of the shear arrivals when unconsolidated or shaley sands are encountered.

APPLICATIONS OF SONIC MEASUREMENTS

Slowness

The primary measurement of the sonic tools has been the velocity of the compressional wave. The tools, as shown in Figure 8.1, measure the velocity between receivers. Rather than present a compressional velocity, it is convenient to present a time measurement, or simply the reciprocal of the compressional velocity, $1/V_p$. This measurement is called the transit time or "slowness" and has units of usec/ft or usec/m. For the tool of Figure 8.1, the slowness is

$$t = \frac{t_2 - t_1}{\text{spacing}} \qquad \text{(Equation 8.7)}$$

The symbol, *t*, is the slowness.

Porosity

A number of different equations may be used for the sonic porosity.[1,6] One is the Wyllie Time-Average equation. Based on laboratory observations in liquid bearing formations, it proposes a weighted average relationship between the bulk transit time and the porosity. It states that

$$t_{\log} = t_f \phi + t_{ma}(1 - \phi)$$

which yields an expression for porosity in the form

$$\phi_S = \frac{t_{LOG} - t_{ma}}{t_f - t_{ma}} \qquad \text{(Equation 8.9)}$$

where

ϕ_S = Sonic log reading for the bulk formation

t_{ma} = Transit time for the matrix

t_f = Transit time of the pore fluid

The transit time for fresh water is 189 microseconds/ft, and 185 us/ft is usually taken for salt water. Transit times for various matrix and other materials are shown on Table 8.1.[3]

The Wyllie equation may be optimistic in some unconsolidated sandstones due to lack of compaction. This is usually not a problem at depths in excess of 7,000 ft (2,133 m) where overburden is large. However, shallower zones may exhibit this error. As a result, a compaction correction may be made. If sands are unconsolidated, so are nearby shales. When the slowness of a nearby shale, t_{sh}, exceeds 100 microseconds/ft, the correction factor, C_p, is approximated by

$$C_p = \frac{t_{SH}}{100} \qquad \text{(Equation 8.10)}$$

and porosity may be approximated by

$$\phi_S = \left(\frac{t_{LOG} - t_{ma}}{t_f - t_{ma}}\right)\frac{1}{C_p} \qquad \text{(Equation 8.11)}$$

TABLE 8.1

	V_{ma}(ft/sec)	Δt_{ma}(µs/ft)	Δt_{ma}(µs/ft) (commonly used)
Sandstones	18,000–19,500	55.5–51.0	55.5 or 51.0
Limestones	21,000–23,000	47.6–43.5	47.5
Dolomites	23,000	43.5	43.5
Anhydrite	20,000	50.0	50.0
Salt	15,000	66.7	67.0
Casing (iron)	17,500	57.0	57.0

Another and somewhat better porosity algorithm is the Raymer-Hunt-Gardner transform. It is

$$\phi_S = \frac{1}{\rho_{ma} - \rho_f}\left(1 - \frac{t_{ma}}{t_{LOG}}\right)$$ (Equation 8.12)

where

$$\rho_{ma} = \text{Density of the matrix}$$

$$\rho_f = \text{Density of the pore fluid}$$

Acoustic porosity measurements respond to well distributed porosity, but miss such things as water filled vugs or fractures. As a result, and especially in carbonates, the estimates of porosity from sonic logs may be lower than true porosity. If ϕ_T is the true porosity, then the sonic porosity differs from true porosity by the contribution of such water filled fractures and vugs. Their contribution is called secondary porosity, and these terms are related as:

$$\phi_T = \phi_S + \phi_{secondary}$$ (Equation 8.13)

where

$$\phi_T = \text{True porosity}$$

$$\phi_S = \text{Sonic porosity}$$

$$\phi_{secondary} = \text{Secondary porosity}$$

If gas is present in the pore space, its effect is to slow the sonic signal as it passes through the rock. As a result, the computation of sonic porosity may be too large over a gas zone.

Elastic Properties

Elastic properties of rocks, including Poisson's Ratio, Shear Modulus, and Bulk Modulus are computable if the compressional and shear slowness are known. Table 8.2 lists the dynamic elastic properties and the equations to compute them from log data.[1,3] This information is relevant to planning hydraulic fractures or evaluation of sand strength to inhibit sand production. Such considerations are beyond the scope of this text.

SONIC EQUIPMENT
Basic Sonic Sonde

The sonde configuration of Figure 8.1 is the basic sonic sonde available from most service companies. The measurement here is the travel time between receivers R1 and R2. As long as the tool is centralized, and the borehole is straight and gauge, this tool is compensated for borehole effects and the transit time is a true transit time between the receivers. The

Table 8.2 Dynamic elastic properties.

v	Poisson's Ratio	$\dfrac{\text{Lateral strain}}{\text{Longitudinal strain}}$	$\dfrac{\frac{1}{2}(t_s/t_c)^2 - 1}{(t_s/t_c)^2 - 1}$
G	Shear Modulus	$\dfrac{\text{Applied stress}}{\text{Shear strain}}$	$\dfrac{e_b}{t_s} \times a$
E	Young's Modulus	$\dfrac{\text{Applied uniaxial stress}}{\text{Normal strain}}$	$2G\,(1+v)$
K_b	Bulk Modulus	$\dfrac{\text{Hydrostatic pressure}}{\text{Volumetric strain}}$	$e_b\left(\dfrac{1}{t_c^{\,2}} - \dfrac{4}{3t_s^{\,2}}\right) \times a$
C_b	Bulk Compressability (with porosity)	$\dfrac{\text{Volumetric deformation}}{\text{Hydrostatic pressure}}$	$\dfrac{1}{K_b}$
C_r	Rock Compressibility (with porosity)	$\dfrac{\text{Change in matrix volume}}{\text{Hydrostatic pressure}}$	$\dfrac{1}{e_g\left(\dfrac{1}{t_{ma}^{\,2}} - \dfrac{4}{3t_{sma}^{\,2}}\right) \times a}$
α	Biot Elastic Constant	$\dfrac{\text{Pore pressure}}{\text{proportionality}}$	$1 - \dfrac{C_r}{C_b}$

Note: coeff. a = 1.34×10^{10} if e_b in g/cm³ and t in µs/ft.

measurement is made by detecting the first significant amplitude arrival at each receiver. When an amplitude above some preset value is detected the arrival time at each receiver is recorded. The slowness measured by this sonde is given in equation 7.

Long spaced sonic logs have become available where a pair of receivers (at 2 ft or 61 mm spacing) are placed 8 or 10 feet (2.44 or 3.05 m) from the transmitter. The purpose of such long spacing is to allow more time for the compressional wave to separate from the shear wave so that they can be discriminated more readily.

Borehole Compensated Sonic Measurement

A borehole compensated sonic tool is described in Figure 8.3. On the left of the figure the conventional borehole compensated measurement is shown.[6] There is an upper transmitter, UT, and a lower transmitter, LT. Midway are two pairs of receivers, R1R2 and R3R4. These can be looked upon as a single pair of receivers straddling the measure point. Transmissions from the upper transmitter are detected between the receivers followed by transmissions from the lower transmitter. This configuration compensates for such effects as sonde tilt and reasonably rugose hole conditions.

The borehole compensated configuration of Figure 8.3, part a, is needlessly long. The service companies have shortened this sonde configuration to that at the right. This is referred to in the figure as the depth-derived measurement. For this configuration, the data must be stored and depth matched for each measurement. The first measurement is made by firing T1 when the receivers straddle the measure point as shown, followed a bit later by the firing of T1 and T2 when they straddle the measure point. The readings are as follows:

$$\text{FIRST } t \text{ READING} = (T_1 \rightarrow R_1) - (T_1 \rightarrow R_2) \qquad \text{(Equation 8.14)}$$

$$\text{SECOND } t \text{ READING} = (T_1 \rightarrow R_2) - (T_2 \rightarrow R_2)$$

and the slowness measurement becomes

$$t = \frac{(\text{MEMORIZED FIRST } t \text{ READING}) + (\text{SECOND } t \text{ READING})}{2 \times \text{SPAN}} \qquad \text{(Equation 8.15)}$$

where the span is the distance between the receivers R1 and R2.

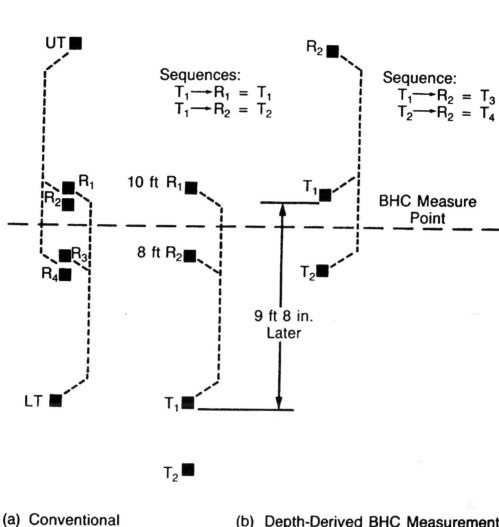

Figure 8.3 Borehole compensated sonic tool (Courtesy Schlumberger, Ref. 6)

Array Tools

The newest tools available are the array type tools shown in Figure 8.4.[3] The main feature of these tools is that they have transmitters and an array of closely spaced (about 6 in. or 15.25 cm) receivers beginning about 8 feet (2.44 m) from the transmitter.[3] The transmitters are monopole, dipole, or both, with appropriate receivers. The tools and service company names are listed below. The * indicates that the name is a mark of the company indicated.

SCHLUMBERGER: Array-Sonic* (Monopole only)
                     Dipole Shear-Sonic Imager*, DSI*
WESTERN ATLAS: Multipole Array Acoustilog*, MAC*
HALLIBURTON: Full Wave Sonic*, FWS*

The sonde in Figure 8.4 is an Array-Sonic tool.[3] It has a monopole source and is capable of cement bond evaluation using the two intermediate receivers.

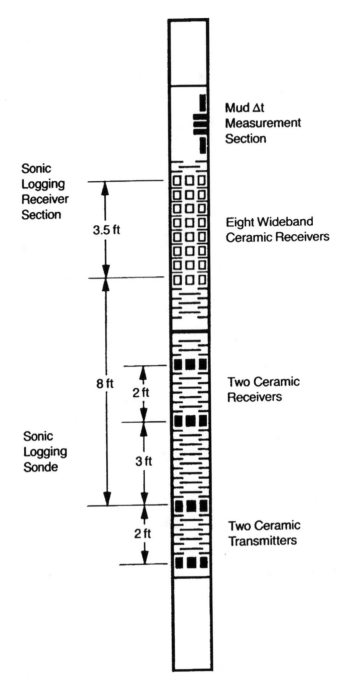

Figure 8.4 Array type tool, Schlumberger Array-Sonic shown (Courtesy Schlumberger, Ref. 3)

MEASUREMENT TECHNIQUE AND REQUIREMENTS

Acoustic Record from an Array Tool

An array type tool records the full acoustic wave signal for processing as it passes each receiver and as shown on Figure 8.5.[7] Figure 8.5, part a, shows the response in a well bonded cased hole while Figure 8.5, part b, shows an array in a poorly bonded section. In the well bonded section, the casing signal is quite low, and the compressional and shear wave arrivals are easily discriminated. When the casing is poorly bonded, the ringing of the casing signal masks the compressive wave, but it appears that the shear can still be detected. The slopes of the lines passing through the onset of each wave are an indication of the velocity of that wave.[7-11]

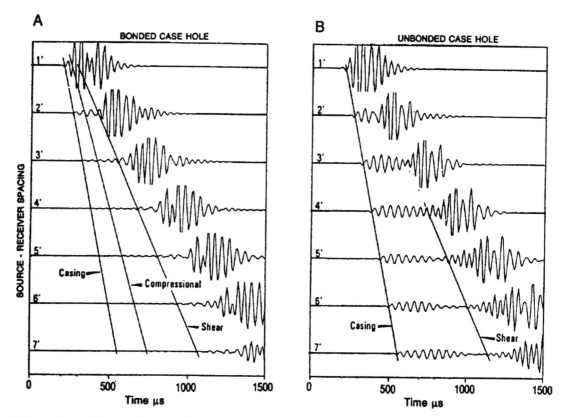

Figure 8.5 Wavetrain recorded at each receiver with an array type tool (Courtesy SPWLA, Ref. 7)

Monopole and Dipole Transmitters

In well bonded carbonates, both the compressive and shear waves are easily detectable. Even older equipment, such as that shown in Figure 8.1, could accurately measure the compressional wave slowness under such circumstances. If the wavetrain could be viewed, even the shear arrivals could often be determined. However, when in poorly consolidated sandstones, the shear wave is often too weak or not present in the wave train. To enhance the shear wave, the array tools listed above use a dipole transmitter. The dipole emits a highly directional pulse perpendicular to the pipe wall whereas the older monopole transmitters are omnidirectional in character. This dipole pulse is analogous to a hammer hitting the wall, and it initiates a strong flexural wave in the borehole wall, which eventually becomes a shear wave by the time it reaches the receiver array. It turns out that these dipole tools are quite effective at arriving at a shear transit time, even in unconsolidated sands and poorly cemented formations.

Note that tools with dipole transmitters also have monopole transmitters for the optimal detection of the compressive wave slowness. Another feature, which is just being understood, is the use of orthogonal dipole transmitters and directional receivers. All of the dipole tools have such directional capability. It is expected that this information will shed light on directions of major and minor stresses in the formation and fracture direction assessment.

ACOUSTIC LOGS THROUGH CASING—APPLICATION EXAMPLES

Early Open Hole-Cased Hole Comparison

The example of Figure 8.6 shows a comparison of open hole to cased hole sonic logs run back in the 1960s.[12] The equipment is like that shown in Figure 8.1. On the left is the open hole log with the transit time recorded on a 40-80 us/ft scale. The tool configuration is

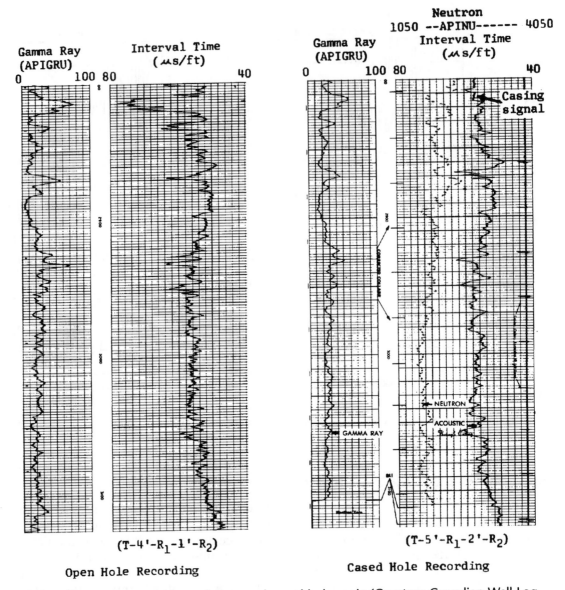

Figure 8.6 Early comparison of open and cased hole sonic (Courtesy Canadian Well Logging Society, Ref. 12)

listed at the bottom of the log, with four feet (122 cm) between the transmitter and R1 and one foot (30.5 cm) between R1 and R2. On the right is a cased hole comparison. The cased hole log was run by another service company and the tool configuration, listed at the bottom of the log, is quite different. The overall agreement is excellent. The main discrepancy is near the top of the logged section where "casing signal" is indicated on the cased hole log. Over this interval, the open hole transit time increases to about 75 us/ft while the cased hole version stays at about 57 us/ft. 57 us/ft is the acoustic slowness in steel, and therefore this is taken to be uncemented pipe.

Effects of Cement Coverage

The log example of Figure 8.7 shows the open hole (dashed line) and cased hole (solid line) interval transit times for the compressive wave overlaying each other over an interval with varying amounts of cement coverage.[6,13] The cement coverage is presented as a cement map from a Schlumberger Cement Evaluation Tool (CET). The black indicates cement, white no cement. Clearly, where the cement coverage is good, both acoustic signals overlay. When

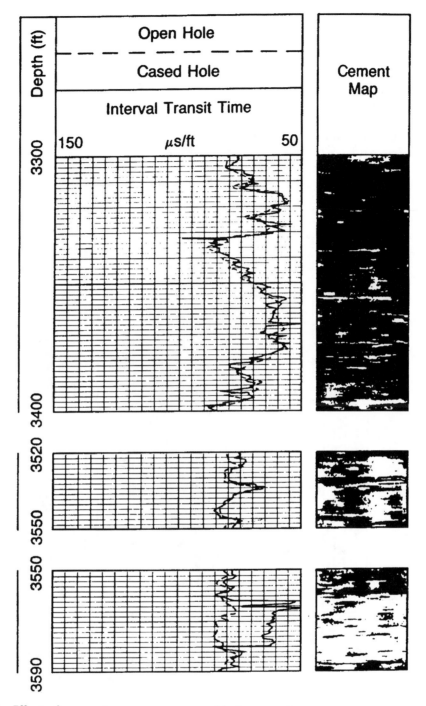

Figure 8.7 Effect of cement coverage on cased hole acoustic measurement (Courtesy Schlumberger, Ref. 6)

cement coverage is reduced to about 50% as shown in the middle section, the correlation still appears to be excellent. However, when the cement is gone altogether as in the lower section of the log, the cased hole transit time drops to a value close to that of steel, 57 us/ft. This example shows that good sonic data is achievable even with only 50% cement coverage.

While the above applies to the compressive transit time, it appears that with dipole sonic tools the shear transit time may be even less sensitive to cement coverage and may even provide good formation shear information in free pipe. Microannulus may play a significant role and may mask the compressional arrivals.[14]

Detection of Hydrocarbons in High Porosity Sands

The example of Figure 8.8 shows an overlay technique in which porosity is computed from both the compressive and shear transit times.[17] When overlayed, the compressive porosity reads greater than the shear wave porosity in hydrocarbon zones. This is an effect of fluid compressibility and its effect on the slowness of the compressional wave, causing it to be too slow in hydrocarbon intervals.[15-17] Such an overlay is shown on track 4 and the separation is shaded when the compression wave porosity reads greater than the shear porosity. The water, oil, and gas legs are clearly evident. For comparison, the open hole density and neutron porosity are shown in track three, with the crossover similarly shaded. The method of computing the shear porosity is discussed in reference.[17]

This technique does not work for carbonates and is limited to clastics with porosities greater than 25% to 30%.

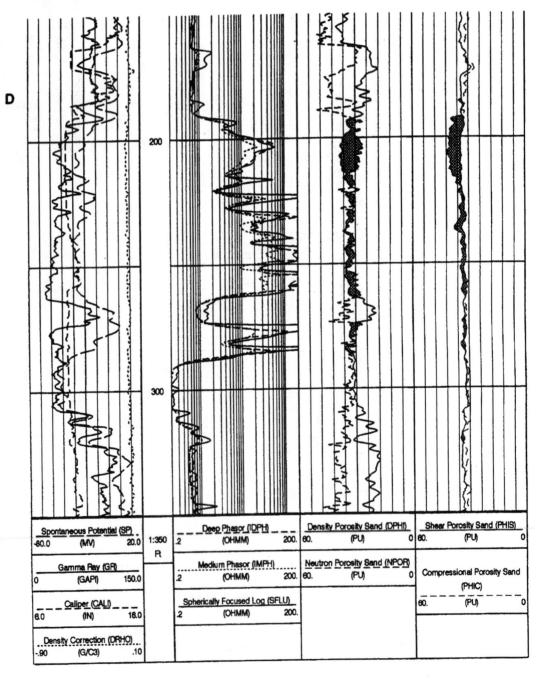

Figure 8.8   Detection of hydrocarbons in high porosity sands (Courtesy Schlumberger, Ref. 17)

Other Applications of Acoustic Wave Logging

The mechanical properties derivable from the compressive and shear slowness are useful for prediction of induced fracture characteristics and migration,[18] and for estimates of the likelihood of a formation producing sand.[6] Other techniques have been published regarding the estimates of permeability from Stoneley waves,[19] the detection of open fractures with tube-waves (Stoneley and Pseudo-Rayleigh waves).[20] The tools having orthogonal dipole transmitters (all dipole tools) also have sets of orthogonal receivers and as a result are measuring the shear anisotrophy of formations.[21,22] This is expected to improve seismic models and provide information useful for natural and hydraulic fracture evaluation.

Vertical Seismic Profiles

Vertical seismic profiling (VSP) is a technique in which a single or an array of geophones is placed downhole to detect acoustic reflections from formation interfaces in the vicinity of and below the measure depth. Such reflections provide information on deeper formations, possible fluid content, faults, salt domes, and other large reservoir features. Seismic sections are routinely run from the surface, but frequently leave questions unanswered. The downhole VSP technique puts the geophones close to the features to be measured and therefore provides locally more accurate and deeper information. A schematic of such an operation is shown on Figure 8.9.[6] Here the tool is comprised of a number of geophones spaced along the wellbore, each recording the waves arriving from the source at the surface.

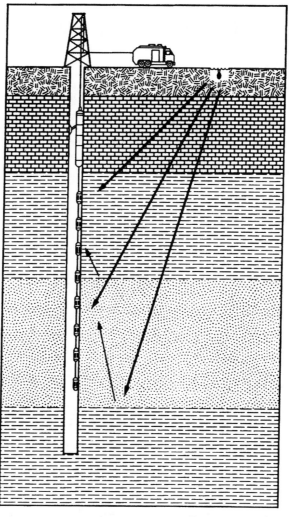

Figure 8.9 Vertical seismic profiling technique (Courtesy Schlumberger, Ref. 6)

POINTS TO REMEMBER

- There are three main types of waves in sonic logging, the compressional, shear, and Stoneley.
- Acoustic waves propagate from the transmitter to receiver by being refracted at the formation interface in a direction parallel to the wellbore.
- While acoustic velocity is the key measurement, it is expressed as its inverse in units of usec/ft or usec/m.
- The transit time is also called slowness.
- Formation porosity can be evaluated from the compressional and shear travel times.
- Both the compressional and shear transit times are required to evaluate formation mechanical properties.
- Sondes may be borehole compensated or of the array type tool.
- Most Array tools have both monopole and dipole transmitters on the same sonde.
- The dipole transmitter is a directional transmitter and is needed for good shear slowness measurement.
- In well bonded casing, both compressional and shear wave slowness can be made easily.
- Cement coverage as low as 50% is adequate for compressional and even less may do for shear transit time measurements.
- Shear measurements in high porosity and unconsolidated sands require the use of a dipole source.
- Overlay techniques are available using both the compressional and shear transit times to detect hydrocarbons in soft sands.
- Cased hole sonic logging has many potential applications, including traditional open hole techniques and cross plots.

References

1. Halliburton, "FWST Training Notes," Document No. 770.06583, Houston, Texas, 1991.
2. Atlas Wireline, "Full Wave Acoustic Logging with the Multipole Array Acoustilog," Houston, Texas, 1992.
3. Schlumberger, "Log Interpretation Principles/Applications," Document SMP-7017, Houston, Texas, 1987.
4. Minear, J.W., and Fletcher, C.R., "Full-Wave Acoustic Logging," 24th Annual CWLS-SPWLA Symposium, Calgary, Alberta, Canada, June, 1983.
5. Minear, J.W., "Full Wave Sonic Logging: A Brief Perspective," 27th Annual SPWLA Symposium, Houston, Texas, June, 1986.
6. Schlumberger, "Cased Hole Log Interpretation Principles/Applications," Document SMP-7025, Houston, Texas, 1989.
7. Georgi, D.T., Heavysege, R.G., Chen, S.T., and Eriksen, E.A., "Application of Shear and Compressional Transit-Time Data to Cased-Hole Carbonate Reservoir Evaluation," *The Log Analyst*, pp. 129-143, March-April, 1991.
8. Schlumberger, "The Technical Review—SONIC LOGGING," Vol. 33, No. 1, Houston, Texas, 1985.
9. Chang, S., van der Hijden, J. and Orton, M., "Acoustic Waveforms Explained," *The Technical Review*, Vol. 35, No. 1, pp. 16-33, 1987.
10. Aron, J., Murray, J., and Seeman, B., "Formation Compressional and Shear Interval-Transit-Time Logging by Means of Long Spacings and Digital Techniques," Paper SPE 7446, 53rd Annual SPE Conference, Houston, Texas, October, 1978.
11. Chang, S.K., and Everhart, A.H., "A Study of Sonic Logging in a Cased Borehole," Paper SPE 11034, 57th Annual SPE Conference, New Orleans, Louisiana, September, 1982.
12. Fons, L., "Acoustic Logging Through Casing," CWLS Second Formation Evaluation Symposium, Calgary, Alberta, Canada, May, 1968.
13. Bettis, F., Borden, C., Rowe, W., and Schwanitz, B., "Sonic Logging in Cased Wells: Opportunities for Enhancing Old Fields," *The Technical Review*, Vol. 35, No. 4, 1987.
14. Streeter, R.W., and Hunt, E.R., "Prefracture Analysis of Red Mountain Sands: A Integrated Analysis of Open Hole and Cased Hole Logs," Paper SPE 19003, SPE Joint Rocky Mountain Regional/Low Permeability Reservoirs Symposium, Denver, Colorado, March, 1989.
15. Jirik, L.A., Howard, W.E., and Sadler, D.L., "Identification of Bypassed Gas Reserves Through Integrated Geological and Petrophysical Techniques: A Case Study in Seeligson Field, Jim Wells County, South Texas," Paper SPE 21483, SPE Gas Technology Symposium, Houston, Texas, January, 1991.
16. Williams, D.M., "The Acoustic Log Hydrocarbon Indicator," 31st Annual SPWLA Symposium, June, 1990.

17. Crary, S., and Badry, R., "Applications of Dipole Sonic Imager Measurement for Formation Evaluation," 13th CWLS Formation Evaluation Symposium, Calgary, Alberta, Canada, September, 1991.

18. Corley, B.H., and Klautt, W.R., "Prediction of Fracture Migration Using Elastic Rock Properties," International Symposium on Petroleum Exploration in Carbonate Areas, Nanjing, People's Republic of China, November, 1986.

19. Saxena, V., "Permeability Quantification from Borehole Stoneley Waves," 35th Annual SPWLA Symposium, June, 1994.

20. Madlin, W.L., and Schmitt, D.P., "Fracture Diagnostics with Tube-Wave Reflection Logs," *JPT*, pp. 239-248, March, 1994.

21. Cengiz, E., Koster, K., Williams, M., Boyd, A., and Kane, M., "Dipole Shear Anisotropy Logging," SEG Annual Symposium, Los Angeles, California, 1994.

22. Mueller, M.C., Boyd, A.J., and Esmersoy, C., "Case Studies of the Dipole Shear Anisotropy Log," SEG Annual Symposium, Los Angeles, California, 1994.

CHAPTER 9

FORMATION EVALUATION: OTHER SERVICES

INTRODUCTION

The remaining available through-casing formation evaluation services are discussed in this chapter. These services are either relatively uncommon or still in the process of development. That is not to say that these services are unimportant. Sometimes a small technological breakthrough propels a service from obscurity into the limelight. Sometimes no other combination of measurements will provide the needed information.

The services covered in this section include the following. An * indicates that the designation is a mark of the company indicated.

- **Borehole Gravity Survey:** Supplied by EDCON Corp., this tool is called the Borehole Gravity Meter* (BHGM*) (*These are marks of EDCON Corp.) and measures the bulk density of the formation.
- **Density Log:** The density is a commonly run open hole logging tool and is available from most logging companies.
- **Through Casing Formation Resistivity:** The Through Casing Resistivity Tool™ (TCRT®)* was developed by ParaMagnetic Logging, Inc. of Woodinville, Washington. This development was funded in part by Gas Research Institute and the U.S. Department of Energy (Bartlesville Project Office).
- **Chlorine Log:** This tool was run by Atlas and McCullough before being taken over by Atlas. The availability of this service is unknown.
- **Wireline Formation Tester:** This type of tool is available through the major service companies. It provides a shaped charge perforator to shoot through the casing, retrieve a sample of formation fluid, and measure accurate formation pressure.

BOREHOLE GRAVITY SURVEY

Applications, Advantages, and Limitations

The borehole Gravity survey measures the bulk formation density.[1,2] This measurement is done by determining the gravitational pull in the wellbore both above and below the formation of interest. The difference in the gravitation pull effect across the formation is primarily due to its mass and hence bulk density. It is especially well suited to the cased hole environment since it is virtually unaffected by casing. These are its advantages in this respect:

- The measurement is not affected by casing, even multiple strings of pipe.
- It is unaffected by rugose hole, washouts, poor cement distribution, fluid invasion, and the like.

* Through Casing Resistivity Tool™ is a Trademark and TCRT® is a Registered Trademark of ParaMagnetic Logging, Inc.

- The radius of investigation can extend 50 feet or more into the formation.
- Salinity and type of fluid, such as gas, oil or water have no detrimental effect on the measurement.
- The measurement is unaffected by borehole fluid type.

The service does offer some disadvantages. Many of these relate to the equipment. However, improved and upgraded equipment may be available since the time of this writing. Some limitations are:

- Measurements are taken with the tool stationary, with reading taking 5 to 15 minutes at a station.
- The tool is limited to a deviation angle of 14°.
- Not available in thru-tubing sizes. Smallest tool is 3.875 in. (9.84 cm).
- Vertical resolution not better than 5 or 10 ft. (1.5 or 3.0 m).

The applications for this measurement are quite interesting, with some provided information unobtainable by any other means. Some applications include the following:

- Good deep formation density measurements regardless of hole or fluid conditions.
- Deep measurements of porosity.
- Monitor fluid saturations and gas/oil/water contact during production.
- Due to large depth of investigation, the service is helpful for mapping of salt flanks, flares and overhangs, mapping reef structures, and detecting fracture or porosity structures away from the well. If a significant density difference exists in a nearby structure, it may be detected.
- Readily monitor gas caps, steam flood fronts, gas storage reservoirs.
- Discriminate tight from producible gas zones.

While a number of tool models exist, some are rated to as high as 20,000 psi (1379 bars) and 200°C (392°F). The tool uses a miniature version of the LaCoste and Romberg land gravity meter.

Principle of Operation

The measurement of the BHGM is made using a spring balance mechanism as illustrated in Figure 9.1.[3] The beam is maintained in a stationary horizontal position by adjusting the

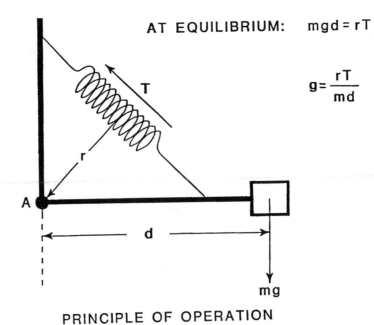

$$\text{AT EQUILIBRIUM:} \quad mgd = rT$$

$$g = \frac{rT}{md}$$

PRINCIPLE OF OPERATION

Figure 9.1 Spring balance gravity meter (Courtesy Edcon, Ref. 3)

spring tension. At equilibrium, the torques or moments about point A must balance, and so the equation may be written[3]

$$mgd = rT$$ (Equation 9.1)

where

m = mass at end of arm d

g = gravitational acceleration

d = distance from pivot point to mass

r = Distance between pivot point and spring

T = Spring tension

As the gravitational acceleration, g, changes, so does the weight on the beam. Since g, d, and r are known, and T is measured when arm d is leveled, then the value of g can be determined. The beam is clamped to immobilize it for transporting to the wellsite and into the well. Meters are limited to a certain range of gravity. It turns out that gravity varies with latitudes since the earth is not a true sphere and is rotating. As a result, each meter can only be used in a certain range of latitudes. Some new meters, however, have a world wide range.

Formation Bulk Density Determination

The force of attraction, F, between two bodies of masses m_1 and m_2 separated by a distance r is given by

$$F = Gm_1m_2/r^2$$ (Equation 9.2)

where G is the gravitational constant. If m_1 and m_2 are in grams and r is in cm, the F is in dynes and $G = 6.67 \times 10^{-8}$. For a unit mass (1 gram) at the earth's surface, assuming the earth is a point mass, M, and the unit mass is a distance, R, from M, the mass would experience a pull or an acceleration g, given by

$$g = GM/R^2$$ (Equation 9.3)

The mass of the earth, M, is 5.968×10^{27} grams and the radius, R, from the center of the earth to the surface is 6.37×10^8 cm.

The model used is not that of the attraction between two point masses, but instead that of the gravitational attraction of a horizontal infinite slab on a unit mass. The model is illustrated on Figure 9.2.[2] Assuming that this slab models a formation rock of uniform density, the gravitational pull on a unit mass at its upper surface is

$$g = 2 \pi \rho G \Delta Z$$ (Equation 9.4)

where

ρ = slab density

ΔZ = slab thickness

If measurements are taken at both the top and bottom of the slab, the effect is to pull down an amount g at the top and pull up an amount g at the bottom. Hence the difference in gravitational force on the unit mass is[3]

$$\Delta g = -4\pi \rho G \Delta Z$$ (Equation 9.5)

However, there is also another very important effect called the Free Air Effect. The equation for this effect states that there is a natural gravitational gradient arising from the mass of the earth and its distribution. This Free Air Effect is a function of the elevation or depth relative to mean sea level and is also a function of the latitude, since the earth is not a perfect sphere. The Free Air Effect gradient is

$$F = .308768 - .000440 \, Sin^2 \o - .0000001442h$$ (Equation 9.6)

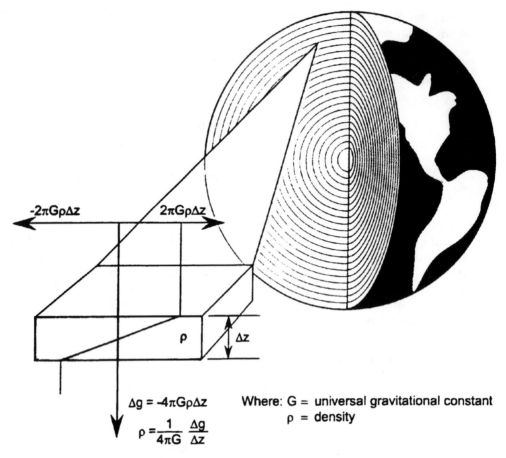

Figure 9.2 Gravitational attraction model (Courtesy Schlumberger, Ref. 2)

where

$$h = \text{elevation above the mean sea level, meters}$$

$$\emptyset = \text{latitude}$$

The measured change of gravitational pull must, therefore, be a function of both the Free Air Effect and Slab Density Effect, and so the equation may be written

$$\Delta g = (F - 4\pi\rho G)\,\Delta Z \qquad \text{(Equation 9.7)}$$

Solving this equation for the slab density, ρ, yields

$$\rho = \frac{1}{4\pi G}\left(F - \frac{\Delta g}{\Delta Z}\right) \qquad \text{(Equation 9.8)}$$

Equation (9.8) may be expressed in engineering units as follows:

$$\rho = 3.68237 - 0.005247\,sin^2\phi + 0.00000172Z - 11.92006\frac{\Delta g}{\Delta Z} \qquad \text{(Equation 9.9)}$$

where Δz = meters

$$\rho = 3.68237 - 0.005247\,sin^2\phi + 0.000000524Z - 39.1273\frac{\Delta g}{\Delta Z} \qquad \text{(Equation 9.10)}$$

where ΔZ = feet

Milligals is the unit for g and gm/cc for the density, ρ, in the above two equations.

For example, if a gravity gradient of 100 microgals/m is detected with a BHGM survey at 45° latitude and at a depth of 1,500 meters, then the density measured using equation 9.9 is 2.49 gm/cc.

While the Δg measurement is obviously critical and difficult to get with a high degree of accuracy, the Δz is also required to be accurate. For larger intervals, say over 10 ft. (3 m) wireline depth control techniques will usually suffice. However, when greater detail is required, a special shuttle sonde is used. This sonde is set up to provide accurate control over the vertical distance between measurements by moving the sensor within the tool an accurately measurable distance. The sensor is moved by a motor driven winch and its position can be measured to an accuracy of better than .04 in. (1 mm).

Porosity Determination

The density log may be used for porosity or saturation determination.[4,5,6] The bulk density of a clean formation, ρ_B, is given by the equation

$$\rho_B = (1 - \phi) \rho_M + \phi(S_W \rho_W + (1 - S_W) \rho_H) \qquad \text{(Equation 9.11)}$$

where

$$\phi = \text{Porosity}$$

$$\rho_M = \text{Density of Matrix}$$

$$\rho_W = \text{Density of Water}$$

$$\rho_H = \text{Density of Hydrocarbon}$$

$$S_W = \text{Water Saturation}$$

Solving the above equation for porosity,

$$\phi = \frac{\rho_M - \rho_B}{\rho_M - S_W \rho_W - \rho_H(1 - S_W)} \qquad \text{(Equation 9.12)}$$

which, if $S_W = 1.0$, becomes the familiar form

$$\phi = \frac{\rho_M - \rho_B}{\rho_M - \rho_W} \qquad \text{(Equation 9.13)}$$

If in a water zone, the porosity can be evaluated from equation 9.13. If open hole porosity is known and gas density can be estimated, then the gas saturation can be evaluated by solving equation 9.12 for S_w. Table 9.1 shows the densities of some common materials.[7] A deep investigating resistivity log can also be used with this density log to measure saturations.

Borehole Gravity Tool Applications

Since this measurement is deep investigating, it offers the opportunity to confirm open hole log indications which might otherwise be affected by filtrate invasion. It can be run by itself for density measurements when no open hole data is available. Often, comparison with the open hole density shows differences. These may be due to invasion or other borehole effects but may also be due to large masses of higher or lower density nearby. These types of examples follow.

The example in Figure 9.3 shows an open hole neutron and gamma ray log.[3] The neutron shows high counts over the interval 230 – 246. This gas zone indication in confirmed by the low BHGM density over this interval.

Figure 9.4 shows a pulsed neutron capture (PNC) log along with open hole resistivity logs.[8] The PNC log clearly indicates a gas from 2,690 to about 3,005 with the sigma curve reading low and the N/F overlay showing the characteristic gas zone crossover shaded black. The open hole resistivity logs appear to confirm this gas indication with high resistivity indications over this zone. The BHGM shows this zone to have a high bulk density and hence is a low porosity tight zone and would not be productive if completed.

Figure 9.5 shows how density would vary due to adjacent structures of significantly different density.[3] Figure 9.5, part a, shows a wellbore through a sedimentary interval having a density of 2.5 gm/cc. This is the value which would be seen by a conventional open hole

Table 9.1

Compound	Formula	Density e_b
Quartz	SiO_2	2.654
Calcites	$CaCO_3$	2.710
Dolomite	$CaCO_3MgCO_3$	2.870
Anhydrite	CaSO4	2.960
Sylvite	KCI	1.984
Halite	NaCL	2.165
Gypsum	$CaSO_42H_2O$	2.320
Anthracite Coal		$\left\{\begin{matrix}1.400\\1.800\end{matrix}\right\}$
Bituminous Coal		$\left\{\begin{matrix}1.200\\1.500\end{matrix}\right\}$
Fresh Water	H_2O	1.000
Salt Water	200,000 ppm	1.146
Oil	$n(CH_2)$	0.850
Methane	CH_4	e_{meth}
Gas	$C_{1.1}H_{4.2}$	e_g

density log. Nearby to this well 500 ft (150 m) away is a salt cap having a density of 2.2. The BHGM survey is clearly affected by the salt dome. Overlaying this conventional density and the BHGM survey would show the bulge as a separation between the two logs. Figure 9.5, part b, shows the affect of a gas sand located 100 ft. (30 m) from the wellbore. Here, the density difference between the BHGM and the open hole density log is shown. Again, a bulge is detected as a result of the large nearby gas sand.

POINTS TO REMEMBER

- The Borehole Gravimeter (BHGM) evaluates formation bulk density.
- The BHGM is insensitive to the borehole environment.
- The radius of investigation is 10s of feet.
- Measurements are taken stationary, 5 to 15 minutes per reading.
- Can be used to evaluate porosity or saturation, and detect gas, oil, and water contacts.
- Detects large density anomalies in the vicinity due to salt domes, fractures, gassy stringers, etc.

DENSITY LOG

The density log is primarily an open hole log used to measure formation density and hence porosity as indicated in the previous section. This tool uses a Cesium source at the borehole wall. The Cs source emits medium energy gamma rays and detects returned Compton scattered gamma rays induced from the formation. There are two gamma ray detectors which measure these scattered gamma rays. The counts at the near and far detectors, along with environmental information, provide a measurement of formation bulk density. The log presentation is in units of bulk density, gm/cc. It is a shallow reading tool and is sensitive to hole conditions. Since it is a nuclear tool, it has been run in cased holes with mixed results.

A study was done to determine the effects of casing wall thickness, cement density and thickness, and tool standoff from the casing.[9,10] These studies conclude that density logs are obtainable through casing under certain conditions. To run such a log, the casing thickness must be known and the casing to formation annulus must be less than 1 in. (2.5 cm) thick for cement and .6 in. (1.5 cm) for mud. These must also be fairly uniform along the

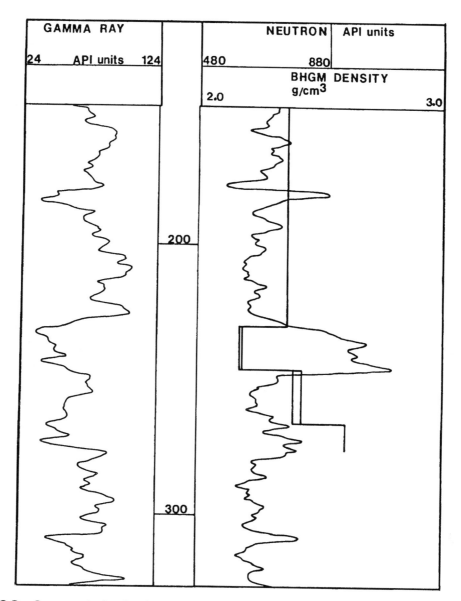

Figure 9.3 Gas zone indication by neutron and gravity meter density (Courtesy Edcon, Ref. 3)

logged interval. There must also be good contact between the pad and the casing wall. The pad radius of curvature must closely match the inside of the pipe.

It would appear that the best conditions to run the density would be in cases where the well is deviated and the density log is run along the low side of the pipe where the cement is thinnest. The density log in cased hole may also be calibrated to a neutron porosity. However, recall that there is typically no invasion in a cased hole and so such a calibration should begin in a wet zone.

FORMATION RESISTIVITY THROUGH CASING

Status of Measurement

The Through Casing Resistivity Tool™ (TCRT®) is currently under development by Para-Magnetic Logging, Inc. (PML). The concept for this device began in the late 1980's [11] and tools have been built by PML for the evaluation of this concept. Results have been quite encouraging to date. The service is not commercial at this time, although major service companies have taken an interest in this technology. A short description of how the technology works plus an example of its performance follow.

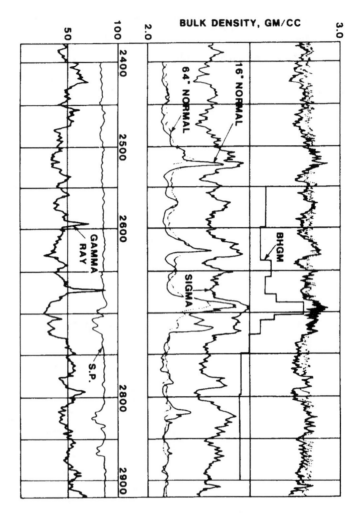

Figure 9.4 Gravity log shows low porosity where PNC log indicates gas (Courtesy SPWLA, Ref. 8)

Principle of Measurement

Referring to Figure 9.6, the schematic shown is used to illustrate the tool's operation.[11,12] First done is a nulling procedure. Switch SW1 is placed in the up position, thereby placing the low frequency AC power source across electrodes A and F. Current then passes through the casing between those electrodes with little loss to the formation. The voltage drop across electrodes C and D is amplified by amplifier A1, and across D and E by amplifier A2. In principle, the gain of amplifier A2 is adjusted so that the output of amplifier A3 becomes zero, thereby balancing the measurement system to a null for any currents flowing along the casing. Null is achieved despite thickness variations of the casing and electrode placement errors.

The switch position is now moved to the down position, thereby placing the AC Power Source across electrode A and a remote electrode B in contact with the earth. Low frequency current introduced to the casing at A will flow out of the casing and through the formation. Now, measurements between C-D and D-E indicate voltages which measure the currents $\Delta I1$ and $\Delta I2$, respectively, leaking into the formation. The net current between C and E, $\Delta I1 + \Delta I2$, is called the total differential current conducted into the formation, ΔI.

In order to calculate the resistance of the formation, the voltage difference between the tool location and a remote electrode must also be known. A measure of such a voltage difference is made between the electrode G in contact with the earth's surface and electrode J which is on the tool in the vicinity of C, D, and E. The measured voltage is called V_o, the "potential voltage."

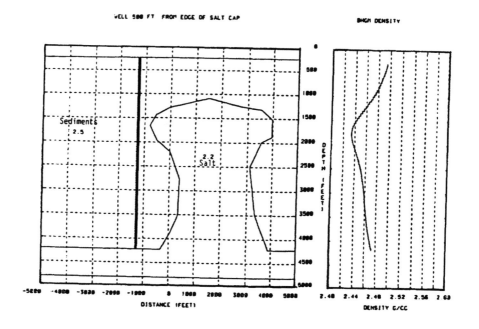

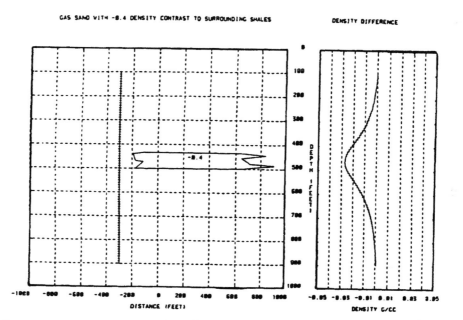

Figure 9.5 Effects on gravity meter density due to adjacent structures—a. Edge of salt cap b. gas sand near the wellbore (Courtesy Edcon, Ref. 3)

If the earth between electrodes C and E is modeled as a resistor, its resistance is given by the equation

$$r_c = V_o / \Delta I \qquad \text{(Equation 9.14)}$$

r_c is measured in ohms. The actual resistivity of the formation, in ohm-meters, may then be related to this resistance value by a constant which is dependent upon casing size and other factors relating to the borehole environment. This equation is

$$R_a = K r_c \qquad \text{(Equation 9.15)}$$

where R_a is the apparent formation resistivity in ohm-meters, which should be equivalent to true formation resistivity, R_t. Here, K is the "calibration constant" that is approximately given by the electrode spacing between electrodes C and D in meters.

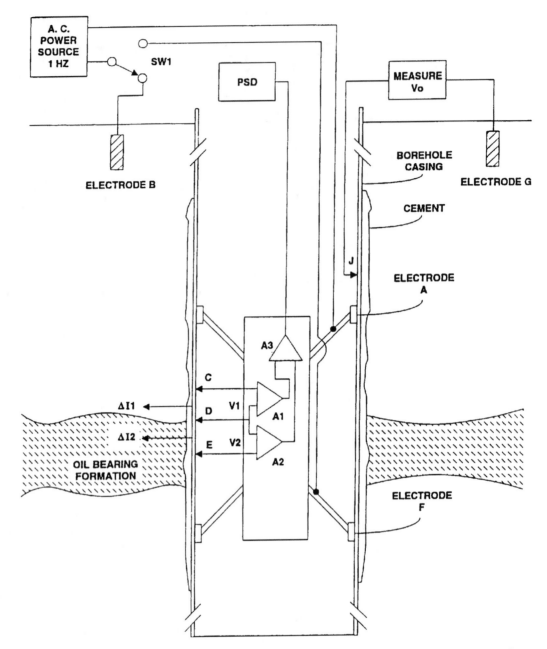

Figure 9.6 Concept of the Paramagnetic Logging Inc. Through Casing Resistivity Tool™ (TCRT®) (Courtesy SPWLA, Ref. 10)

Operational and Environmental Considerations

The tool is a contact device and in its present configuration is a stop and measure tool. It is not run continuously across the logged interval. At present, each station is quite time consuming, but it is anticipated that later generation tools can reduce the time at a station to the order of a few minutes at most, or even be run continuously. The vertical resolution of the tool is approximately two to three times the spacing of the electrodes (currently 5 ft. [1.5 m])

The tool is affected by borehole environmental factors such as cement and casing.[13] It turns out that if formation resistivity is higher than the cement resistivity, the influence does not exceed 10% of the measurement. If the formation resistivity is low and the cement resistivity is high, the error may be somewhat larger. Problems arising from scale and corrosion appeared to be non-existent during studies in a research well. There is an effect arising when measurements are taken within about 100 ft (30 m) of the bottom of casing which must be compensated for.

Measurement Results

The early measurements for this tool were made at the Gearhart test well in Fort Worth, Texas. A section of the test interval is shown on Figure 9.7, part a.[10] This shows a deep induction, a smoothed Laterolog 3 (LL3), and data from the Through Casing Resistivity Tool (TCRA in this figure). The TCRA old and new data were separated in time by about one year. This logged section clearly shows that the TCRA repeats quite well and responds much like the LL3, and not reading as low as the deep induction. The section of Figure 9.7, part b, also shows correlation of the TCRA with the LL3 over a resistivity range of 7 to about 150 ohm-meters.

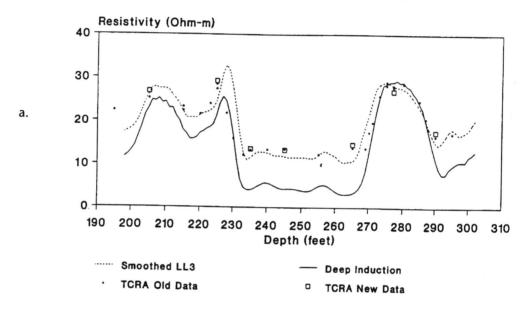

a.

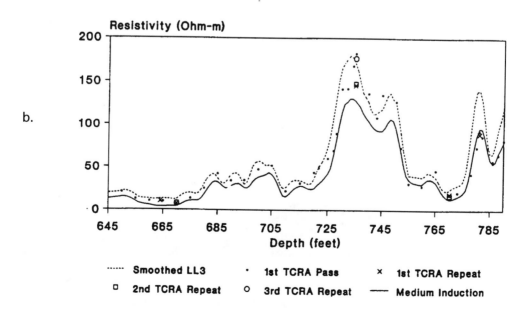

b.

Figure 9.7 Comparison of Paramagnetic Logging Inc. TCRT resistivity measurements with open hole data in test well (Courtesy SPWLA, Ref. 10)

Application of Resistivity Through Casing

This technology offers the opportunity to measure resistivity when no open hole logs are available. This may occur in old wells, where no logs have been run, or where well conditions dictated that no logs could be run before casing the well. The TCRT offers a means to evaluate zones which would have been bypassed or could not have been logged otherwise. The gamma ray, neutron, and sonic logs may all be effective through casing, and so the basic open hole combination or something closely equivalent may be available through casing in the future. The TCRT is an excellent means of monitoring producing or waterflooded wells by detecting water breakthrough. With other measurements, it may offer a means of discriminating formation from injected flood waters.

POINTS TO REMEMBER

- Formation resistivity can be measured through casing.
- The measurements are taken at stationary positions.
- The tool appears to respond like a LL3.
- The tool is not commercially available.

CHLORINE LOG

Chlorine Log Tool Operation

Chlorine logs have been used to discriminate water from hydrocarbon bearing sands with little effect of lithology. These tools have been particularly helpful in low resistivity apparently shaley pays. The chlorine log carries a 20 Curie Am-Be neutron source and a scintillation gamma ray detector set to evaluate formation gamma rays in two spectral windows. These are the Hydrogen and Chlorine windows as shown in Figure 9.8.[14,15]

Chlorine Log Example

By properly selecting the hydrogen and chlorine windows, an overlay technique may be used for zonal evaluation. In a salt water zone, both hydrogen and chlorine counts are expected to increase or decrease together, typically with changes in porosity. When in a

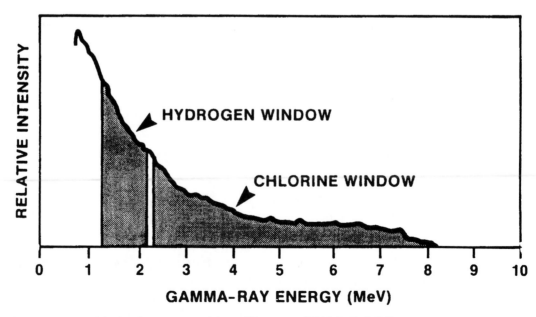

Figure 9.8 Chlorine log gate positions (Courtesy SPWLA, Ref. 14)

hydrocarbon zone, the chlorine counts disappear. As a result, the hydrogen and chlorine curves, if properly scaled to overlay in a water zone, separate in a hydrocarbon zone. Such a separation is shown in the log of Figure 9.9.[14] This measurement, when supported by lab data, is susceptible to quantitative saturation evaluations. This technique is relatively insensitive to shale.

CASED HOLE WIRELINE FORMATION TESTER

Overview

Cased hole wireline formation testers are used for taking and retrieving formation fluid samples to the surface. During this process, measurements of formation pressures and permeabilities are also made. The procedure entails perforating the casing, taking the sample, performing the measurements, and returning the sample to the surface. One major problem with this type of tool is that the perforation hole remains after the test.

The cased hole testers are typically equipped with one or two shaped charges, and therefore can perform only one or two tests per trip in the hole. The minimum hole size is rather large, with most tools requiring at least a 5 1/2 in. (14 cm) inside diameter casing. A slim hole version (Schlumberger RFTT-N*) is capable of running in casing as small as 4.5' 13.5#/ft with nominal I.D. of 3.92 in. (10.0 cm). These tools have high quality quartz pressure gauges which offer high accuracy and resolution.

Tool Operation

The cased hole formation tester is offered by most of the major wireline companies. A schematic of such a tool is shown on Figure 9.10. This particular tool is the Schlumberger Cased Hole RFT* (Repeat Formation Tester*). (* indicates that these are marks of Schlumberger.) These tools are run with a gamma ray or CCL for depth control. Once zones are identified for testing, the tester is run to the requisite depth. To set the tool for a test, the

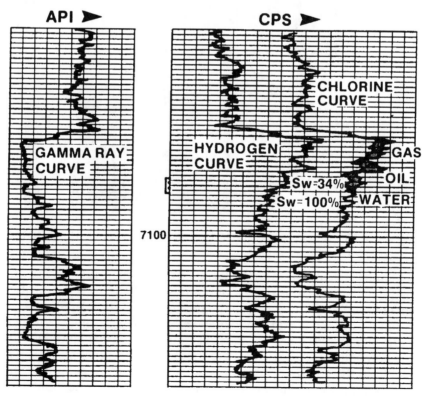

Figure 9.9 Overlay of Hydrogen and Chlorine count rate curves (Courtesy SPWLA, Ref. 14)

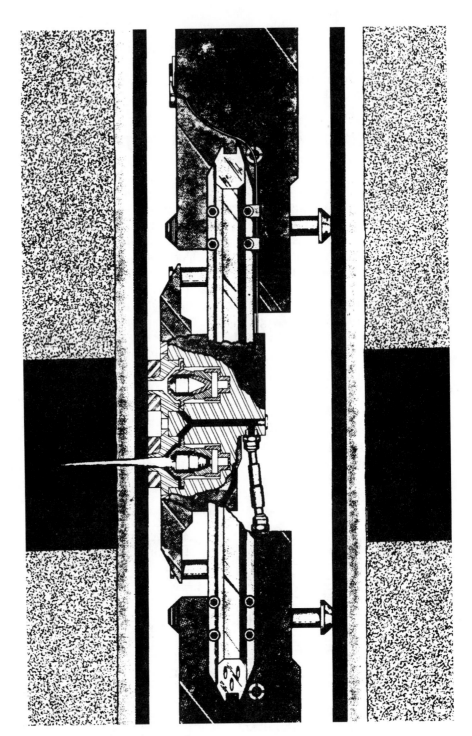

Figure 9.10   Schlumberger Cased Hole Repeat Formation Tester (RFT) (Courtesy Schlumberger)

hydraulic system within the tool causes the circular packer to be pressed against the casing by the extension arms on the opposite side of the tool. At this time, the tool draws in a small internal piston to reduce the pressure in the flow line to the packer. This is called the "pretest" and the piston would draw only about a 10 cubic centimeters (cc) volume. This is a critical part of the test, for the packer must hold the reduced pressure without a leak from the borehole fluids. If the pressure, recorded at all times at the surface, drops, it indicates a seal from the borehole fluids and is the "go ahead" signal to continue the test sequence.

To sample the formation fluids, the shaped charge is now initiated. It perforates through the casing and several inches into the formation to establish communication between the formation and sample chamber of the tester tool. The sample from the formation is then allowed to flow into a sample chamber contained at the bottom of the tool. These sample jugs are typically measured in 1, 2.5, or 5 gallon sizes (4k, 10k or 20k cc). (See Figure 9.11.[12]) Sometimes, the sample jugs contain a water cushion device to reduce the flow rate and pressure drop at the formation. After flow into the sample chamber is completed, the tool is retracted, the pretest pistons return to their initial position, and the test is completed. Some tools have the capability of taking a second test by having a second shaped charge and packer assembly. From the beginning and throughout this test sequence, the pressure is recorded at the surface.

After the test is completed, the perforation hole remains. At present, there is no provision to repair this hole and a packer or cement squeeze must be performed if this communication is not desired. Some older systems no longer in use included a chamber filled with cement. After perforating and sampling, the differential pressure from the borehole was used to squeeze this cement into the perforation. Unfortunately, cement tends to develop strength if not moved frequently, and so the cement would often set up within the tool and not go into the perforation.

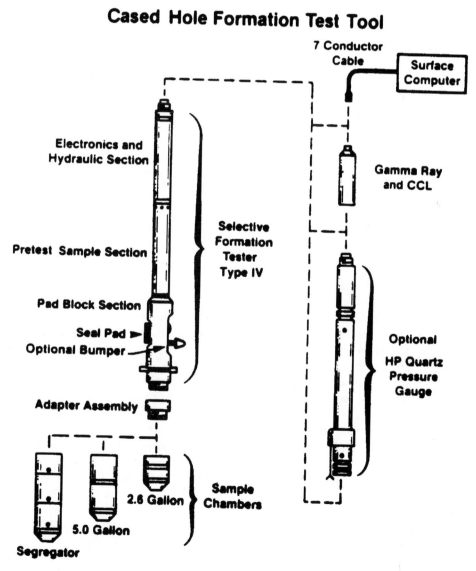

Figure 9.11 Typical components of a cased hole formation tester (Courtesy Halliburton Co. and NAPE, Ref. 16)

The Pressure Record

A typical pressure record for a cased hole formation tester is shown on Figure 9.12.[16,17] Pressure is recorded vs. time, with time increasing toward the bottom of the figure. On the left-hand track is shown the analog pressure record detected by a pressure gauge within the tool flow system. Since the pad is open to the wellbore, the initial pressure recorded is the pressure of wellbore fluid, referred to as "hydrostatic" pressure in the figure. When the pad is set, fluid is trapped within the pad and flow system of the tool while the rubber pad continues to be pushed against the casing, causing an increase in recorded pressure. The decline of pressure following is caused by the retraction of pretest pistons within the tool, thereby increasing the volume of the tool flow system and causing a pressure reduction. The new stable pressure recorded is low and nearly constant, indicating an effective seal. This is the signal to proceed with the test.

The shaped charge is fired, establishing communication, and the sample jug is opened. The large drawdown is due to the large volume of the sample jug at low pressure. The formation fluid flows into the sample chamber, filling it and causing the pressure detected to gradually build up toward formation pressure. Finally, the sample jug is closed, the equalizing valve is opened to equalize the pressure between the wellbore and tool flow system, the tool is retracted, and the pretest piston(s) return to the initial position. The total time for this test is about three and one half minutes. Times for such a test typically range from about two minutes to as long as one is willing to wait for pressure build up to be satisfactorily completed.

On the right-hand tracks is shown a digital record of the pressure. There are four small tracks, each representing one of the four digits from 0000 to 9,999 psi. The digital recording in each track is in units of 0 to 9. For example, at the top of the log, the hydrostatic pressure is indicated from the tracks as 1-1-2-3, or 1,123 psi. Notice on the last digit, the recording is bouncing between 2 and 3 for a short time before the pad contacts the casing, whereupon the pressure increases rapidly. At the end of the test, the pressure recorded is 1-1 9-5, 1,195 psi, and increasing. This indicates that pressure is building up to formation pressure but has not yet reached it. Notice that the values of pressure are also recorded in the depth track.

Interpretation of the Formation Tester

The main feature of this type of tool is that it offers a real sample from a formation which may not have been evaluated or completed earlier. It is critical that a good cement job exists over the interval tested and that the casing is in good condition. Without good cement, the sample retrieved did not necessarily come from the zone tested and therefore is of little value. The casing condition is critical for achieving seal with the packer and if it is too poor, it may be damaged when the tool is set. A good sample can be analyzed at the well site for hydrocarbons and/or gas or may be sent to a lab for more accurate analysis.

The data may be analyzed further than simply for sample analysis.[18,19] The formation pressure may be evaluated. Formation pressure may be taken as the last recorded pressure if the pressure appears to have stabilized. The last pressure of the log in Figure 9.12 is 1,195 psi. It is near its stable value since the recorded pressure has changed only one psi during the previous 10 seconds. Actual formation pressure will be slightly larger. Modeling the build-up using either a spherical (early time) or cylindrical (late build-up) flow model provides a more accurate measure, provided the correct model is selected. The technique uses a Horner type build-up plot of pressure vs. the appropriate function of time. Such plots are shown on Figure 9.13 for the data of the log.[12] Clearly, the cylindrical model applies and yields a final formation pressure of about 1,207 psi.

Permeability may also be evaluated from the slope of the build-up data. The numbers generated for permeability are somewhat controversial,[20] since they may vary by orders of magnitude depending on which model is selected. Suffice it to say that if the build-up occurs rapidly, the formation, at the depth of the perforation, has good producing characteristics. If build-up occurs slowly, it is tight. The ideal formation will have a high pressure, and the sample will be retrieved quickly, indicating high or at least reasonable permeability. The

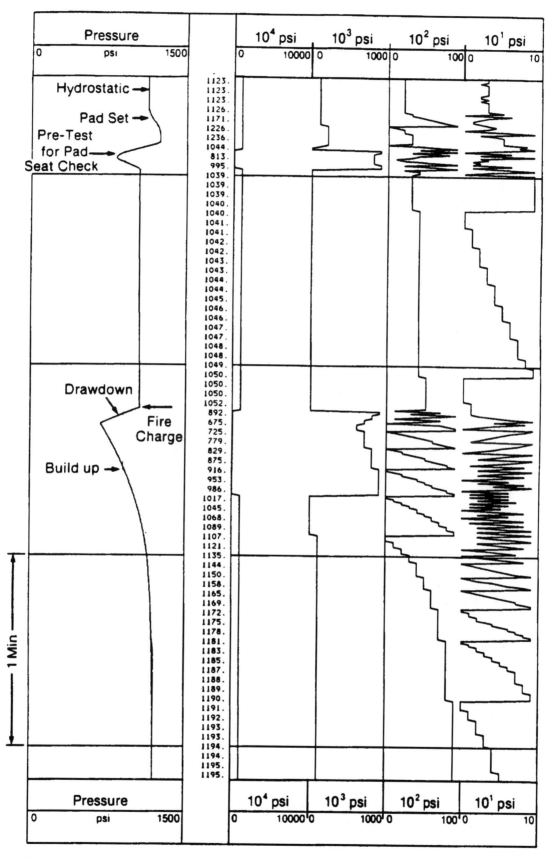

Figure 9.12 Cased hole formation tester pressure record (Courtesy Halliburton Co. and NAPE, Ref. 16)

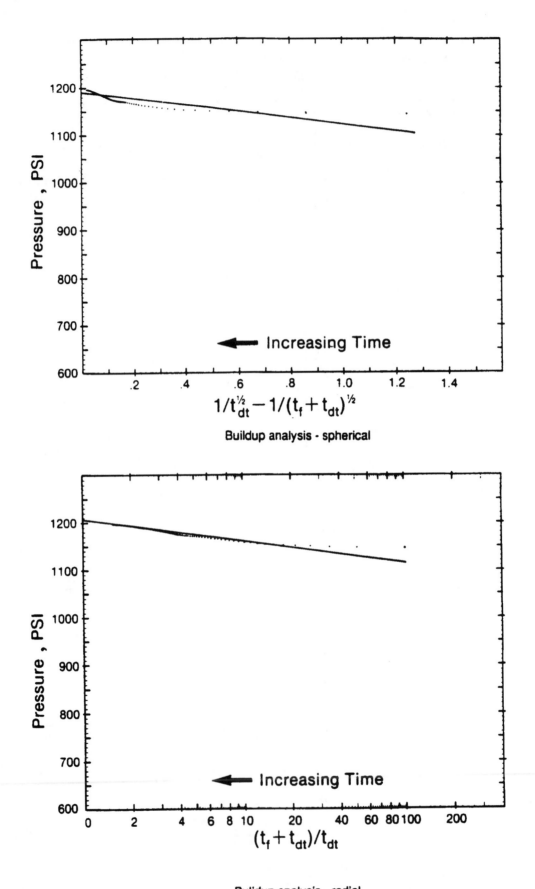

Figure 9.13 Pressure buildup plots from the formation tester pressure record—a. Spherical buildup, b. Radial buildup (Courtesy Halliburton Co. and NAPE, Ref. 16)

sample should contain hydrocarbons. Problems with filtrate, as are incurred with open hole testers, are usually not a problem unless the tests are run shortly after setting casing.

Figure 9.14 shows an example application of such pressure data.[5] On this figure, a number of zones are being tested for pressure. Each zone shows both the final gauge pressure and the pressure extrapolated from buildup. Clearly, there are at least four separately pressured intervals, at least one of which looks to be seriously depleted. Furthermore, as a result of taking a number of test points across an interval, the gradient of the fluid in that interval can be determined thereby indicating water, oil, or gas in an interval. The hydrostatic pressure is also recorded and is an excellent quality control test since the gradient that it defines should be the gradient of the well bore fluid, provided that the fluid column is continually being filled and the interface is not allowed to drop. While this is an open hole example, the same principles apply to cased hole tester data. The main difference would be that the cased hole tester requires a trip out of the hole for each two tests, and therefore it is unlikely that this quantity of data would be gathered.

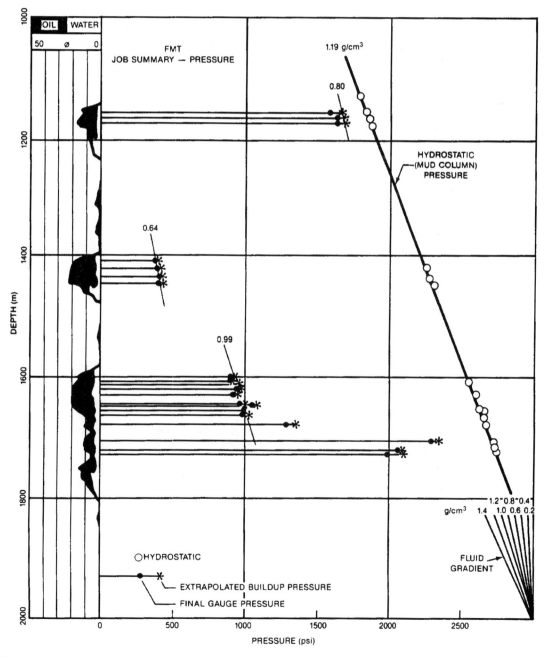

Figure 9.14 Pressure summary of a well from tester data (Courtesy Western Atlas, Ref. 15)

POINTS TO REMEMBER

- The cased hole formation tester takes formation sample by perforating through casing and producing the formation fluid into a sample jug.
- The sample jug may be evaluated at the surface or sent to a lab for PVT analysis.
- The tester can also measure formation pressure with great accuracy.
- Testers can evaluate permeability, but this is a highly questionable calculation in terms of accuracy.
- When a sample is taken, a hole is left in the casing.
- Good cement isolation is necessary for accurate measurements and representative samples.
- Casing condition should be good to assure good sealing by the packer.

References

1. Black, A.J., "Borehole Gravity Surveying, Current Instrumentation, Capabilities and Applications," *Borehole Geophysics for Mining and Geotechnical Applications,* ed., P.G. Killeen, Geological Survey of Canada, Paper 85-27, pp. 181-187, 1986.
2. Popta, J.V., and Adams, S., "Gravity Gains Momentum," Schlumberger, *Middle East Well Evaluation Review,* pp. 6-11, November 12, 1992.
3. Edcon, Inc., *Borehole Gravity Density Logging,* In-house Publication, Denver, Colorado.
4. Gournay, L.S., and Lyle, W.D., "Determination of Hydrocarbon Saturation and Porosity Using a Combination Borehole Gravimeter (BHGM) and Deep Investigating Electric Log," 25th Annual SPWLA Symposium, June 10-13, 1984.
5. Popta, J.V., Heywood, J.M.T., Adams, S.J., and Bostock, D.R., "Use of Borehole Gravity for Reservoir Characterization and Fluid Saturation Monitoring," Paper SPE 20896, Presented at Europec 90, The Hague, Netherlands, October 22-24, 1990.
6. Adams, S.J., "Gas Saturation Monitoring in North Oman Reservoir Using a Borehole Gravimeter," Paper SPE 21414, SPE Middle East Oil Show, Bahrain, November 16-19, 1991.
7. Schlumberger, "Log Interpretation Principals/Applications," Schlumberger Document SMP-7017, Houston, Texas, 1987.
8. Gournay, L.S., Maute, R.E., "Detection of Bypassed Gas Using Borehole Gravimeter and Pulsed Neutron Capture Logs," *The Log Analyst,* pp. 27-32, May-June, 1982.
9. Jacobson, L.A., and Fu, C.C., "Computer Simulation of Cased-Hole Density Logging," Paper SPE 19613, 64th Annual SPE Technical Conference, San Antonio, Texas, October, 1989.
10. Cigni, M. and Magrassi, M., "Gas Detection from Formation Density and Compensated Neutron Log in Cased Hole," 28th Annual SPWLA Symposium, 1987.
11. Vail, W.B.III, "Methods and Apparatus for Measurement of the Resistivity of Geological Formations from within Cased Boreholes," U.S. Patent No. 4,820,989, April 11, 1989.
12. Vail, W.B., Momii, S.T., Woodhouse, R., Alberty, M., Peveraro, R.C.A., and Klein, J.D., "Formation Resistivity Measurements Through Metal Casing," 34th Annual SPWLA Symposium, Calgary, Canada, June, 1993.
13. Tabarovsky, L.A., Cram, M.E., Tamarchenko, T.V., Strack, K.M., and Zinger, B.S., "Through-Casing Resistivity (TCR): Physics, Resolution and 3-D Effects," 35th Annual SPWLA Symposium, June, 1994.
14. Tripathi, S.N., Domangue, E.J., and Murdoch, B.T., "Low-Resistivity Sand Evaluation with the Chlorine Log," 25th Annual SPWLA Symposium, June, 1984.
15. NL McCullough, "Chlorine Log. It Unmasks Oil In Shaly Sands," NL Industries, Houston, Texas, June, 1984.
16. Kessler, C., Hampton, D., and Harness, P., "Application of the Cased Hole Formation Tester," Paper NAPE-001, Nigerian Association of Petroleum Explorationists 7th Annual Conference, Lagos, Nigeria, November, 1989.
17. Smolen, J.J., and Litsey, L.R., "Formation Evaluation Using Wireline Formation Tester Pressure Data," *Journal of Petroleum Technology,* pp. 25-32, January, 1979.
18. Schlumberger, "RFT—Essentials of Pressure Test Interpretation," Document M-081022, Schlumberger ATL-Marketing, 1981.
19. Atlas Wireline Services, "Formation Multi-Tester (FMT) Principles, Theory, and Interpretation," Document 9575, October, 1987.
20. Cooke-Yarborough, P., "Reservoir Analysis by Wireline Formation Tester: Pressures, Permeabilities, Gradients, and Net Pay," *The Log Analyst,* pp. 36-46, November, 1984.

CHAPTER 10

WELL INTEGRITY: ACOUSTIC CEMENT EVALUATION SURVEYS

OVERVIEW

Cement and Isolation

The main purpose of cement over the production interval is to provide isolation between neighboring zones. Hydraulic isolation allows the well operator to selectively complete certain zones and assures that fluids will not move into or from neighboring zones through the borehole behind casing. Failure of isolation can cause a myriad of problems such as water production, depletion of gas drive mechanism, loss of production to neighboring zones, contamination of fresh water sands, and the like. The remedy for such problems is an expensive squeeze cement job which may only have a marginal chance of success.

Acoustic Bond Logs—What They Measure

The intent of running cement bond logs is to evaluate hydraulic seal. Other reasons exist for running bond logs, such as determining cement coverage, compressive strength, locate cement tops, and possibly others. **Acoustic cement bond logs do not measure hydraulic seal!** Instead they measure the loss of acoustic energy as it propagates through casing. This loss of acoustic energy can be related to the fraction of the casing perimeter covered with cement. If the cement compressive strength is constant over the logged interval, this fraction of cement annular fill is called Bond Index.

Referring to Figure 10.1, acoustic bond logs are suitable to detect annular fill of cement and therefore can detect channels within the cement of the types I and II. These channels

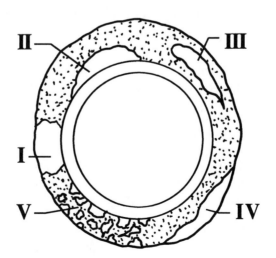

Figure 10.1 Types of channels in cement

161

directly affect the fraction of the annulus contacting cement. Types III and IV are virtually invisible to the acoustic bond logs unless they are large and the cement sheath contacting the pipe becomes very thin. Type V, the condition called "gas cut cement," would be detected but not recognized as such.

Types of Acoustic Bond Logs

There are at least four types of acoustic bond logs. There is the conventional cement bond log (CBL). This is the most common bond log. An improvement to this tool is the borehole compensated bond log. Neither of these tools has the capacity for azimuthal discrimination of cement. However, in recent years bond logs have incorporated multiple directional receivers for some azimuthal resolution. One company has placed the transmitters and receivers on pads contacting the casing wall. These tool variations are discussed in this chapter.

CEMENT BOND LOG (CBL) TOOL CONFIGURATION AND OPERATION

Tool Configuration

The CBL tool is shown schematically on Figure 10.2. The tool typically has a single omnidirectional acoustic transmitter and two receivers. Most common is a receiver at three feet and another at five feet from the transmitter. Other variations may exist, including tools having a single receiver at a four foot spacing. This tool has no azimuthal capability.[1-5] Instead the received signal is an average from all around the pipe. This tool must be accurately centralized and cannot be run with gas or gas bubbles in the wellbore. CBL transmitters typically operate in the 15,000 to 30,000 hertz range and pulse at rates from 15 to 60 pulses per second, depending on the tool and service company.

Acoustic Signal Path

Figure 10.2 shows a CBL tool in a borehole.[6] The acoustic signal from the transmitter can get to the receiver by way of a number of different paths. One, of course, is through the tool. Slots or other means are usually incorporated in the tool housing so that this signal is not detected at the receiver during the periods of interest. Four other paths are shown on Figure 10.2. For these paths, the acoustic signal propagates through the wellbore fluid, the casing, the cement, and the formation. The signal observed at the receiver is a composite of these signals.[6] The receiver signal is the basis of all interpretation of this log.

THE RECEIVED SIGNAL AND LOGS PRESENTED

Wavetrain Display

Each path is a different material having a unique acoustic velocity. Table 10.1 shows the acoustic velocities of various materials encountered downhole.[2] A typical composite wavetrain is shown at the top of Figure 10.3.[6,7] The transmitter signal is indicated. After a certain time, the earliest arrival is the casing or pipe signal. Its velocity through the casing is 57 microseconds/ft. This arrival is followed by the formation signal, which usually overwhelms and contains the slower cement signal. The acoustic velocity of formation materials, with the exception of limestone or dolomite, is usually slower than or close to the velocity through steel and the acoustic path is longer. These factors account for the later arrival of the formation signal. Lastly, the arrivals from the mud or borehole fluid appear.

The effect of the cement is to cause the acoustic energy in the casing to dissipate into the formation. This is caused by the acoustic signal propagating like a compression wave along the casing from the transmitter to the receiver. Good cement contact, called good bond, provides shear support and causes that wave to die down or be greatly attenuated by the time it gets to the receiver. If contact is poor or nonexistent, as in free pipe, there

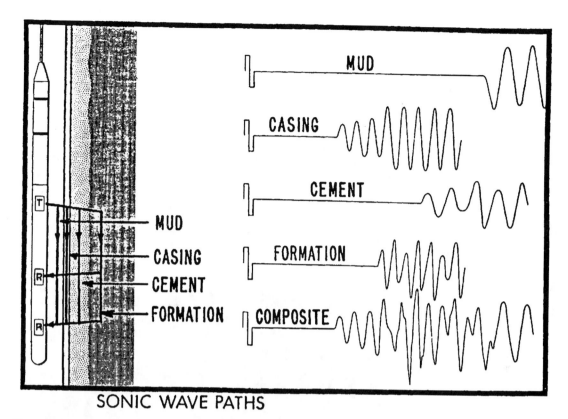

SONIC WAVE PATHS

Figure 10.2 Cement Bond Log tool configuration and sonic wave paths (Courtesy Schlumberger, Ref. 6)

TABLE 10.1 Travel Time through Various Materials

Material	Travel Time, microsecond/ft
Sandstone	55.5
Limestone	47.6
Dolomite	43.5
Salt	67.0
Anhydrite	50.0
Polyhalite	57.5
Water (Fresh)	200.0
Water (100,000 ppm NaCl)	189.0
Water (200,000 ppm NaCl)	182.0
Oil	222.0
Air	919.0
Steel Casing	57.0
Mud	167.0
Cement	90.0–160.0

is no shear support on the casing and the signal is attenuated very little.[8,9] This is poor board. The contrast between these two conditions is shown in the casing arrival period of the wavetrain on Figure 10.3.[7]

The effects on the wavetrain for a variety of conditions downhole is shown in the "signature" track of Figure 10.4.[2] The free pipe condition at the top shows the acoustic signal "ringing" loudly. Notice that this signal totally dominates, and the formation arrivals are non-existent. This is due to the great acoustic impedance (density x acoustic velocity) mismatch between the pipe and the fluid outside. The acoustic signal is trapped within the

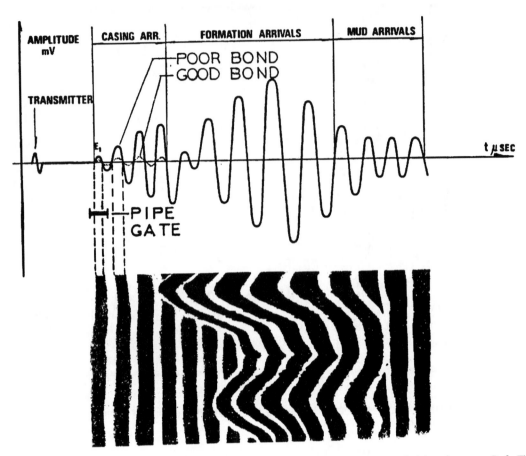

Figure 10.3 Sonic wavetrain at receiver and Variable Density Log (VDL) (Courtesy Schlumberger, Ref. 7)

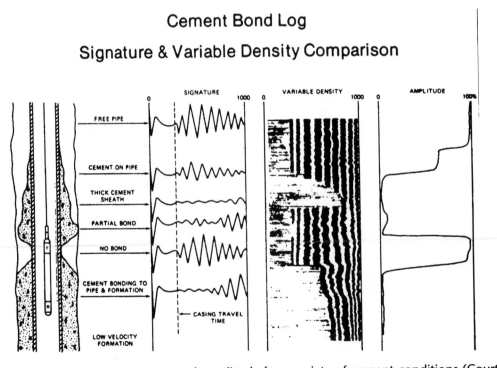

Figure 10.4 Comparison of wavetrain, VDL, and amplitude for a variety of cement conditions (Courtesy Western Atlas, Ref. 2)

pipe wall and little energy leaves the pipe. In contrast, the bottom section shows the best possible bond. Here, the pipe signal is low, and the formation signal comes in loud and clear. If the pipe is bonded in some parts and not in others, then both pipe and formation signals may be seen.

Notice the condition of cement adhering to the pipe, but not contacting the formation. In this case, the amplitude of the casing arrival decreases (attenuation increases) until the cement thickness reaches about 3/4 in. (1 cm).[9] Thicker cement will have no added effect and the signal is as low as it will ever be. Notice also that the formation signal is weak or non-existent. This is again a result of the acoustic signal not jumping the liquid gap easily and the energy remains trapped in the casing and cement. While this case of unsupported cement is unlikely, and since the smooth and impermeable casing would seem more difficult to bond to than the rough and permeable formation, it has been observed that thin cement sheaths may act as if unsupported in terms of the attenuation of the pipe signal.[10,11] As a result, thin cement sheaths may have a somewhat higher pipe signal amplitude.

It is important at this time to be sure to understand the different signals comprising the wavetrain. All measurements made with the CBL tool are based on this wavetrain.

CBL Amplitude Curve

The CBL amplitude curve is a measure of the amplitude of the casing or pipe signal at the 3 foot (.9 m) receiver. Notice that when a bond log is run, the acoustic path from the transmitter to the receiver should not change. If the borehole fluid, casing pipe size and weight, and pipe material do not change, and if the tool stays centralized, as it should, then the acoustic signal through the pipe will always arrive at the same time with only its amplitude affected by the presence of cement. An electronic gate or window is set which is open during the time that the casing signal should be arriving at the receiver. The maximum amplitude is detected within the gated period. Generally, the gate opens at a specific time after the transmitter fires and is open for a fixed period of time.

The gate may be set to measure the amplitude of a specific wave, i.e., the first or second positive arrival or the first negative arrival of the wavetrain. Some equipment opens a somewhat broader gate and looks at the first few waves. In this regard, focusing on the first positive or negative arrival would appear to be best since it minimizes contamination from later formation signals appearing in the pipe gate.

There are both fixed and floating gates. Fixed gates are as described above. Floating gates are not set to open at a specific time, but are triggered when a certain minimum amplitude is detected.[12] This floating gate should never be run without the fixed gate. The following discussion refers to fixed gates unless otherwise specified.

A fixed pipe gate, set to focus on the amplitude of the first arrival (referred to as E1) is shown on the wavetrain of Figure 10.3. Generally, the highest amplitudes are associated with wholly uncemented or "free" pipe. The lowest amplitudes are associated with the best bond. The amplitude curve is simply the maximum amplitude recorded within the gate and recorded as a log. It is usually expressed in millivolts, but may be expressed in other units by the service companies. Figure 10.4 shows the amplitude curve on the right-hand track. Notice that this curve is scaled from zero to 100 percent. This amplitude log shows free pipe at the top and just below the middle of the interval where the schematic indicates no bond. The lowest amplitude at the bottom of the logged interval shows good bond. Notice that the middle interval shows similar low amplitude and good bond over the interval where the cement sheath is thick and not contacting the formation.

The Variable Density Log (VDL)

The variable density log, VDL, is derived directly from the wavetrain.[13] Looking at Figure 10.3, imagine that each discrete wavetrain corresponds to a flash causing a film to be exposed. The positive amplitude waves expose the film while the negative amplitudes do not. With 15 to 60 pulses per second, the sequence of such exposures while logging is to create a continuous map of the positive wavetrain peaks as shown on the lower portion of this figure. This is a contour map with positive values (black) and negative values (white).

Looking carefully at the figure, notice that the pipe signal, which should always arrive at the same time, forms straight lines. The formation signal, however, must pass through varying thicknesses of cement and formations whose acoustic properties vary with depth. As a result, the formation signal shows up as the non-straight or wavy set of lines on the VDL display. The mud arrivals may show up as straight lines, but usually long after the business at hand is finished. The VDL is recorded from the 5 foot (1.5 m) receiver to allow greater separation in time of the pipe and formation arrivals.

The center track of Figure 10.4 shows the VDL over a variety of intervals. The strong straight lines characterizing free pipe are shown at the top. Good bond is shown at the bottom. Because the pipe signal is low, it does not expose the film adequately to darken it. As a result, the pipe arrivals have disappeared, but the wavy or ratty formation signal comes in loud and clear. Where partial bond exists, both the pipe and formation signals may be seen. The case of a thick cement sheath with no contact to the formation shows neither pipe nor formation signal on the VDL.

Travel Time Curve

The travel time is the time it takes for the signal to travel from the transmitter to the receiver. It is measured using a preset amplitude detection level or bias. This level is set to be high enough to not trigger on tool road noise but to trigger on the first wave, E1 in Figure 10.5, part a.[7] When the transmitter is pulsed, a clock is started and when an amplitude greater than the detection level is detected, the clock is stopped. Travel time is measured in microseconds.

Travel time does some odd things.[14] When the amplitude gets low, the detection level is reached somewhat later. This is solely a result of the decrease in amplitude since the wave frequency and hence zero crossings do not change. This phenomenon is called "cycle stretch" and is shown on Figure 10.5, part b. Some equipment computes the zero crossing travel time and does not exhibit this cycle stretching condition.

When the amplitude gets very low, it is possible that the detection level may be higher than the first arrival. When that happens, the detection is made at the second or even the third positive arrival. When this occurs, the travel time is said to "cycle skip" as shown on Figure 10.5, part c. Such lengthening of travel time, either by stretching or skipping, is a result of good bond, and so increases of travel time are considered good indicators of cement fill.

In the past, many service companies did not run a travel time curve. This curve is critical to a properly run and centered CBL log. While service company charts are available for travel time, it may be approximated by using 57 microsecond/ft (187 microsecond/m) as the acoustic velocity through steel pipe, and 200 microsecond/ft (656 microsecond/m) as the velocity through fresh water. The length of steel travel is approximately the transmitter receiver spacing, and the length of travel through fluid is twice the annular clearance between the tool and the pipe.

CBL LOG PRESENTATION

Basic Presentation

The CBL log presentation of Figure 10.6 is typical.[2] On the left track, the gamma ray and the travel (SRT) time are presented. In this case, the travel time is scaled from 240 to 340 microseconds, i.e., to read 100 microseconds across the track, a highly desirable scale. Travel time is also often presented with a range of 200 microseconds across the track, also acceptable. The collars are shown in the narrow depth track. The middle track contains the amplitude (CBL%) and 5x amplitude. In this example the amplitude is scaled from 0 to 100% of receiver amplitude. A presentation of amplitude in millivolts is more common, scaled from 0 to 100 mv. The VDL is shown on track three. The other curves listed on the log heading are not usually shown. They are the tension on the cable (TEN) and the predicted pipe travel time (PPT).

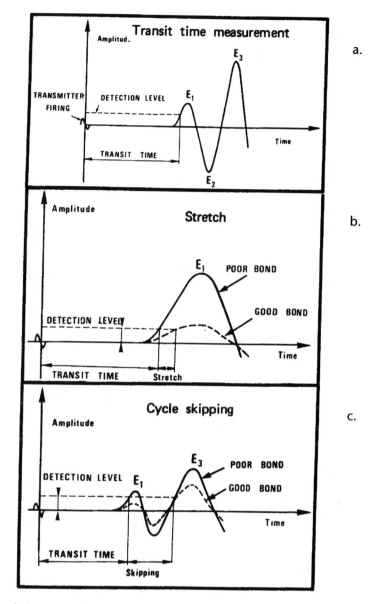

Figure 10.5 Travel time measurement—a. Normal, b. Stretch, and c. Cycle Skipping (Courtesy Schlumberger, Ref. 7)

This log example shows a free or nearly free pipe over the top 75 ft (22 m). The amplitude is high at about 90% signal, the travel time is about 263 microseconds, and the VDL shows strong straight lines for the pipe signal. The rapid decline in amplitude shows a transition from free pipe to good bond and is referred to as the top of cement. Below the cement top, the travel time skips at low amplitudes to a value greater than 340 microseconds, i.e., off scale. The VDL pipe arrival disappears while the formation signal is received very well. Note that the 5x amplitude shows that the amplitude has dropped to less than 2% of the maximum value.

FACTORS AFFECTING TOOL PERFORMANCE

Centralization

Centralization is critical for a quality CBL measurement. Figure 10.7 puts centralization into perspective.[7] It indicates that a tool eccentralization of only 1/4 inch (.64 cm) is enough to cause the amplitude to be reduced by 50% from what it should be based on the

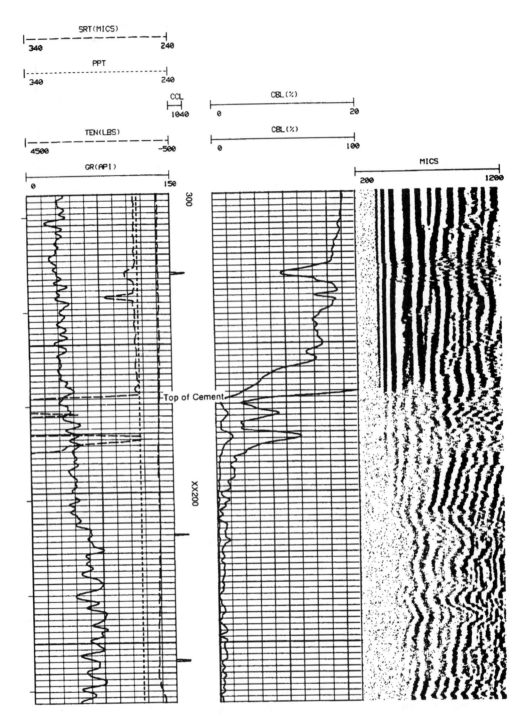

Figure 10.6 Bond log showing cement top (Courtesy Western Atlas, Ref. 2)

cemented condition. Clearly, if a tool is not centered, there is a serious likelihood of saying there is good bond when in fact there is not.

The main way to monitor tool centralization is through the travel time measurement. The drawing at the upper right of Figure 10.7 shows an eccentered tool. On the close side, there is a short acoustic path. The travel time detected from this signal will be shorter than if the tool was properly centered. The generally accepted criterion for a good log is that the travel time should never be shorter than 4 to 5 microseconds less that of the centralized free pipe signal. The 4–5 microseconds corresponds to about a 1/8 in. (.32 cm) eccentering and a loss of amplitude of about 30%.

Figure 10.8 shows the effects of eccentralization on a free pipe signal.[15] Curve C, the free pipe signal, is centralized with a travel time of 312 microseconds. The recorded free pipe

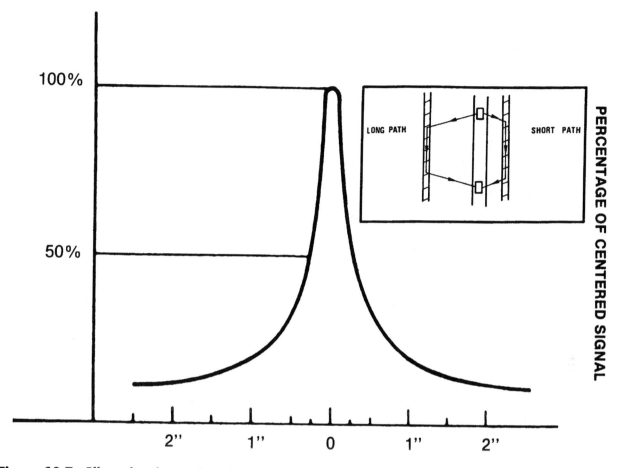

Figure 10.7 Effect of tool centralization or receiver amplitude (Courtesy Schlumberger)

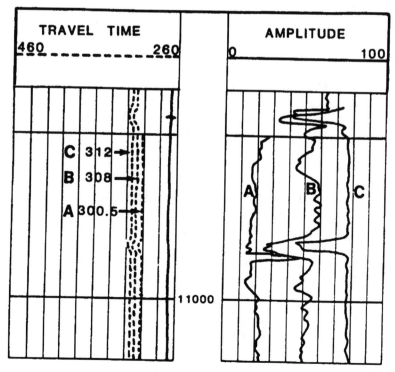

Figure 10.8 Log example showing effects of tool decentralization (Copyright2 SPE, Ref. 15)

amplitude is about 78 mv. A 4 microsecond eccentralization (1/8 in. or .32 cm) to 308 microseconds reduces the amplitude to about 50–60 mv, a 23-35% loss of signal. A nearly 12 microseconds (3/8 in. or 1.0 cm) eccentralization causes an amplitude reduction to less than 20 mv, for a nearly 75% signal loss.

Remember, for eccentralization a simultaneous shortening of travel time and reduction of amplitude will occur.

While shortening of travel time is undesirable, recall that lengthening of travel time was good news, since it was associated with improved bond and lower amplitude. However, cycle skipping should be looked at carefully. Such skipping should not occur at amplitudes greater than about 5 mv. If skipping occurs at significantly higher values, then the travel time curve loses its diagnostic capability regarding centralization when the bond is fairly good.

Fast Formation

A fast formation signal occurs when the acoustic signal propagating through the formation actually arrives at the receiver before the pipe arrival. From the acoustic velocities listed on Table 10.1, it is clear that this can only occur when the formation is a dense limestone or dolomite. When the formation signal arrives earlier than the pipe signal, the travel time detector will be triggered early and a shortened travel time will be detected. Even if the bond is quite good and the pipe signal is low, the formation signal will now appear in the fixed gate and the amplitude will be high. Remember, in a fast formation a shortening of the travel time and an increase of the amplitude will occur.

The log of Figure 10.9 shows a fast formation below the depth of about 6,940.[16] Above that depth, there is essentially free pipe with an amplitude in excess of 50 mv. The VDL shows strong straight lines and the travel time is reading about 240 microseconds. Below the cement top, the travel time has shortened considerably. This shortening of travel time is also evident on the VDL since the formation signals are obviously arriving before the pipe signal over most of this interval. The amplitude reads high in many parts of this interval, especially over the intervals where the travel time has shortened. This example shows a cement top in a limestone or dolomite section. Compare this to Figure 10.6.

Do we have good bond if we can detect a fast formation? Clearly, there must be cement in the annulus to detect the formation signal. However, to get a formation signal, only an acoustic path is required. The presence of cement only on one side of the pipe can provide such a path, and formation signals, even from a fast formation, can be detected with less than 100% annular fill. The best that we can say is that the pipe signal must be less than or equal to the measured amplitude in the pipe gate.

Notice that in free pipe the travel time and VDL both show a marked effect at the collars. These are also evident, but not as obvious, in Figure 10.6. The travel time lengthens across the collar due to a longer path. The VDL shows reflections of the acoustic signal from the collars. These reflections are called "W" or "chevron" patterns.

Microannulus

A microannulus occurs when the pipe contracts slightly relative to its size when the cement cured, thereby causing a small microseparation from the cement.[15-18] Such a microseparation may be partially or completely around the pipe. The presence of this separation reduces or eliminates the ability of the cement to support the pipe in shear, and the acoustic signal is free to travel through the pipe with little loss of acoustic energy to the surroundings. Some causes of microannulus are listed below:

1. The pressure inside the casing is less at the time of logging that at the time of cement curing. This may be caused by:
   - Reducing the casing fluid weight
   - Lowering of the fluid level in the casing
   - Holding pressure on the casing while cement cures
2. Thermal expansion of the casing while the cement is curing followed by a return to normal temperatures

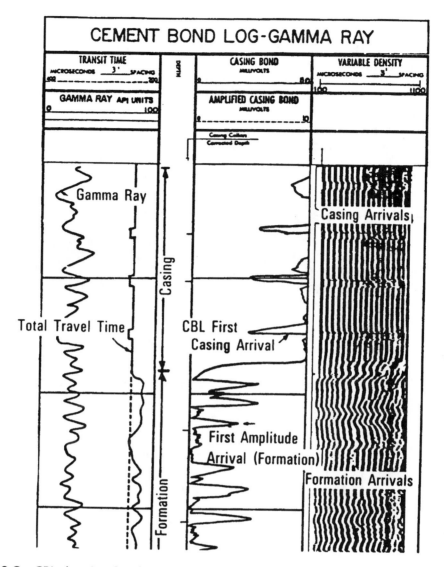

Figure 10.9 CBL showing fast formations (Courtesy Schlumberger, Ref. 16)

3. Cool fluid circulated shortly before running the CBL
4. Pressure testing or squeezing after cement has cured

It has been reported that 90% of all new wells exhibit some microannulus effects.

The general consensus appears to be that even though a microannulus is present, hydraulic isolation is still adequate. However, it would appear that some concern may be valid when the pressure gradient through the microannulus is large or the fluid flowing through is a gas. The main problem with a microannulus is that the bond log will indicate little or no cement when the annulus is full of cement. The bond log is indicating that a corrective cementing operation (squeeze) should be performed, but in fact it is not necessary and not possible. To rectify this problem, the CBL must be run under pressure. Pressure eliminates the microannulus by swelling the casing back to the position it had when the cement cured. Now the cement is in contact with the pipe and the bond log may be representative of the annular fill.

Programs to test under pressure vary. Some companies in certain areas run bond logs under zero and 1,000 psi (7 mPa), or some other pressure standard. The standard should take the pressure to at least what it was at the time of curing. Another approach suggested by the service companies is to increase the pressure from zero in increments of, say, 500 psi (3.5 mPa), until the amplitude curve no longer changes. At that point, the bond log is as good as it can be and is representative of the cement job. Such pressuring up should, of course, be consistent with ratings of pressure equipment and tubulars downhole. In the example shown on Figure 10.10, the left log is run at zero wellhead pressure, and the amplitude curve indicates poor bond.[15] Notice that this example has a wavetrain rather than a VDL display. This well was pressured to 3,300 psi (22.75 mPa) and logged with the result shown on the right. The amplitude is low, consistent with good bond, the pipe signal of the wavetrain is very low, and the travel time is skipping over much of the interval.

Just how large is the microannulus? Once the pressure necessary to restore contact with the cement is ascertained, the size of the microannulus may be determined from Figure 10.11.[2] If the casing is 5.5 in. 17 lb/ft (14 cm 25.3 kg/m), and it takes 1,000 psi (7 mPa)

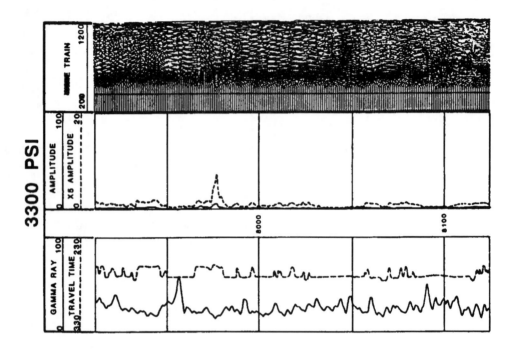

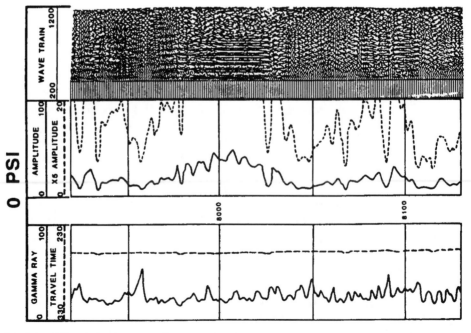

Figure 10.10 Effect of microannulus on CBL response (Copyright SPE, Ref. 15)

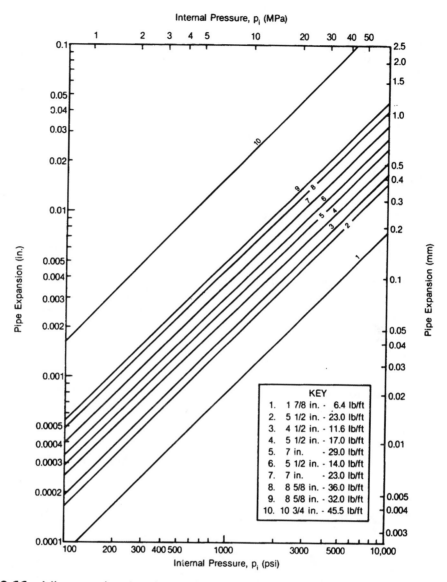

Figure 10.11 Microannulus size chart (Courtesy Western Atlas, Ref. 2)

internal pressure to close the microannulus, then the diametral expansion of the pipe is .0024 in. (.06 mm). If the annulus is uniform around the pipe, its width is .0012 in. (.03 mm). The flow, in b/d, can be estimated for a microannulus using the following equation:[19,20]

$$Q = 1.337 \times 10^6 \times (Dh^3/u) \times (delP/L)$$

where

$$Q = \text{Leak rate, B/D}$$

$$D = \text{Pipe outside diameter, in.}$$

$$h = \text{Microannulus width, in.}$$

$$u = \text{Viscosity, cp}$$

$$delP = \text{Pressure differential acting between zones, psi}$$

$$L = \text{Length of cemented section between zones, ft}$$

In the above example, if water is to leak across such a microannulus over a 10 foot (3.5 m) interval with a 100 psi (.7 mPa) differential, this equation indicates a flow rate of .13 B/D (.02 m³/D), certainly insignificant relative to probable production rates.

Cement Curing Time

Uncured cement is essentially a liquid slurry and affords no support to the casing in shear. Therefore, its presence around the pipe looks like poor bond to a CBL log prior to curing. The key to amplitude attenuation is the development of compressive strength. While this may vary with the cement mix used, a rule of thumb is to wait 72 hours after circulating cement before running any bond log. The example of Figure 10.12 shows how the amplitude was affected in a well 4, 18, 28, and 33 hours after cementing.[21] Clearly, uncured cement does not attenuate the acoustic signal appreciably.[21]

Borehole Fluids

The fluids in the borehole may affect the ultimate amplitude detected at the receiver.[4,22] Charts, such as that shown for a Gearhart CBL on Table 10.2, are typically indicative of expected amplitudes when water is the borehole fluid. However, fluids of greater density may cause the amplitude to be larger than the expected values. Figure 10.13 shows the necessary multiplying factor to bring the excessive values into conformance with chart values.[4] For example, in 7 in. casing filled with 10.5 lb/gal CaCL2, free pipe measured would be 92 mv. Using the multiplying factor of .63 would correct the free pipe reading to .63 x 92 = 58 mv, which is approximately that of free pipe with fresh water in the borehole.

Other Factors

Other factors which may affect the measured amplitude include the cement compressive strength, foamed cement, casing size, and weight.[23-25] These factors are discussed in the section on quantitative bond log evaluation. The presence of cement or excessive scale on the pipe may cause centralization problems. The presence of coatings on the pipe outer surface

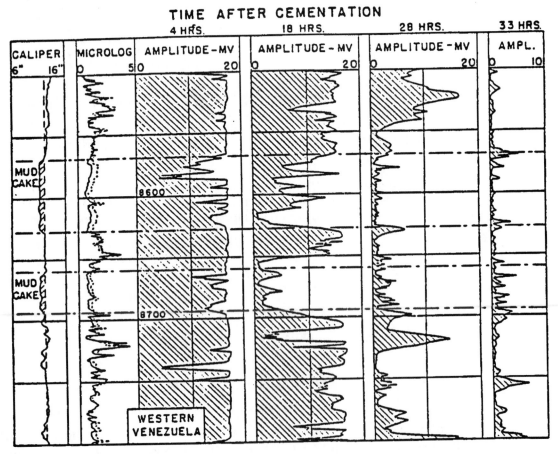

Figure 10.12 Effects of cement curing time on CBL amplitude response (Copyright SPE, Ref. 21)

TABLE 10.2

| Casing Size | Wt. | Travel Time u-sec | Free Pipe Signal | —Class H Cement— | | Interval For Isolation |
				3000 PSI 100% Cement	60% Bond Cut off	
4½"	9.5			0.2 mv	2.3 mv	
	11.6	254	81 mv	0.6 mv	4.6 mv	5 Feet
	13.5			1.0 mv	7.0 mv	
5"	15.0			0.9 mv	5.5 mv	
	18.0	258	76 mv	2.2 mv	10.0 mv	5 Feet
	21.0			3.6 mv	15.0 mv	
5½"	15.5			0.7 mv	4.8 mv	
	17.0	269	72 mv	1.0 mv	6.0 mv	6 Feet
	20.0			2.1 mv	9.0 mv	
	23.0			3.5 mv	13.0 mv	
7"	23.0			1.0 mv	5.5 mv	
	26.0			1.7 mv	7.5 mv	
	29.0			2.4 mv	9.3 mv	
	32.0	289	62 mv	3.3 mv	13.0 mv	11 Feet
	35.0			4.0 mv	14.0 mv	
	38.0			5.0 mv	15.0 mv	
	40.0			6.0 mv	17.0 mv	
7⅝"	26.4			1.1 mv	5.5 mv	
	29.7	302	59 mv	1.8 mv	7.5 mv	12 Feet
	33.7			2.6 mv	10.0 mv	
	39.0			3.5 mv	13.0 mv	
9⅝"	40.0			1.8 mv	6.8 mv	
	43.5			2.2 mv	8.5 mv	
	47.0	332	51 mv	2.7 mv	9.0 mv	15 Feet
	53.5			4.0 mv	12.0 mv	
10¾"	40.5			1.2 mv	5.1 mv	
	45.5			1.8 mv	6.5 mv	
	48.0			2.1 mv	7.6 mv	
	51.0	352	48 mv	2.5 mv	8.0 mv	18 Feet
	54.0			2.7 mv	8.4 mv	
	55.5			2.8 mv	8.8 mv	

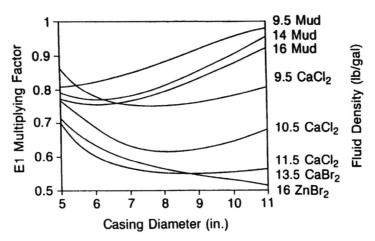

Figure 10.13 Effect of borehole fluids on CBL amplitude (Courtesy Schlumberger, Ref. 4)

may have an effect like a microannulus.[26] If the cement sheath is thinner than 3/4 in. (1.9 cm), then the amplitude measured will be larger since the this sheath cannot attenuate as effectively as the thicker sheath. Oftentimes, the cement mix is changed during the cementing process or the cement job may be staged with different cements used for each stage. The change in cement properties may affect the resulting amplitude. Gas cut cements may have the affect of decreasing the attenuation causing a higher amplitude over the gas cut interval. Concentric casings, such as in a liner overlap situation, may be difficult to evaluate since the outer pipe signal is appearing the pipe gate for the inner pipe, causing the amplitude to be inaccurate.

QUANTITATIVE CEMENT BOND LOG EVALUATION

Bond Index—Definition

Bond Index (BI) is a calculatable value which is the key to the quantitative interpretation of cement bond logs.[27] Bond Index is defined as follows:

$$BI = \frac{\text{Attenuation rate in zone of interest (db/ft)}}{\text{Attenuation rate in well cemented zone (db/ft)}}$$

If the BI = 1.0, the bond is considered perfect with 100% cement coverage of the pipe. If the BI is less than one, the bond is not perfect, but hydraulic seal may still exist. Figure 10.14 shows that a "reasonable assurance" of isolation may be achieved with a long enough interval of less than perfect bond.[27] This chart shows the length of continuous interval required with a bond index greater than or equal to .8 to achieve the "reasonable assurance" criterion. For example, for 7 in. (17.8 cm) casing, a 10 foot (3.3 m.) cemented interval with BI=.8 or greater is required to predict isolation.

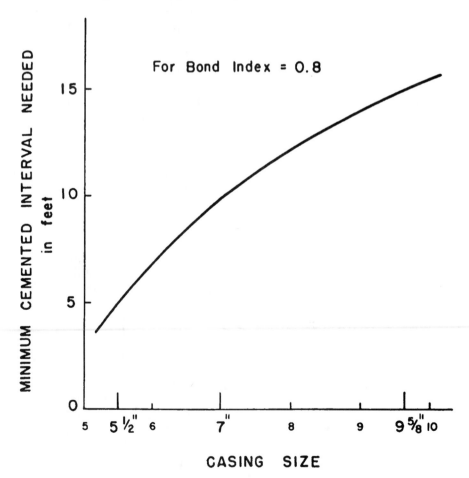

Figure 10.14 Interval required for isolation with bond index = 0.8 (Courtesy SPWLA, Ref. 27)

Bond Index—Computation Using Schlumberger Chart

Consider the log example of Figure 10.15.[15] This log shows the cement job in question with a free pipe section inserted at the top. The free pipe travel time and cemented interval travel times are shown. The cemented section travel time shows both stretching and skipping effects. The amplitude and 5x amplitude are shown in the middle track. First, scan the logged interval for a well bonded (BI = 100%) section. This will usually correspond to the lowest sustainable amplitude, in this case about .8 mv at the interval marked "100% cement 'A'."

To compute the BI, we must convert the amplitude reading in mv to the db/ft equivalent. Attenuation in db/ft may be related to the measured amplitude in mv through the following equation:[8]

$$\text{Attenuation} = \frac{20}{d} \log_{10}\left(\frac{A}{A_0}\right)$$

where

$$\text{Attenuation} = \text{Rate of attenuation, db/ft}$$

$$A_0 = \text{Amplitude of signal at transmitter, mv}$$

$$A = \text{Amplitude of signal at receiver, mv}$$

$$d = \text{Transmitter to receiver spacing, ft}$$

Rather than use the equation, it is common practice to use service company charts to do this, such as the Schlumberger chart of Figure 10.16.[24] For example, if the amplitude

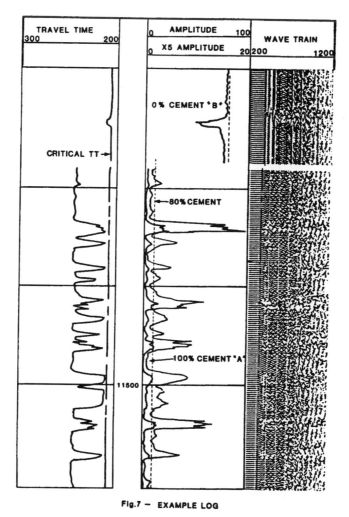

Fig.7 – EXAMPLE LOG

Figure 10.15 CBL amplitude log showing BI = .8 level (Copyright SPE, Ref. 15)

CBL INTERPRETATION CHART

CENTERED TOOL ONLY

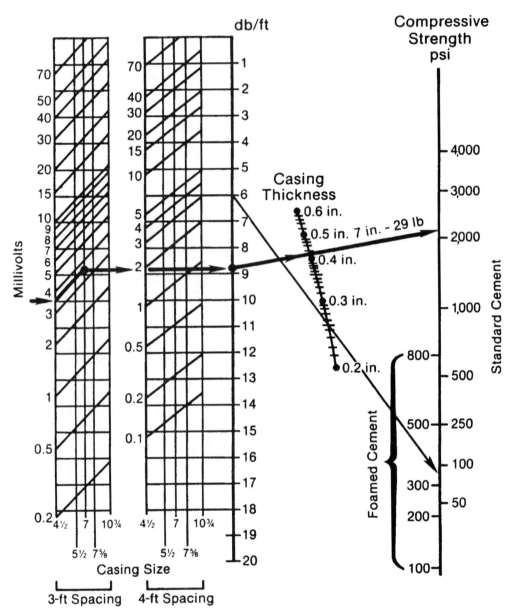

Figure 10.16 Schlumberger CBL Interpretation Chart (Courtesy Schlumberger, Ref. 24)

detected at the 3 ft. spaced receiver was 3.5 mv and the casing was 7 in. (17.8 cm.), then enter 3.5 mv, move diagonally to the 7 in. casing line, then horizontally to the db/ft value of 8.9 db/ft. For the example log of Figure 10.15, and 5.5 in. (14 cm) casing, the .8 mv amplitude reading corresponds to an attenuation of 13.5 db./ft. The attenuation corresponding to BI=.8 is determined by multiplying (13.5 db/ft) x .8 = 10.8 db/ft. Going backwards on the chart to compute the amplitude reading in mv corresponding to BI = .8 yields an amplitude of 2.0 mv. Drawing a line corresponding to 2.0 mv corresponds to the 80% cement (BI=.8) line in Figure 10.15. From Figure 10.14, five continuous feet (1.5 m) of BI=.8 or better are required for a reasonable assurance of isolation. On this basis, the only isolating intervals on this log are 11,398-11,418, 11,428-11,438, 11,447-11,453, 11,480-11,492, and 11,538-bottom.

Bond Index—Tenneco/Fitzgerald Technique

A short technique is available which can be used with bond logs from almost any service company.[28] The only requirement is that the received signal is not processed or operated on prior to amplitude log presentation. The amplitude presented must be linear with respect to the detected signal amplitude. A semilog plot is made as shown on Figure 10.17.[28] Bond Index is scaled on the linear horizontal axis while the amplitude is on the vertical logarithmic axis. The amplitude units here are millivolts, although this technique

CBL INTERPRETATION

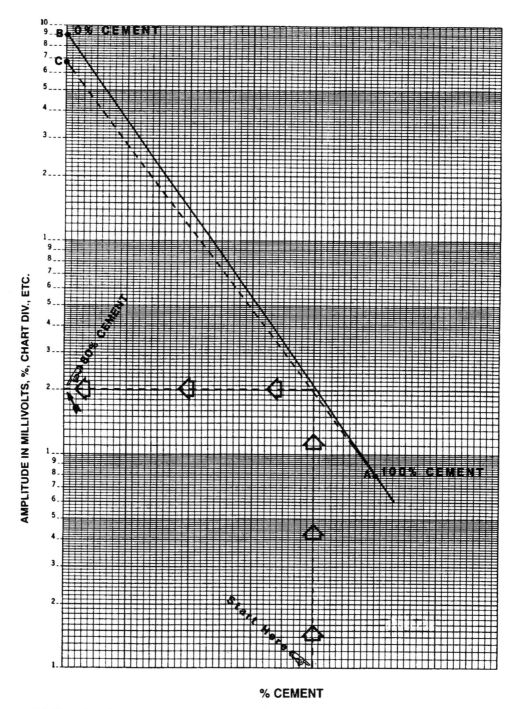

Figure 10.17 Graphical technique for determination of BI amplitude (Courtesy Donald Fitzgerald, Copyright 1983 Tenneco Oil and Exploration Co., Ref. 28)

will work regardless of units if the linearity of the measurement is not compromised. The 100% cement point is marked "A" at an amplitude of .8 mv and the free pipe is labeled "B" at about 90 mv. A straight reference line drawn between points A and B can then be used to calculate the amplitude in mv corresponding to any bond index value. For example, the BI=.8 value corresponds to an amplitude of 2.1 mv, just as earlier computed.

If the free pipe amplitude value is not well known, then an estimate can be made with little fear of significant error. For example, a second line is drawn from the well cemented point A in Figure 10.17 to an erroneous value for free pipe of 68 mv. Even so, the effect on the calculation of the pipe amplitude at BI=.8 is to change the value from 2.1 to 2.0.

Bond Index—Annular Fill vs. Compressive Strength

Up to this point, bond index was treated as being dependent on annular fill. If the cement compressive strength is constant, then BI is a direct measurement of such fill, or at least the fraction of the pipe perimeter in contact with cement. While certain parameters mentioned earlier may have an effect on the amplitude and hence the bond index, compressive strength is far and away the dominating factor after annular fill. For example, the chart in Figure 10.16 was used to obtain the attenuation rate in db/ft and BI. Beginning with 3.5 mv in 7 in. (17.8 cm), pipe an attenuation was determined to be 8.9 db/ft (29.2 db/m). However, if the pipe wall thickness was .41 in. (1.04 cm) thick, this attenuation required a cement compressive strength of about 2,000 psi (14 mPa). Going backwards on this same chart, if the cement compressive strength was 4,000 psi (28 mPa), the attenuation for a .41 in. (1.04 cm) wall pipe would be 10.5 db/ft (34.4 db/m) and the amplitude for the 7 in. (17.8 cm) pipe would be 2.0 mv. If only 1,000 psi (7 mPa) cement was on the outside, then the amplitude would be about 7 mv.

The interpretation process requires that a well bonded section be located, and this is usually taken as the interval having the lowest consistent amplitude. If that value happens to be 2 mv in 7 in. (17.8 cm) casing, then a BI of .8 will have a higher amplitude of 4.1 mv. While this may correspond to 80% annular fill, it also may correspond to 100% annular fill with a compressive strength of about 1,800 psi (12.6 mPa). In the first possibility, there is a good possibility of a channel, in the second case there is almost certainly isolation. This is the main problem with the idea of Bond Index and the CBL logs of this type. We essentially have a situation where there are two major unknowns affecting a single measurement, and we cannot discriminate the effect of one from the other. The other factors mentioned earlier are details by comparison.

Using Bond Index

The foregoing discussion does not mean that the CBL and "Bond Index" are useless.[29] 90% accuracy regarding squeeze decisions has been claimed using the bond index concept and a BI criterion of .8. In some areas, the BI criterion has been reduced to .6. Clearly, this allows what appears to be worse cement to be considered satisfactory. Why the difference? In certain areas, the formation may tend to slough in or shales may tend to swell and cause isolation even when the cement job is only marginal. In such areas, a BI=.6 may be acceptable. In hardrock areas, where holes are competent, the burden of isolation falls more heavily on the cement, and hence a BI=.8 may be more appropriate.

Another way of rationalizing this reduction of BI is to look at this bond index concept as a squeeze/no squeeze guideline. If the more stringent BI=.8 criterion is used, and it is discovered that too frequently the squeeze is not necessary because the cement cannot be displaced into the channel, then clearly this criterion is too tough. It is causing a loss of money to too many unnecessary squeezes. However, if the criterion is reduced to BI=.7 or to BI=.6, then an economic balance is formed where expensive squeeze jobs are performed less and those performed have a greater likelihood of having been necessary.

When the service companies compute BI, the basic 100% bond point is usually taken from a chart and is the amplitude detected by their tool at a 3,000 psi compressive strength. The chart of Table 10.2 is a set of parameters used for the Gearhart (now Halliburton Energy Services) CBL tool. Most service companies have such charts available. In the approach advocated herein, the best bond is evaluated from the apparently 100% bonded section of the log and the BI=.8 amplitude calculated relative to that well bonded section. The resulting differences in the threshold amplitudes are usually small, but may at times be significant. The service company has the advantage, by using chart values, of having a fixed standard for comparison. It is, however, subject to error due to borehole fluids, calibration, and in general the factors which systematically affect the amplitude. The approach of calculating BI relative to a well bonded section is independent of many corrupting factors, not calibration sensitive, and independent of the service company running the tool.

SPECIAL AND NON-STANDARD CBL EXAMPLES

Foamed Cement

The log of Figure 10.18 shows a transition in a well cemented interval from conventional to foamed cement in 4.5 in. (11.4 cm) casing.[24] Below about 4,580 ft (1,396 m) the conventional cement shows good annular fill and an amplitude of about 1 to 2 mv. Above this cement is a column of foamed cement having an amplitude of about 8 mv. Foamed cements have less compressive strength and less attenuation. Notice also that the pipe signal is still strong in the foamed cement section, even though there is good annular fill.

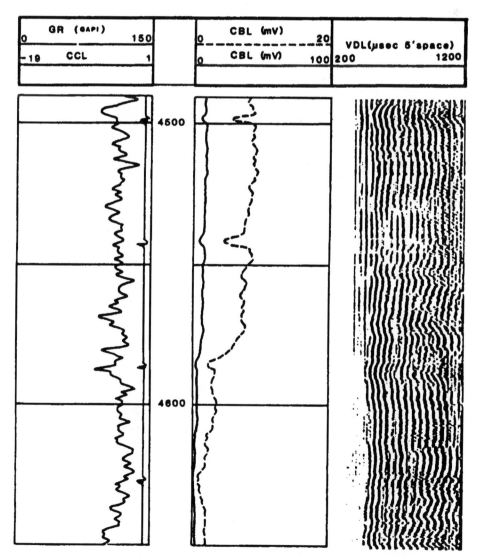

Figure 10.18   Transition from neat to foamed cement (Courtesy Schlumberger, Ref. 24)

Open and Cased Hole VDL

The example in Figure 10.19 shows an open and cased hole VDL over the same interval.[30] The amplitude curve on the right indicates a good cement job. The formation character is clearly seen in both the cased hole and open hole logs. Comparisons such as this are frequently not as good because of either poorly centralized pipe or a poor cement job.

Detection of Natural Fractures and Laminations

Figure 10.20 shows a VDL and dipmeter over the same interval. The formation signal shows numerous discontinuities which are interpreted to be either natural fractures (F) or laminations (L) in the formation.[31] These discontinuities are caused by reflections of the acoustic signal from the fractures or laminations.[31]

Evaluation of Squeeze Effectiveness

Figure 10.21 shows the VDL and amplitude curves both before and after a cement squeeze.[30] Before squeeze, the cement job shows high amplitude and strong pipe signals. Perforations were placed at 9,016 and 9,070 through which the squeeze was performed. The after squeeze amplitude and VDL curves show that the squeeze cement improved the bond over the whole interval from 8,950 down.

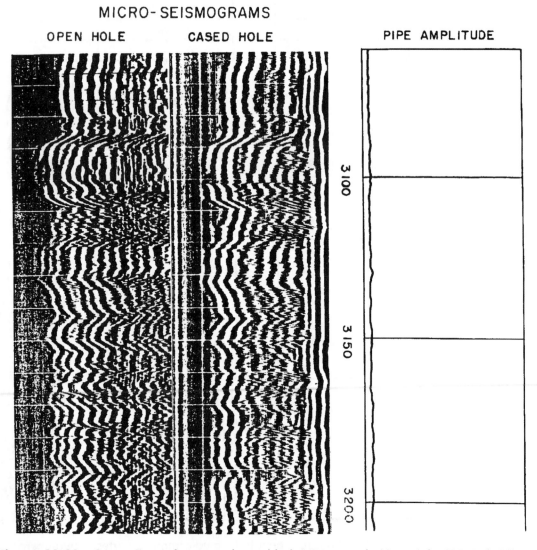

Figure 10.19 Comparison of open and cased hole VDL records (Copyright SPE, Ref. 30)

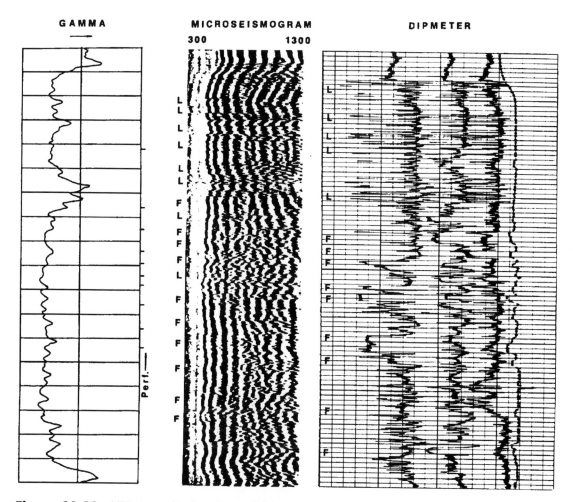

Figure 10.20 VDL type display shows information fractures (F) and laminations (L) (Reproduced with permission of Halliburton Co., Ref. 31)

BOREHOLE COMPENSATED CEMENT BOND LOGS

Availability of the Service

The borehole compensated cement bond logs are available from the major service companies. Their trade designations are listed below. The * indicates that name is a mark of the company associated with it.

Schlumberger: Cement Bond Tool* (CBT*)
Western Atlas: Bond Attenuation Log* (BAL*)
Halliburton: Compensated Cement Attenuation Tool* (CCAT*)
Computalog: Compensated Sonic Attenuation System*

Tool Configuration and Operation

The borehole compensated CBL (BCCBL) is comprised of two omnidirectional acoustic transmitters separated by three receivers.[32-34] Figure 10.22 shows a schematic of the Schlumberger CBT tool.[32] All of the BCCBL tools are similar, having two receivers, labeled R2 and R3, separated from each other by one foot (.3 m) and symmetrically located between the transmitters as shown on the figure. Receiver R1 is spaced five feet (1.5 m) from the lower transmitter for recording of the VDL. The main difference in terms of tool configuration is that the CBT transmitters are 5.8 ft (1.77m) apart while those on the BAL and CCAT are spaced 6 ft (1.83m).

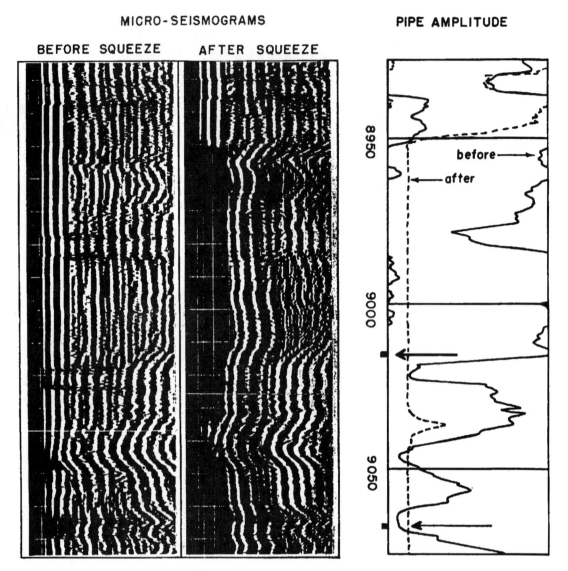

MICRO-SEISMOGRAMS   **PIPE AMPLITUDE**

BEFORE SQUEEZE   AFTER SQUEEZE

Figure 10.21 Bond log shows improvement in cement after cement squeeze (Courtesy SPE, Ref. 30)

The transmitters fire in sequence, and the amplitudes are detected at receivers R2 and R3. The attenuation between these receivers is computed using receiver amplitudes in both directions with the equation

$$\text{Attenuation} = (10/S) \times \log_{10}((AUR3 \times ALR2)/(AUR2 \times ALR3))$$

where

    Attenuation = Attenuation between receivers R2 and R3, db/ft

        AUR3 = Amplitude from upper transmitter measured at R3, mv

        ALR2 = Amplitude from lower transmitter measured at R2, mv

        AUR2 = Amplitude from upper transmitter measured at R2, mv

        ALR3 = Amplitude from lower transmitter measured at R3, mv

           S = Distance between receivers, ft (1 ft for BCCBL tools)

The advantage of this system is that by taking a ratio of amplitudes, the sensitivity of the tool to variations in transmitter strength or receiver sensitivity cancels out in the ratio. If such variations introduce a factor, such as a transmitter which is putting out 1.2 times

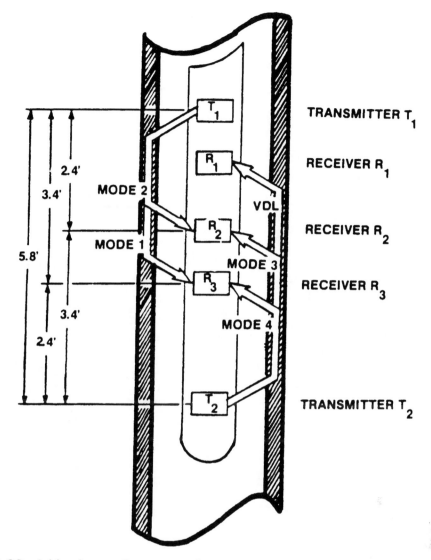

Figure 10.22 Schlumberger Cement Bond Tool (CBT) diagram (Courtesy Schlumberger, Ref. 32)

the proper signal strength or a receiver measuring .88 of the actual signal, then this factor cancels out in the above equation. Furthermore, since each receiver is exposed to the same environment and reduction of signal due to eccentralization, the tool is not sensitive to centralization.

So, the advantages of this tool over the conventional CBL are that it is not sensitive to variations in transmitter strength or receiver sensitivity, no signal calibration is required, the BCCBL is less sensitive to tool eccentralization, and the tool is not sensitive to borehole fluid. However, this tool continues to share some disadvantages with the CBL. It is sensitive to microannulus, fast formations, cement curing time, cement coating, and most of all it cannot discriminate compressive strength effects from annular fill.

Log Presentation

The log presentation for the BCCBL is shown on Figure 10.23.[5] This is an Atlas Bond Attenuation Log (BAL). A short section of free pipe is inserted at the top of the log. The left track shows the usual Gamma Ray and Travel Time curves. The middle track shows attenuation on a 0-20 db/ft (0-66 db/m) scale. An amplitude from the 2.5 ft (.76 m) receiver is shown on a 0-130 mv scale, along with a 5x amplitude curve (confined to the left half of the middle track). Notice that the attenuation curve is shaded at bond index values greater than

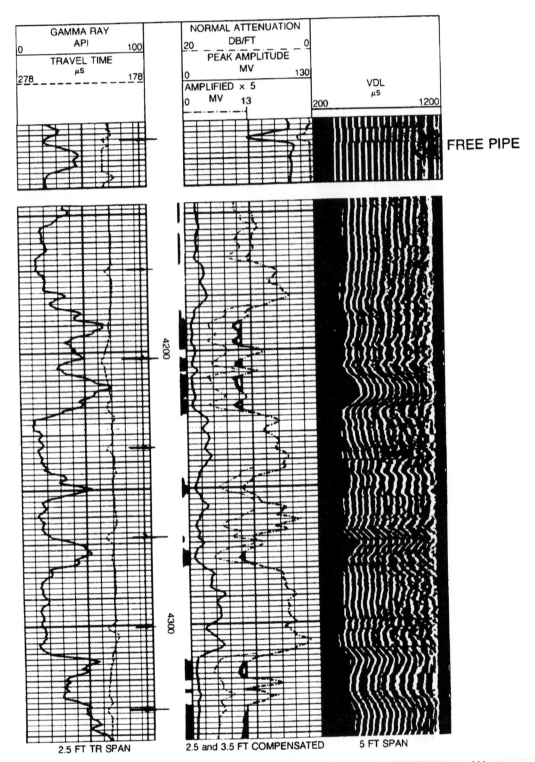

Figure 10.23   Example of Western Atlas Bond Attenuation Log (BAL) (Courtesy Western Atlas, Ref. 5)

.8. A flag to highlight such intervals is also shown in the depth track. The VDL is recorded from the third receiver spaced five feet (1.5 m) from the lower transmitter.

This tool, as mentioned earlier, is sensitive to fast formations. However, if the casing size is about 7 1/2 in. (19 cm) or less, then the amplitude from the short spaced receiver (.8 or 1.0 ft [24.4 - 30.5 cm] from the upper transmitter) may be helpful for fast formation evaluation. This short signal will always reach the receiver before the signal going through the formation and so could be monitored across a fast formation section.

PAD TYPE CBL, THE SEGMENTED BOND TOOL (SBT)

SBT Tool Configuration and Operation

The Segmented Bond Tool*, SBT*, (*Mark of Western Atlas) is based on a pad type bond measurement.[35-37] It is designed with six pads, each spring loaded on a centralizer-like arm and held against the wall during logging operations. Each pad contains both an acoustic transmitter and receiver. A cross section of the tool in a borehole is shown on Figure 10.24, part a. The pads are numbered 1 through 6. If the casing around the tool is unwrapped to show the footprints of the pads, the footprints would appear as shown in Figure 10.24, part b. Again the pads are numbered 1 through 6. Each pad has a transmitter, T, and a receiver, R.

When a transmitter emits an acoustic signal, the signal propagates along the casing wall and is detected by the receivers of neighboring pads. For example, if the transmitter of pad 4 emits an acoustic pulse, that pulse is detected by the receivers of pads 3 and 2 and of pads 5 and 6. When pad 1 pulses, its pulse is detected by pads 6 and 5 and by pads 2 and 3. This configuration is the same as that used in the BCCBL tools in that the acoustic signal passes two detectors in both directions for measurement of attenuation. This has the advantage of being insensitive to transmitter and receiver variations. By firing the pad pulses in an appropriate sequence, each of the six 60° sectors of the borehole can be examined. To a great extent, this removes the basic uncertainty between compressive strength and annular fill common to previous bond logs.

SBT Log Presentation

Figure 10.25 shows a simplified schematic of the SBT log presentation. In the middle, a cement sheath is shown with 100% annular fill at the bottom, free pipe at the top, and intermediate amounts of fill in between. On the left is a log with six tracks numbered to correspond to the sectors between pads. These tracks show the attenuation in each sector and are typically scaled from 0 to 20 db. On the right is a simplified cement map. This presentation maps the inside of the casing, unwrapping it and shading the segments covered with cement in black. In reality, there are shades of gray not shown on this simplified schematic. Both presentations show good cement coverage at the bottom of the interval, with free pipe at the top. The cement map give a sense of how the cement is distributed over the casing.

Figure 10.26 a and b show a SBT log.[5] Both primary and segmented data are shown. Figure 10.26 a shows a gamma ray and collar locator in the left track, with a VDL on the right. The middle track shows the average attenuation of the six sectors (heavy continuous line) with the minimum sector attenuation also shown. The region is shaded white between these. When attenuation is fairly high in all of the sectors, separation may indicate a small channel. The amplitude and 5x amplitude is computed from the average sector attenuation and is also shown. It is scaled in mv and is similar to CBL amplitude. Figure 10.26 b

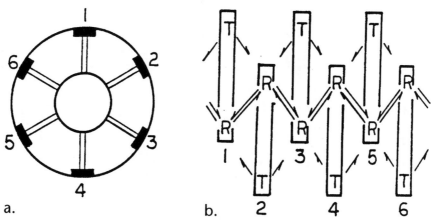

Figure 10.24 Pad array for the Western Atlas Segmented Bond Tool (SBT)

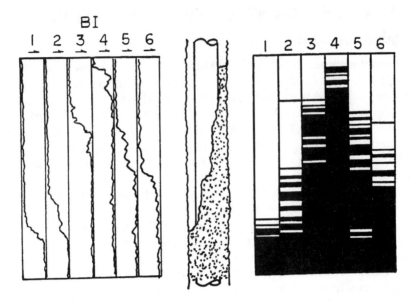

Figure 10.25 Basic presentation of measurements for the SBT

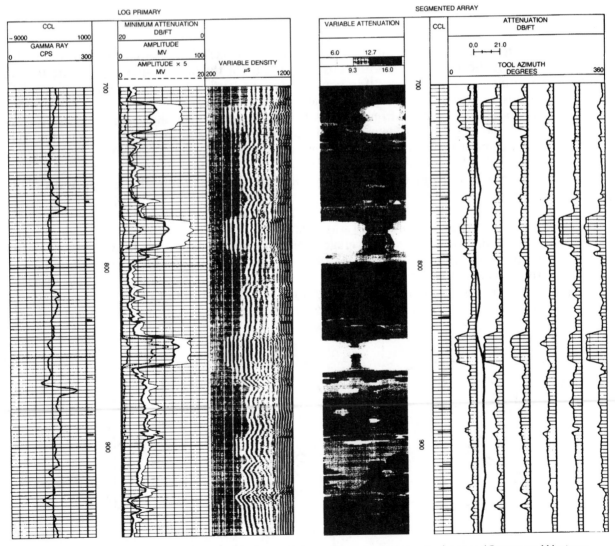

Figure 10.26 Western Atlas SBT presentation a. Primary log, b. Segmented array (Courtesy Western Atlas, Ref. 5)

shows the segmented array information. Each sector is displayed on the right with attenuation between pad scaled here on a 0-21 db scale. The heavy line wiggling around in the second small track is the tool azimuth and indicates the location of the low side of the hole. The cement map, sometimes called the Variable Attenuation Log (VAL) is shown on the left. It is shaded corresponding to five levels of attenuation. Three very poorly cemented intervals are evident on this log.

The advantages of the SBT include those of the BCCBLs. Furthermore, this tool measures annular fill and to a great extent removes the uncertainty of whether the apparent poor bond is a result of weak cement or poor annular fill. Due to the short distance the acoustic signal travels, it is not affected by fast formations unless the cement is quite thin. Since it is a pad type device, it is not affected by gas bubbles in the wellbore. This tool is, however, still sensitive to microannulus, cement curing time, and coatings on the outside of pipe. This tool has come a long way to overcome the problems inherent in acoustic CBL logging.

BOND LOGS WITH DIRECTIONAL RECEIVERS

Another approach to eliminate the uncertainty between compressive strength and annular fill is to use a conventional CBL type of tool with directionally sensitive receivers for the amplitude measurements. This service is offered by the following companies and * indicates that the name is a mark of the company indicated:

Computalog: Sector Bond Log*
Wedge Wireline: Radial Analysis Cement Bond Log* (RAL*)

The tool configuration typically consists of an omnidirectional transmitter, a near and far receiver for amplitude and VDL measurements, and an array of eight directional receivers in close proximity to the near receiver.[38-40] Presentations include the usual CBL curves plus data from the individual directionally oriented receivers. The oriented receiver amplitudes are presented individually and as a map much like the sectored data from the SBT discussed earlier. Fancier processed presentations are also available on computer disc for viewing on a personal computer.

POINTS TO REMEMBER

- Bond logs do not measure isolation.
- At best, bond logs measure the annular fill of cement contacting the pipe on the outside.
- Bond logs are acoustic devices and measure the attenuation of an acoustic signal as it moves along the pipe wall.
- The conventional CBL presents an amplitude of the pipe signal, the signal travel time, and the VDL or wavetrain log.
- The poorest bond is called "free pipe" and corresponds to the highest receiver amplitude and lowest attenuation rate.
- The best bond corresponds to the lowest amplitude and highest attenuation rate.
- There are fixed and floating amplitude measurement gates available with some tools. Never run the floating gate without the fixed gate.
- The Variable Density Log (VDL) is a contour map of the received signal as the tool is logged uphole.
- The pipe arrivals on the VDL are strong straight lines, while the formation arrivals tend to be wiggly or ratty.
- In the best bonded intervals, the pipe portion of the VDL tends to disappear since it is too low to expose the film.
- Travel time is the main quality control curve, and is used to detect tool eccentralization.
- An eccentralization of 4-5 us corresponds to about a 1/8 in. (.32 cm) and causes about a 305 loss of signal. Any greater eccentralization is considered to be a poor log.
- Shortening of travel time with a decrease of amplitude is a sign of eccentralization.

- Shortening of amplitude with an increase of amplitude is a fast formation.
- Lengthening of travel time is caused by cycle stretching or skipping, and is considered a good indication.
- Fast formation occurs when in a limestone or dolomite interval.
- Microannulus occurs when the casing diameter is less than at the time of cement curing and is remedied by pressuring up on the casing.
- Microannulus does not detrimentally affect isolation, but only the CBL response.
- Wait until the cement cures before running the bond log. A good rule of thumb is 72 hours after cementing.
- Denser borehole fluids cause an increase of amplitude at the receiver.
- A thin cement sheath will cause a larger amplitude to be detected.
- Foamed cements do not attenuate as strongly as conventional cements.
- Bond Index (BI) represents annular fill of cement if compressive strength is constant.
- A certain distance of less than perfect bond can provide a "reasonable assurance" of isolation.
- Bond Index is affected by both annular fill of cement and cement compressive strength.
- Bond Index may be viewed as a "squeeze/no squeeze" criterion.
- Borehole compensated cement bond logs have two major advantages over conventional CBL logs. They are not so sensitive to centralization, and the variations in the transmitters and receivers are tolerated.
- BCCBL remains sensitive to microannulus and cannot discriminate between the effects of annular fill and compressive strength.
- The Pad type CBL has all the advantages of the BCCBL plus it can discriminate annular fill from cement strength to some extent.
- The pad type tool can operate with gas bubbles in the wellbore.
- Tools with directional receivers can discriminate annular fill from cement strength factors.
- Pad type tools and tools with directional receivers are sensitive to microannulus.
- Easy to understand cement map type presentations are available with the pad and directional receiver CBLs.

References

1. Gearhart Industries (Now Halliburton Energy Services), "Basic Cement Bond Log Evaluation," Document No. WS-597, Fort Worth, Texas, 1982.
2. Atlas Wireline, "Acoustic Cement Bond Log," Document 2206, Houston, Texas, 1085.
3. NL McCullough (Now Atlas Wireline), "Cement Bond Log Interpretation Manual," Publication No. Mc L-011, Houston, Texas, 1985.
4. Schlumberger, "Cased Hole Log Interpretation Principles/ Applications," Document No. SMP-7025, Houston, Texas, 1989.
5. Atlas Wireline Services, "Cement Evaluation Guidelines," Document No. 9618, Houston, Texas, 1991.
6. Schlumberger, "Cement Bond Variable Density Log," Central Cased Hole Division document, 1985.
7. Schlumberger, *The Essentials of Cement Evaluation*, Houston, Texas, March, 1976.
8. Pardue, G.H., Morris, R.L., Gollwitzer, L.H., and Moran, J.H., "Cement Bond Log—A Study of Cement and Casing Variables," *JPT*, pp. 545-555, May, 1963.
9. Hayman, A.J., Gai, H., Toma, I., "A Comparison of Cementation Logging Tools in a Full-Scale Simulator," Paper SPE 22779, 66th Annual SPE Technical Conference, Dallas, Texas, October, 1991.
10. Jutten, J.J., and Corrigall, E., "Studies with Narrow Cement Thickness Lead to Improved CBL in Concentric Casings," *JPT*, pp. 1158-1192, November, 1989.
11. Pilkington, P.E., "Cement Evaluation—Past, Present, and Future," *JPT*, pp 132-140, February, 1992.
12. Fertl, W.H., Pilkington, P.E., and Odd, R.A., "Cement Bond Logging in the North Sea," *Petroleum Engineer*, pp. 80-88, May, 1975.
13. Welex (Now Halliburton Energy Services), "An Illustrated Brochure on MSG Acoustic Cement Bond Logs," Document No. CL-2004, Houston, Texas, 1980.
14. Bigelow, E.L., "A Practical Approach to the Interpretation of Cement Bond Logs," *JPT*, pp. 1285-1294, July, 1985.
15. Fitzgerald, D.D., McGhee, B.F., and McGuire, J.A., "Guidelines for 90% Accuracy in Zone Isolation Decisions," Paper SPE 12141, 58th Annual Technical Conference, San Francisco, California, October, 1983.
16. McGhee, B.F., and Vacca, H.L., "Guidelines for Improved Monitoring of Cementing Operations," 21st Annual SPWLA Symposium, July, 1980.
17. Fertl, W.H., Pilkington, P.E., and Scott, J.B., "A Look At Cement Bond Logs," *JPT*, pp. 607-617, June, 1974.

18. Killion, H.W., "Acoustic Cement Evaluation-ACE—Operation— Calibration—Interpretation," Southwestern Petroleum Short Course Association, Texas Tech University, Lubbock, Texas, April, 1980.
19. Halliburton Energy Services, "Cement Evaluation Manual," 1994.
20. Amyx, J.W., Bass, D.M. Jr., and Whiting, R.L., *Petroleum Reservoir Engineering*, pp. 84-85, McGraw Hill Book Company, New York, New York, 1960.
21. Grosmangin, M., Kokesh, P.P., and Majani, P., "A Sonic Method for Analyzing the Quality of Cementation of Borehole Casings," *JPT*, pp. 165-171, February, 1961.
22. Nayfehm T.H., Wheelis, W.B. Jr., and Leslie, H.D., "The Fluid Compensated Cement Bond Log," Paper SPE 13044, 59th Annual SPE Technical Conference, Houston, Texas, September, 1984.
23. Harms, W.M., and Febus, J.S., "Cementing of Fragile- Formation Wells with Foamed Cement Slurries," Paper SPE 12755, SPE California Regional Meeting, Long Beach, California, April, 1984.
24. Masson, J.P., and Bruckdorfer, B., "CBL Evaluation of Foam-Cemented Casings Using Standard Techniques," 24th Annual SPWLA Symposium, June, 1983.
25. Bruckdorfer, R.A., Jacobs, W.R., and Masson, J.P., "CBL Evaluation of Foam Cemented and Synthetic-Cemented Casing," Paper SPE 11980, 58th Annual Technical Conference, San Francisco, California, October, 1983.
26. Carter, L.G., and Evans, G.W., "A Study of Cement-Pipe Bonding," *JPT*, pp. 157-160, February, 1964.
27. Brown, H.D., Grijalva, V.E. and Raymer, L.L., "New Developments in Sonic Wave Train Display and Analysis in Cased Holes," 11th Annual SPWLA Symposium, May, 1970.
28. Tenneco Oil Exploration and Production Co., *Cement Bond Logging Field Reference Manual*, Houston, Texas, 1983.
29. Wydrinski, R., and Jones, R.R., "Integrated Cement Evaluation Saves Time and Money," Paper SPE 28441, 69th Annual SPE Conference, New Orleans, Louisiana, September, 1994.
30. Walker, T., "Utility of the Micro-Seismogram Bond Log," Paper SPE 1751, SPE Regional Meeting on Mechanical Engineering Aspects of Drilling and Production, Fort Worth, Texas, 1967.
31. Walker, T., and Kessler, C. "Detection of Natural Fractures with Well Logs," WELEX (Now Halliburton Energy Services) Document No. L-38, 1980.
32. Meisenhelder, J., *Cement Bond Tool Interpretation*, Schlumberger, Houston, Texas, September, 1986.
33. Gollwitzer, L.H., and Masson, J.P., "The Cement Bond Tool," 23rd Annual SPWLA Symposium, July, 1982.
34. Computalog, "Cement Bond Attenuation System," Brochure, Fort Worth, Texas, 1990.
35. Lester, R.A., "The Segmented Bond Tool: A Pad-Type Cement Bond Device," Trans. Canadian Well Logging Symposium, 1989.
36. Bigelow, E.L., Domangue, E.J., and Lester, R.A., "A New and Innovative Technology for Cement Evaluation," Paper SPE 20585, 65th Annual SPE Conference, New Orleans, Louisiana, September, 1990.
37. Atlas Wireline Services, "SBT Segmented Bond Tool," Document No. 9616, Houston, Texas, 1990.
38. Schmidt, M.G., "The Micro CBL—A Second Generation Radial Cement Evaluation Instrument," 30th Annual SPWLA Symposium, June, 1989.
39. Computalog, "Sector Bond Cement Bond System," Document 0593, Fort Worth, Texas, 1993.
40. Wedge Wireline, Inc. "Seeing Is Believing," Brochure, Fort Worth, Texas, 1992.

Other Suggested Reading

Allen, S.L., and Wood, M.W., "Cement Bond Quality Control Through Simultaneous Recording of Fixed Gate and Sliding Gate Amplitudes with Transit Times," 26th Annual SPWLA Symposium, June, 1985.

Leslie, H.D., de Selliers, J., and Pittman, D.J., "Coupling and Attenuation: A New Measurement Pair in Cement Bond Logging," Paper SPE 16207, SPE Production Operations Symposium, Oklahoma City, Oklahoma, March, 1987.

Jutten, J.J., Parcevaux, P.A., and Guillot, D.J., "Relationship Between Cement Slurry Composition, Mechanical Properties, and Cement Bond Log Output," Paper SPE 16652, 62nd Annual SPE Conference, Dallas, Texas, September, 1987.

Norel, G., Dubois, C., and Georges, G., "Test Bench Checks Cement in Horizontal Holes," *Petroleum Engineer International*, pp. 54-59, November, 1988.

Broding, R.A., and Buchanan, L.K., "A Sonic Technique for Cement Evaluation," 27th Annual SPWLA Symposium, June, 1986.

CHAPTER 11

WELL INTEGRITY: ULTRASONIC PULSE ECHO CEMENT EVALUATION

OVERVIEW

Why the Need?

Conventional cement bond logging techniques provide a Bond Index (BI) indication. This measurement, when compressive strength is constant, is an indication of annular fill. If this was all there was to it, a BI of 70% would indicate a channel occupying 30% of the annular space. Unfortunately, this is not necessarily so since the BI measurement is also affected by cement compressive strength. A BI less than 100% can be caused by a channel or by a reduction of cement compressive strength. Conventional bond logs offer no way to resolve the issue. Furthermore, such an indication may be a continuous channel or a series of non-connected gaps in the cement. That is why the directional and pad type devices discussed earlier were developed.

Another approach to this issue is highlighted by the development of the pulse echo bond log tools. This technology was first introduced in 1981 by Schlumberger.[1,2] It provides an estimate of cement coverage and a direct measurement of the cement compressive strength. It indicates the geometry of the cement job adjacent to the pipe and hence is excellent for finding potential channels. Like the conventional bond logs, this tool only sees cement contacting the pipe, and therefore cannot detect channels of types III or IV of Figure 11.1.

The Pulse Echo Tools

As of this time, the pulse echo tools available from the industry include the following. The * indicates that the name is a mark of the company with which it is associated:

Schlumberger: Cement Evaluation Tool* (CET*)
Ultrasonic Imager* (USI*)
Halliburton: Pulse Echo Tool* (PET*)

The CET and PET use fixed transducers and are discussed in the sections following. The USI uses a single rotating transducer and is discussed later in this chapter. These tools are also discussed in the chapter on casing inspection for such applications.

Applications

These tools provide a number of new and unique perspectives on the casing and cement annulus. They:

1. Directly measure and map the cement distribution around pipe,
2. Determine acoustic impedance and possibly compressive strength of annular materials,
3. Discriminate gas from liquid behind pipe,

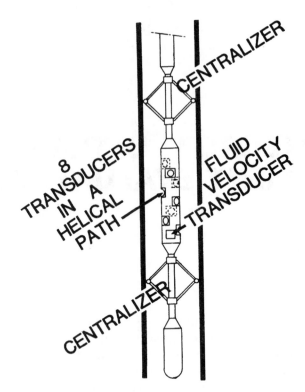

Figure 11.1   Tool diagram, Schlumberger Cement Evaluation Tool (CET) (Courtesy Schlumberger, Ref. 16)

4. Assess the presence of gas cut cement, and
5. Provide an acoustic measurement of the pipe ID, inner wall surface roughness and wall thickness, as well as ovality and tool centralization.

FIXED TRANSDUCER PULSE ECHO TOOL CONFIGURATION AND OPERATION—THE SCHLUMBERGER CET

Tool Configuration

The configuration for the fixed transducer pulse echo tool is shown on Figure 11.1.[16] The tool is comprised of eight transducers (these act as both transmitter and receiver) oriented in a helical array. The transducers are separated by 45° and each transducer samples an area of the wall about 1 in. (2.5 cm) in diameter. Coverage is not complete, varying from about 65% in 4.5 in. (11.4 cm) casing to about 29% in 9-5/8 in. (24.4 cm) casing. A ninth transducer is included in the tool. It is oriented a fixed distance from a target plate within a cavity accessible to wellbore fluid. This transducer is used to measure the acoustic velocity of the wellbore fluid.

Each transducer emits an ultrasonic pulse of about 500 kilohertz. Figure 11.2, part a, shows the orientation of each transducer with respect to the wellbore environment.[16] Figure 11.2, part b, shows a detailed diagram of the transducer relative to the casing, cement and formation. The acoustic wave travels from the transducer to the wall and returns. This first wave is shown as wave A1 and is the primary reflection. It is followed by the wave series arising from resonance within the casing, and these are designated B1, B2, B3, etc. These waves essentially rattle around within the casing wall, losing energy every time they reflect from an interface. This wave series, A1 plus B1, B2, . . . is the basis of all interpretation of the pulse echo tools. The time of arrival of the A1 wave is used for measurement of the casing diameter, and its strength is used as an indication of inner surface roughness. The B series is used to measure the casing wall thickness and the acoustic impedance of whatever is behind pipe.

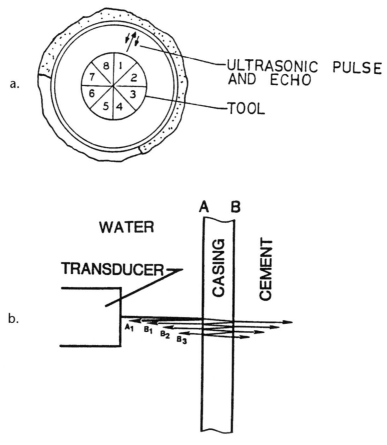

Figure 11.2 Pulse echo signal and reflections—a. Borehole arrangement, b. Reflected signal detail (Courtesy Schlumberger, Ref. 16)

Signal Analysis, Schlumberger CET

The wavetrain arriving back at a transducer is shown in Figure 11.3.[4] For the Schlumberger CET, the A1 wave is monitored with a gate positioned at W1. The B series of resonant casing waves are detected in gates W3 and W2.[1-4] Note that the B series is shown with the vertical scale amplified about 10 times. In free pipe, the energies or amplitudes in both W3 and W2 are high. In well bonded casing, energy is lost to the cement and formation, and all of the B series is reduced. The amplitude of W2 reduced somewhat more than that of W3. Occasionally, a reflection from the formation/cement interface returns and interferes with the decay in gate W2. In this case, the W3 energy is high relative to W2.

The CET gate 2 and 3 energies are normalized according to the following equations:

$$\text{WiN} = \frac{\text{Wi}}{\text{W1}} \times \frac{1}{\text{WiFP}} \qquad \text{(Equation 11.1)}$$

where

    W1 = Energy in Gate 1

    Wi = Energy in Gate i

  WiFP = Energy in Gate i in Free Pipe with Water Inside and Outside of the Pipe.

      i = Gate 2 or 3

Note in the above equation that the free pipe condition is the condition with water both on the inside and outside of the pipe.

If the normalized values W2N and W3N are cross plotted, the data would look like that of Figure 11.4.[5] Both W2N and W3N would each read 1 in the free pipe condition. In well bonded pipe, both W2N and W3N are less than 1 with W2N lower than W3N.[5] The scatter

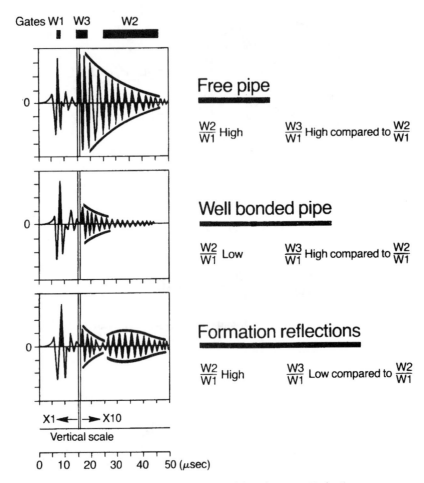

Figure 11.3 CET gate positions and signal (Courtesy Schlumberger, Ref. 4)

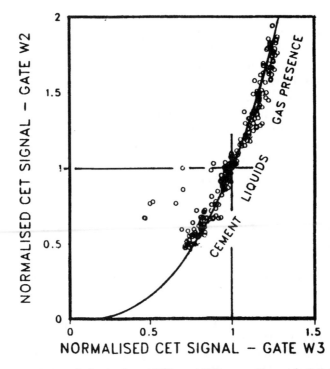

Figure 11.4 Normalized CET signals from Gates W2 and W3 pass through 1,1 (Courtesy Schlumberger, Ref. 5)

of points to the left of the plot corresponds to formation reflection points where the W3N is too low relative to the W2N detected. When gas exists on the outside of pipe, the response of both W2N and W3N are each greater than 1. The cluster of data points passing through W2N=1 and W3N=1 as shown would normally indicate that the log was properly calibrated and normalized and appear on the log tail. Such a plot is called a "banana plot."

Computation of Acoustic Impedance, Cement Compressive Strength, and Geometric Parameters, CET

The W2N and W3N energy levels can be used to compute both the acoustic impedance and compressive strength of whatever is outside of the casing. Acoustic impedance, Z, is defined as the product of the density and acoustic velocity of a medium.[6] It is expressed in MRayl (10^6kg/m^{2}sec).

$$Z = rho \times v \qquad \text{(Equation 11.2)}$$

where

Z = Acoustic Impedance, MRayl (10^6kg/m^{2}sec)

rho = Density, kg/m^3

v = Acoustic Velocity, m/sec

In general, the greater the mismatch of acoustic impedances at an interface, the stronger the reflection and the less energy passes through the interface. The reflection coefficient, C_r is given as

$$C_r = \frac{Z_2 - Z_1}{Z_1 + Z_2} \qquad \text{(Equation 11.3)}$$

where

C_r = the reflection coefficient

Z = Acoustic Impedance

1,2 = Subscripts of the media on each side of the interface

Hence the B series of waves discussed earlier die out more rapidly when cement is outside of pipe than if liquid or gas is outside. A list of acoustic impedance values for common downhole materials is shown in Table 11.1.[7]

Figure 11.5 shows a series of charts (A, B, and C) useful for the purpose of computing the acoustic impedance and compressive strength.[5] Chart A is similar to that of Figure 11.4 and shows the correspondence of W2N and W3N. With either W2N or W3N, proceed to chart B and determine the outside acoustic impedance for the appropriate casing wall thickness. Next proceed downward to the appropriate cement correlation curve on chart C, such as the neat cement line, to get cement compressive strength. For example, if W2N=.5 (W3N=.8) and the casing wall thickness is .4 in. (1.0 cm), then the acoustic impedance is

TABLE 11.1 Acoustic Properties of Materials (Courtesy Schlumberger, Ref. 7)

Material	Density (kg.m^{-3})	Acoustic velocity (m.s^{-1})	Acoustic impedance (MRayl)
Air (1–100 bar)	1.3–130	330	0.0004–0.04
Water	1000	1500	1.5
Drilling fluids	1000–2000	1300–1800	1.5–3.0
Cement slurries	1000–2000	1800–1500	1.8–3.0
Cement (Litefil)	1400	2200–2600	3.1–3.6
Cement (class G)	1900	2700–3700	5.0–7.0
Limestone	2700	5500	17
Steel	7800	5900	46

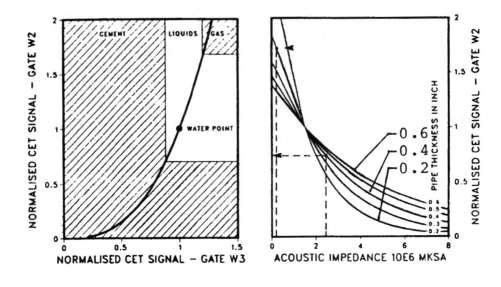

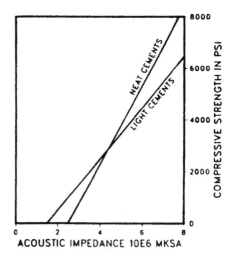

Figure 11.5 Chart set to compute cement compressive strength from CET signal in gates W2 or W3 (Courtesy Schlumberger, Ref. 5)

about 3.5 MRayl and the compressive strength for neat cement is about 2,400 psi (16.5 Pa). Note that the curve for neat cement is actually not linear and will be discussed in more detail later.

Geometric parameters are also presented with the CET. Figure 11.6 shows how the geometric parameters are related to the transducer measurement of distance.[16] Recall that the time of travel of the acoustic pulse is measured as well as the acoustic velocity from the ninth transducer. Hence, the distance from transducer to the wall, or alternatively the radius from the center of the tool to the wall, can be accurately measured. This radius is given by the equation:

$$\text{Radius} = (.5 \times 12 \times 4TT/FV) + RadT \qquad \text{(Equation 11.4)}$$

where

Radius = Radius from the center of the tool to the wall, inches

TT = Acoustic travel time, transducer to wall and return, microseconds

FV = Fluid velocity, microseconds/foot

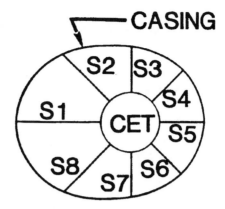

```
Caliper 1  =  S₁  +  S₅  +  Tool O.D.
Caliper 2  =  S₂  +  S₆  +    "      "
Caliper 3  =  S₃  +  S₇  +    "      "
Caliper 4  =  S₄  +  S₈  +    "      "


Mean Caliper  =  (C₁  +  C₂  +  C₃  +  C₄)  ÷  4
Ovality  =  Caliper max  -  Caliper min
Eccentering  =  S_max  -  S_min
```

Figure 11.6 CET geometrical measurements (Courtesy Schlumberger, Ref. 16)

RadT = Radius of tool, inches

12 = Factor to change from feet to inches

.5 = Converts TT to one way travel time

The log typically presents the mean caliper, ovality, and tool eccentering. With special equipment and computational techniques, the pipe wall thickness may also be measured based on the resonant frequency of the B series of waves. This thickness may be calculated from the equation:[8]

$$\text{Casing thickness} = \frac{6.0}{52.55 \times f} \qquad \text{(Equation 11.5)}$$

where the casing thickness is measured in inches and the frequency is in MegaHertz (MHz).

Log Presentation, CET

Figure 11.7 shows a CET log presentation. The geometric parameters are generally located in the left track.[10] Here is shown the CCL, mean diameter, ovality, and tool eccentering. Two parameters shown in the middle track are also sometimes shown in the left track, and these are the well deviation angle and the relative bearing. The latter gives an indication of tool rotation and orientation downhole. Also in the middle track are the minimum and maximum compressive strengths calculated from the acoustic impedances seen by the

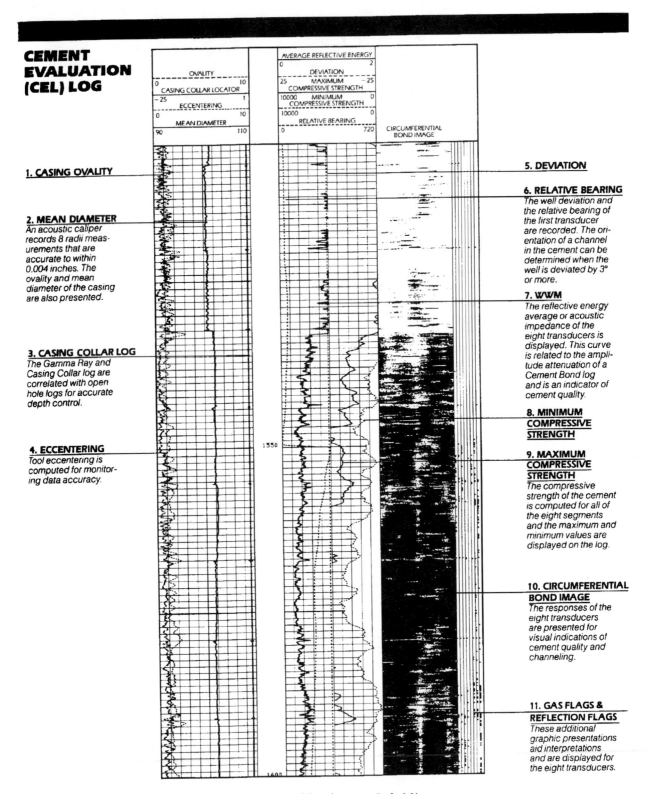

Figure 11.7 CET log presentation (Courtesy Schlumberger, Ref. 10)

eight detectors. In this case, the cement top is located at about 1,533 m (5,028 ft) and both the minimum and maximum compressive strengths go to zero at the free pipe condition. The remaining parameter, called average reflective energy or WWM, is the average of the eight transducer readings. Referring back to chart A of Figure 11.5, W2N in good cement should read about .5 or less while in free pipe with water on the outside it should read about 1.0. Indeed, this example reads about .3 to .5 over the well cemented interval, and about 1.0 in free pipe.

The right-hand track shows the cement map. For this map, the data from the transducers are stored and played back on depth with respect to each other. Each 1/8 of the distance across this track corresponds to data from one single transducer. This map is typically presented in terms of acoustic impedance. It is often looked at as a map of cement compressive strengths since there is assumed to be a linear relationship between compressive strength and impedance. White corresponds to acoustic impedances of 2.5 MRayls or less, i.e., fluid in the annulus. Impedances of 3.2 MRayls or larger are shaded black to indicate that whatever is behind pipe, it is not a liquid. In the first years of CET usage, the impedance values (or compressive strengths) used for the black and white shading were not consistent and this map presentation was often quite misleading.

Depending on the processing, there may be shades of gray between the white and black limits, or the map may be processed to smooth the area between transducers as shown in the map. While the processing may make the cement fill appear to continuously and smoothly vary, the transducers do not in fact sample the whole perimeter. This track is also typically oriented to where the center of this track corresponds to the lower side of the hole. This map clearly shows the cement top. It shows some competent cement. Unfortunately, it also shows a possible serious channel at the low side of the pipe, especially over the bottom 30 meters (100 ft).

At the far right are eight fine tracks. These tracks each correspond to a single transducer. If the track has a substantial mark in it, such as shown at various depths in this example, a reflection from the formation back into gate W2 is indicated. The presence of this signal is a good sign, for it confirms the presence of cement between the casing and formation. However, the lack of such a "formation flag" is not bad news. The acoustic impedance of the formation may be close to that of the cement in which case little energy is reflected back. Perhaps the cement sheath is thick, thereby requiring the reflection to traverse a great distance and not return early enough to be detected in the W2 gate. Sometimes a thin vertical line appears within these tiny tracks. Such a line is a "gas flag" and indicates the presence of gas on the outside of pipe. No gas flags are shown on this example.

THE HALLIBURTON PET

Signal Analysis

Overall, the PET operation is similar to the CET. The transducer arrangement is slightly different, forming a double helical array. As with the CET, each transducer examines a one in. (2.5 cm) diameter area in its respective 45° sector. Again, a ninth transducer is used to measure borehole fluid acoustic velocity.

The gating of the PET consists of two gates. The first gate, called the "first arrival window," measures the A1 arrival and roughly corresponds to the W1 gate of the CET. The second gate measures the B1, B2, . . . series. It opens shortly after the closure of the CET W3 gate and stays open until about midway through the CET W2 gate. This gate period is called the "resonance window."

Based on measurements in these two windows, an "area ratio" is computed. The area ratio is given by the following equation:[8]

$$\text{Area Ratio} = \frac{\left(\dfrac{RW}{FA}\right) \text{ at depth}}{\left(\dfrac{RW}{FA}\right) \text{ in free pipe}} \qquad \text{(Equation 11.6)}$$

where

RW = Amplitude in resonance window

FA = Amplitude of first arrival

This Area Ratio is related to the acoustic impedance and compressive strength by using the charts shown in Figure 11.8.[8] For example, an area ratio measurement of .25 indicates an acoustic impedance of 5.6 MRayls outside of the pipe and compressive strength in neat cement of 5,200 psi (36Pa).

Log Presentation, PET

The PET presentation is shown on Figure 11.9.[10] In this schematic, the presentation is quite similar to the CET of Figure 11.7. The left track shows geometrical parameters and the middle track shows deviation and relative bearing.[8,9] In this presentation, the middle track shows maximum, minimum, and average acoustic impedance as seen by the eight transducers. The map at the right shows an acoustic impedance map. This map shows two distinct and continuous channels through which fluid can flow.

Since the assumed relationship between the acoustic impedance and cement compressive strength is linear, the plots or maps of impedance and compressive strength are essentially the same. The only difference is that zero compressive strength, a liquid, corresponds to some non-zero value of acoustic impedance. From an earlier discussion, this value is approximately 2.5 MRayls.

MISCELLANEOUS FACTORS

Microannulus

When the CET was first introduced, it was touted as a tool with no microannulus effect. This was not quite so. While the slightest microannulus will affect the CBL response, Figure 11.10 shows that a pulse echo tool is not affected until the microannulus exceeds a 0.1 mm (.004 in.) change in diameter (the width of the microannulus is one half of the dimension given).[4] For 5.5 in. (14 cm) 17 lb/ft (23.5 Kg/m) pipe, it would take over 2,000 psi (15 Pa) to close this microannulus. So, even with a pulse echo tool, one should be prepared to pressure up on the casing to close off a microannulus.

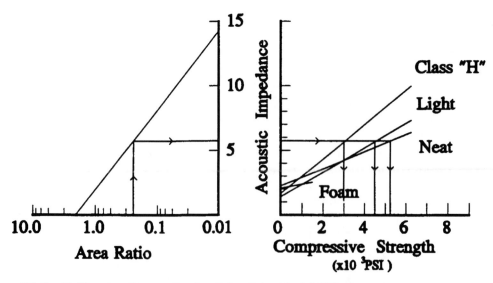

Figure 11.8 Halliburton Energy Service Pulse Echo Tool (PET) charts to compute acoustic impedance and cement compressive strength (Reproduced with permission of Halliburton Co., Ref. 8)

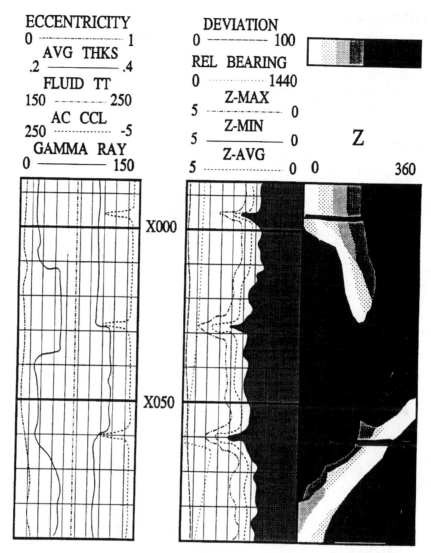

ECCENTRICITY
0 --------- 1
AVG THKS
.2 --------- .4
FLUID TT
150 --------- 250
AC CCL
250 --------- -5
GAMMA RAY
0 --------- 150

DEVIATION
0 --------- 100
REL BEARING
0 --------- 1440
Z-MAX
5 --------- 0
Z-MIN
5 --------- 0
Z-AVG
5 --------- 0

Z

0 --------- 360

X000

X050

Figure 11.9 PET log presentation (Reproduced with permission of Halliburton Co., Ref. 10)

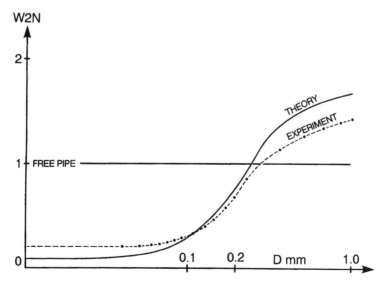

W2N

2

1 FREE PIPE

THEORY

EXPERIMENT

0   0.1   0.2   D mm   1.0

Figure 11.10 Effect of microannulus on CET response (Courtesy Schlumberger, Ref. 4)

Tool Eccentralization

When a pulse echo tool is not centered, certain of its transducers emit an acoustic pulse which strikes the wall obliquely. This reflected wave may partly miss the transducer upon its return and thereby show an erroneously low signal amplitude. Figure 11.11 shows that, as a rule of thumb, the pulse echo tools can tolerate about 1 mm of eccentralization for each inch of casing size.[4]

Mud Weight

The pulse echo tools are affected by mud weight. For water based muds, these tools are limited to mud weights in the neighborhood of 12 lb/gal (1.44 gm/cc). At higher weights, the muds attenuate the signal and may give a false indication of cement. Special transducers are available for higher mud weights, up to about 16 lb/gal (1.9 gm/cc). In oil based muds, the limits on mud weight are even lower.

Gas Cut Cement

When the CET was first introduced, it was always run with a conventional or borehole compensated cement bond log (CBL). A cross plot of this data from the CBL and CET showed some interesting features.[5,11] Figure 11.12 shows such a cross plot.[5] The vertical axis is a plot of normalized CET signal, W3N, appearing in gate W3. Recall that this is not as sensitive as W2N, but is free of formation reflections and should read one at free pipe. On the horizontal axis, and plotted on a log scale, is the CBL amplitude in mv.

There are four clusters of data points on this cross plot. The cluster captioned "liquids" is clearly free pipe for both the CBL and CET, with the CET reading one and showing water on the outside. The cluster captioned "gas" is clearly a free pipe based on the CBL with the CET showing a greater than one response, indicating gas outside the casing. The cluster

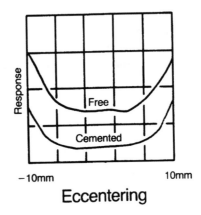

Casing size	Max Ecc
9⅝	10mm
7⅝	7mm
5	5mm

Figure 11.11 Effect of tool centralization on CET response (Courtesy Schlumberger, Ref. 4)

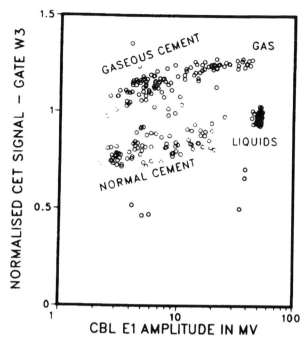

Figure 11.12 CET vs. CBL cross plot (Courtesy Schlumberger, Ref. 5)

indicated as "normal cement" shows a reduced CET signal plus shear support indicated by the CBL, and both measurements confirm that this is cement. What about the cluster at the top? The CBL indicates shear support and reduced amplitude while the CET shows a response greater than one, i.e., a gas. This cluster is interpreted as gas cut cement.

CBL/PULSE ECHO COMPARISON

Channel Recognition

Figure 11.13 shows a PET run over the same interval as a conventional CBL/variable denisty log (VDL) log.[12,13] While the CBL amplitude would indicate intervals of both good and poor cement, the VDL shows low pipe arrivals across the whole interval, implying a somewhat better cement job than the amplitude curve. The cement map presentation of the PET clearly shows that the interval has a number of continuous vertical channels. There are also significant intervals of isolation over the interval shown.

Gas Cut Cement

For accurate analysis of gas cut cement, both a pulse echo and CBL type log are required.[14] The contrary indications of free pipe with the CET and shear support and hence bond with the CBL are indications of gas cut cement. Figure 11.14 shows the CET and CBL in the same well.[15] Notice that the cement map of the CET shows significant regions of poor cement or free pipe while the borehole compensated bond log tool, the CBT, shows low amplitude over much of the same intervals shown. Processing of these two services yields a product shown at the right, the Cement Scan Log (mark of Schlumberger). The cement distribution map shows solid cement (black), liquid mud (white), and gas cut cement (gray) in the track labeled cement distribution. Clearly, the middle and top of the upper interval have good coverage of gas cut cement and would be difficult to squeeze. The liquid bearing channels are white. The relative bearing curve would be helpful to orient perforating equipment to properly intercept the channel for cement squeezing. The track on the far right shows the cement coverage shaded gray and the gas cut cement shaded bubbly. Whiter corresponds to liquid behind the pipe and the dotted area indicates gas.

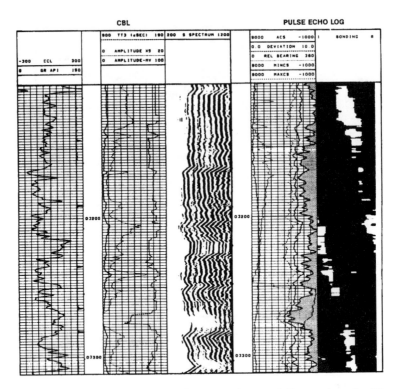

Figure 11.13 Comparison of CBL and PET over the same interval (Reproduced with permission of Halliburton Co., Ref. 12)

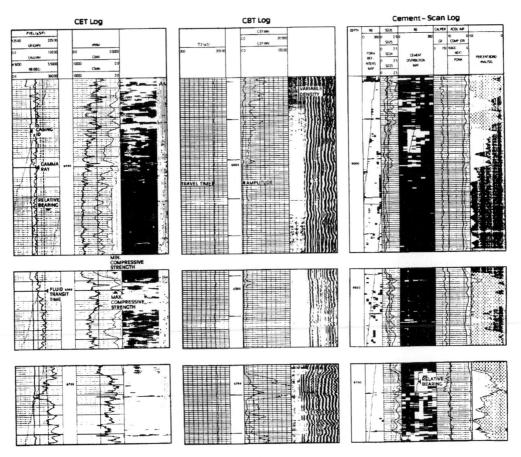

Figure 11.14 CET and CBT over same interval and showing gas-cut cement in all sections (Courtesy Schlumberger, Ref. 15)

THE MOBIL/GOODWIN TECHNIQUE

Acoustic Impedance of Cements and Muds

The cement map, even though shaded based on acoustic impedance, is frequently looked upon as a map of compressive strengths due to the presumed linear relationship between them. Squeeze decisions based on this map often were unsuccessful or not needed. A review of the relationship revealed that it was, in fact, non-linear and a function of cement weight as shown in Figure 11.15, part a.[6] An acoustic impedance of 3 MRayl could correspond to virtually no compressive strength in heavy cements or relatively high compressive strengths in light cements. As a result, the use of the pulse echo tools to indicate compressive strength is inaccurate at best. Figure 11.15, part b, shows the acoustic impedance of muds over a wide range of densities. Heavy muds and some cements clearly could look the same in term of acoustic impedance.

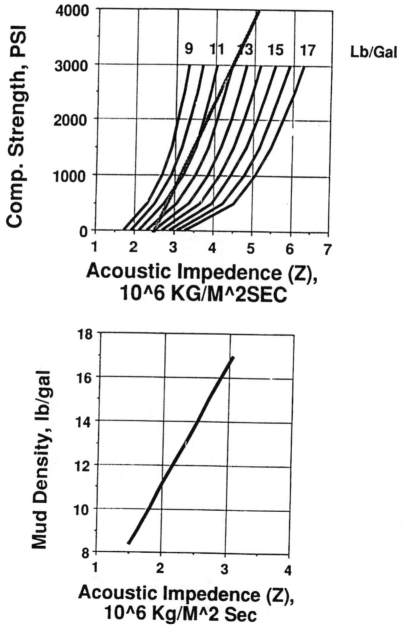

Figure 11.15 Acoustic impedance of cements and muds (Copyright SPE, Ref. 6)

The Mobil-Goodwin Technique

Rather than determine the compressive strength of the cement, Mobil suggests that it is not relevant, as long as it is not zero. The idea is that even weak cement cannot generally be displaced since it has no place to go.[6] For example, fill a soda pop bottle with sand to the top. Try to push your finger into the bottle. The sand has little compressive strength, but enough to keep your finger out of the bottle. The key to channel detection is to detect regions behind the pipe which are liquid filled.

The direct recording of the acoustic impedances from each transducer was reviewed and showed unique signature responses for differing conditions behind pipe. Figure 11.16 shows the response of the acoustic impedance recording for liquid, set cement, and gas cut cement.[6] Each of the four tracks shown records the responses of two adjacent transducers. Sometimes a relative bearing curve is superimposed over the acoustic impedance responses to determine the orientation of the channel. The liquid response is smooth and uniform and consistent with acoustic impedance values of the mud in the borehole. Cement is not perfectly homogeneous. Due to mixing, impurities, water loss, and the like, the cement acoustic impedance is rough and at values consistent within a range of compressive strengths. In gas cut cements, the pulse echo tools detect pockets of gas as well as pockets of good cement. As a result, the acoustic impedance varies from nearly zero to values consistent with cement having substantial compressive strength.

Reducing the evaluation process to a squeeze/no squeeze criterion, the Mobil/Goodwin technique clearly provides insight into whether a squeeze is likely to be successful. If the region behind pipe is a liquid, the channel can be intercepted by perforating and the liquid therein may be displaced with cement slurry. This cannot be done if weak or poor cement fills the region. Gas cut cement poses another problem. It is partially solid, par-

Liquid in the Annulus

Set Cement in the Annulus

Gas Cut Cement in the Annulus

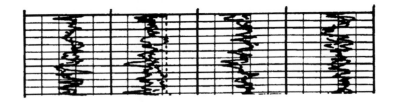

Figure 11.16 Acoustic impedance responses to various cement conditions (Copyright SPE, Ref. 6)

tially riddled with gas channels. If few channels exist and can be identified, then a squeeze may be effective. If many tiny channels exist, it is practically impossible to intercept all of the channels and correct the poor cement condition.

THE SCHLUMBERGER ULTRASONIC IMAGING TOOL (USI)

Tool Configuration and Operation

The Ultrasonic Imager (USI) was introduced by Schlumberger in 1991. It is basically a continuously rotating pulse echo type tool, and is an improvement over the CET in that the coverage is nearly 100% of the casing wall. The processing of the echo is, however, quite different from the CET. The USI is shown schematically in Figure 11.17, part a.[17] The main working element is the rotating transducer indicated as "sensor" on the bottom of the tool string.[17-19] The transducer rotates, emitting and receiving signals reflected back from the casing wall. Sampling can be done in a number of modes, as frequently as each five degrees of rotation and 1.5 in. (3.8 cm) of depth, and more frequently if only corrosion information is desired.

The USI can only be run on the Schlumberger Maxus 500 surface system. The USI is 3 3/8 in. (8.6 cm) diameter. By changing the rotating transducer subassemblies, the tool can operate in casing sizes from 4.5 in. (11.4 cm) to 13 3/8 in. (34.0 cm). Borehole mud densities are limited to less than 1.9 gm/cc (16 lb/gal) for water based muds and 1.4 gm/cc (11.6 lb/gal) for oil based muds. The minimum channel size detectable is about 1.2 in. (3.0 cm).

The rotating transducer is shown schematically at the bottom of Figure 11.17, part b.[17] In the measurement position it is aimed toward the wall and in the fluid properties position it is aimed toward the target plate. The fluid properties are measured when going in the hole.

USI Signal Processing

The transducer emits a short burst of acoustic energy widely scattered around a preselected frequency closest to the expected resonance frequency of the casing wall. Only the very early part of the reflected wave is analyzed. The frequency spectrum reflected back is typically characterized as shown in Figure 11.18, part a.[7] The phase of these reflected ultrasonic waves is also shown. The amplitude and phase of each frequency is computed by best fitting a Fourier Series to the returned signal and plotting the amplitude and phase of each term to get the Figure 11.18, part a. The group delay curve is the derivative of the phase relationship.

From Figure 11.18, part a, the first most apparent dip in the group delay corresponds to the resonant frequency of the casing. This resonant frequency may be related to the wall thickness of the casing as discussed earlier. Note that the second dip is of the first harmonic. The width of this first dip, as shown on Figure 11.18, part b, of the group delay at resonance can be related to the acoustic impedance outside of the pipe. In general, a narrow dip indicates a gas or liquid outside of pipe since the energy is essentially trapped in the pipe wall due to a large acoustic impedance mismatch. Cement on the outside produces a wider dip since the acoustic mismatch is less and the acoustic energy leaks out more rapidly.

The USI Presentation

The USI presentation of Figure 11.19, here shown in black, and white, and gray, is highly colored and computer processed.[20] The parameters shown are similar to the CET. Tracks 1 through 7 relate primarily to casing condition. The left-hand track, track 1, contains the usual geometrical parameters and the gamma ray. The track 2 shows diagnostic signals for the field engineer. These may indicate some problem with the received signal, such as poor returned signal strength or the inability to process it correctly. Notice that collars and the apparent cable groove indicate such a problem. Track 3 measures the A1 amplitude. A high value indicates that the signal is reflecting strongly. If the casing inner surface is rough, the reflected signal will be diffused and return to the transducer weakly. Tracks 4, 5, 6, and 7 show inner radius and wall thickness traces and maps.

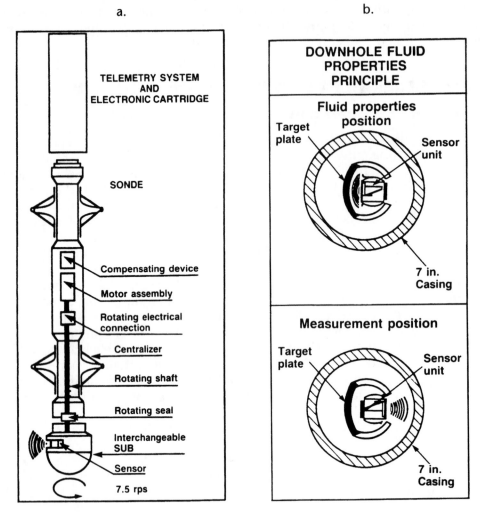

Figure 11.17 UltraSonic Imager, USI: a. Tool diagram, b. Transducer position (Courtesy Schlumberger, Ref. 17)

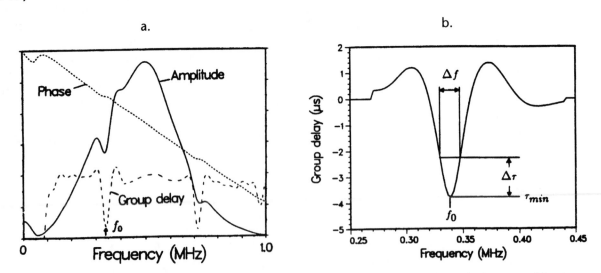

Figure 11.18 USI interpretation: a. Evaluation of group delay from amplitude and phase, and b. determining acoustic impedance (Courtesy Schlumberger, Ref. 7)

USI* IMAGE INTERPRETATION

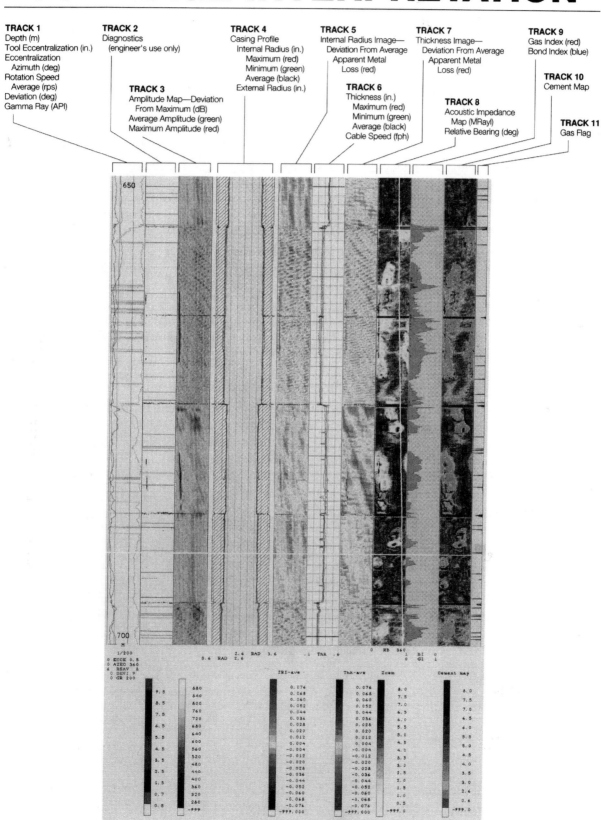

TRACK 1
Depth (m)
Tool Eccentralization (in.)
Eccentralization
Azimuth (deg)
Rotation Speed
Average (rps)
Deviation (deg)
Gamma Ray (API)

TRACK 2
Diagnostics
(engineer's use only)

TRACK 3
Amplitude Map—Deviation
From Maximum (dB)
Average Amplitude (green)
Maximum Amplitude (red)

TRACK 4
Casing Profile
Internal Radius (in.)
Maximum (red)
Minimum (green)
Average (black)
External Radius (in.)

TRACK 5
Internal Radius Image—
Deviation From Average
Apparent Metal
Loss (red)

TRACK 6
Thickness (in.)
Maximum (red)
Minimum (green)
Average (black)
Cable Speed (fph)

TRACK 7
Thickness Image—
Deviation From Average
Apparent Metal
Loss (red)

TRACK 8
Acoustic Impedance
Map (MRayl)
Relative Bearing (deg)

TRACK 9
Gas Index (red)
Bond Index (blue)

TRACK 10
Cement Map

TRACK 11
Gas Flag

*Mark of Schlumberger

Figure 11.19 USI Image Interpretation guide (Courtesy Schlumberger)

The conditions outside of casing are shown on Tracks 8 through 11. Track 8 is a color coded map of acoustic impedance with no interpretation. Track 10 is exactly the same as track 8 except that values of acoustic impedance less than .3 MRayl are specially shaded to indicate gas. Values greater than .3 and less than 2.6 MRayl are shaded to indicate liquid behind pipe. The remainder of the map is shaded identically to Track 8. Track 9 shows the fractional coverage of the casing with cement, liquid, and gas. Track 11 is a gas flag indicator. The color coding may vary from one geographical area to the next. Furthermore, a new presentation to show gas cut cement based on the local standard deviation of acoustic impedance is being introduced.

POINTS TO REMEMBER

- The CET, PET, and USI emit an acoustic pulse perpendicular to the wall and listen for the reflection.
- The primary reflected wave is used to measure time of travel and hence radius from the center of the tool and pipe surface roughness.
- The series of waves following the primary reflection are used to evaluate the pipe wall thickness and acoustic impedance of whatever is outside of pipe.
- The CET and the PET do not cover 100% of the pipe wall.
- The CET W2 and W3 gates must be normalized to read 1,1 with water inside and outside of the pipe.
- If properly normalized, a "banana plot" passing through 1,1 will be on the tail of the log.
- The CET W2N reads about .3-.6 in cemented intervals, and the W3N somewhat higher.
- Both W2N and W3N read greater than 1 if gas is outside pipe.
- Pulse Echo tools measure acoustic impedance, density times velocity, of the material outside of pipe.
- It has been erroneously assumed that acoustic impedance is linearly related to cement compressive strength.
- The log presentation shows geometrical parameters relating to the pipe as well as tool centralization.
- The geometrical parameters form the basis for a high quality casing inspection survey.
- The pulse echo tools must be centralized within about 1 mm of center for each inch of casing diameter.
- The pulse echo tools are less sensitive to microannulus than other types of bond logs, but may require pressuring if a large microannulus is present.
- Special considerations are required if run in heavy muds.
- The cement map presentation is a map of impedance. If a linear relation is assumed, it is also a map of compressive strength.
- The cement map shows channel configuration and connectedness, which could not be seen on most older types of bond logs.
- The orientation of the channel can be determined.
- Formation flags and gas flags are helpful for detection of good cement or gas outside of pipe, respectively.
- When run together, the CBL and CET can locate gas cut cement.
- The Mobil/Goodwin technique focuses on raw acoustic impedance data to make squeeze/no squeeze decisions.
- The USI has a continuously rotating acoustic transducer for nearly 100% coverage of the casing wall.
- The USI uses highly sophisticated processing.
- The USI presentation is color coded, showing both curves and maps for virtually every parameter of interest.
- The USI must be run on Schlumberger's Maxus 500 equipment.

References

1. Benoit, F., Pittman, D., and Seeman, B., "Cement Evaluation Tool—A New Approach to Cement Evaluation," Paper SPE 10207, 56th Annual SPE Conference, San Antonio, Texas, October, 1981.
2. Benoit, F., Dumont, A., Pittman, D., and Seeman, B., "Cement Evaluation Tool: A New Approach to Cement Evaluation," *JPT*, pp. 1,835-1,841, August, 1982.
3. Dumont, A., Patin, JB. and LeFloch, G., "A Single Tool for Corrosion and Cement Evaluation," Paper SPE 13140, 59th Annual SPE Conference, Houston, Texas, September, 1984.
4. Schlumberger, "Cement Evaluation Tool," Document M-086301, Schlumberger ATL Marketing, Montrouge, France, June, 1983.
5. Catala, G.N., Stowe, I.D., and Henry, D.J., "A Combination of Acoustic Measurements to Evaluate Cementations," Paper SPE 13139, 59th Annual SPE Technical Conference, Houston, Texas, September, 1984.
6. Goodwin, K.J., "Guidelines for Cement Sheath Evaluation," Paper SPE 19538, 64th Annual SPE Conference, San Antonio, Texas, October, 1989.
7. Hayman, A.J., Hutin, R., and Wright, P.V., "High Resolution Cementation and Corrosion Imaging by Ultrasound," 32nd Annual SPWLA Symposium, Midland, Texas, June, 1991.
8. Halliburton, "Cement Evaluation Manual," Houston, Texas, 1994.
9. Albert, L.E., Standley, T.E., Tello, L.N., and Alford, G.T., "A Comparison of CBL, RBT, and PET Logs in a Test Well With Induced Channels," Paper SPE 16817, 62nd Annual SPE Conference, Dallas, Texas, September, 1987.
10. Schlumberger, "Schlumberger's Cement Evaluation Tool, "Document SMP 5040, Schlumberger, Houston, Texas, 1989.
11. Rang, C.L., "Evaluation of Gas Flows in Cement," Paper SPE 16385, SPE California Regional Meeting, Ventura, California, April, 1987.
12. Sheives, T.C., Tello, L.N., Maki, V.E. Jr., and Blankinship, T.J., "A Comparison of New Ultrasonic Cement and Casing Evaluation Logs with Standard Cement Bond Logs," Paper SPE 15436, 61st Annual SPE Conference, New Orleans, Louisiana, October, 1986.
13. Halliburton Logging Services, "Pulse Echo Log," Document EL-1046, Houston, Texas, 1989.
14. Schlumberger, "Cased Hole Log Interpretation Principles/Applications," Document SMP-7025, Houston, Texas, 1989.
15. Schlumberger, "Schlumberger's Cement-Scan Log—A Major Advancement in Determining Cement Integrity," Document SMP 5058, Houston, Texas, 1990.
16. Schlumberger, "Cement Evaluation Tool," Marketing Document for Oklahoma, California, and Rocky Mountain Sales, 1984.
17. Schlumberger, "USI—UltraSonic Imager," Document M-090247, Houston, Texas, 1991.
18. Schlumberger, "Ultrasonic Imaging," Document SMP-9230, Sugar Land, Texas, November, 1993.
19. Hayman, A.J., Parent, P., Cheung, P., and Verges, P., "Improved Borehole Imaging by Ultrasonics," Paper SPE 28440, 69th Annual SPE Technical Conference, New Orleans, Louisiana, September, 1994.
20. Schlumberger, "USI Image Interpretation," Document SMP-5518, Houston, Texas, 1991.

CHAPTER 12

WELL INTEGRITY: DOWNHOLE CASING INSPECTION

OVERVIEW

Applications

Casing inspection measurements are used to examine the casing or tubing in a well. Unless the inspection is of the tubing, the assessment of pipe condition is not a through tubing operation, and requires pulling of the tubing and casing and killing the well. Scale and other buildup on the inside of the casing may affect the tool response, and therefore a bit and scraper run is usually required to clean up the hole prior to running the survey.

The applications of these include the location of pits, leaks, and holes in the casing. With some equipment, a pit on the outside can be discriminated from one on the inside wall of the pipe. A caution should be noted about hole detection. Holes smaller than 1/8 in. (.4 cm) diameter are difficult to detect with any of the instruments discussed in this chapter. Furthermore, larger defects such as pits may be easily detectable, but cannot easily be discriminated from smaller holes through the casing. Larger holes are easily detectable. Perforations are detectable since they are expected and are oriented in a predictable pattern. If small holes which are producing an undesirable fluid need to be detected, temperature or other fluid movement logs may be better suited to their detection.

These tools are helpful in assessing the overall condition of pipe, whether it is to evaluate the suitability of changing a producing well to an injector, assessing the condition of intermediate strings during drilling operations, or determining the financial value of pipe before plugging and abandoning a well. Corrosion management is an important application. Certain surveys are necessary for proper cathodic protection planning. Often surveys are run on a time lapse basis to monitor the progress of corrosion with time. Proper combinations of tools discussed herein even afford the opportunity to detect serious damage to the outer of two concentric strings of pipe.[1-6]

Types of Inspection Equipment

Generally, there are a number of approaches to casing inspection, all using vastly differing technologies. They are:

1. **Mechanical Calipers** examine the inner pipe surface.
2. **ElectroMagnetic Tools** examine and discriminate the inner from outer pipe surface and may shed light on concentric casing strings.
3. **Acoustic Devices** assess the pipe ID, surface roughness, and wall thickness.
4. **Borehole Video Cameras** provide an actual view of damage or lost equipment downhole.
5. **Casing Potential Profiles** foresee electrochemical corrosion and are the basis for cathodic protection.

These technologies are each discussed separately in this chapter.

MECHANICAL CALIPERS

Tool Configuration

The mechanical calipers are among the simplest and most accurate measurements of the inside condition of pipe. A schematic of such a tool is shown on Figure 12.1.[8] The tool is centralized with an array of fingers or arms which reach out to the inside casing wall. The arms are tungsten carbide tipped for wear and, when released downhole, spring loaded to press against the pipe wall. Typically, such a tool can have from 40 to 80 fingers, depending on the tool design. The coverage is not 100% since a gap between the fingers will exist. The tool is insensitive to borehole fluids. Such tools are available on electric wireline for real time surface readout and also available with downhole recording devices for slickline or pump down applications.[7-12]

Mechanical Caliper Log Presentation

Figure 12.2 shows a typical log presentation for a 40 finger caliper.[10] The upper part of the figure shows a piece of casing with a slot at A, a swelled casing section at B, and perforations at C. The right-hand tracks of the log show the traces across this casing interval for

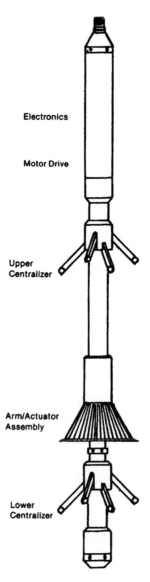

Figure 12.1 Mechanical multifinger caliper (Courtesy Western Atlas, Ref. 8)

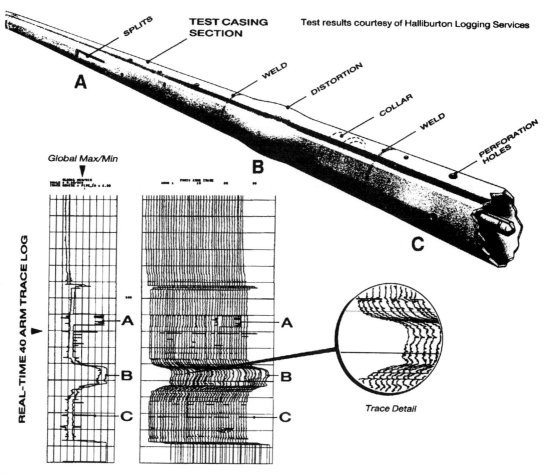

Figure 12.2 Multifinger caliper with min/max and individual 40 finger recording (Reproduced with permission of Sondex and Halliburton Co.)

each of the individual 40 fingers, with radius increasing toward the right. The split at A shows up on a single finger, the swelled casing at B shows up on all fingers, and the perforations at C show up on four fingers. On the left-hand track, the minimum and maximum as seen among the 40 fingers is recorded. Such min/max presentations help highlight the gross number of defects on a piece of pipe. Prior to the development of tools capable of recording each finger and sending that data to the surface, the min/max presentation was all that was available (some equipment could do a min/max measurement over discreet segments of the perimeter). Indeed, sometimes only the max was shown.

ELECTROMAGNETIC CASING INSPECTION TOOLS

Two Types of Tools Available

There are two basic types of electromagnetic casing inspection tools available. These are the pad type and phase shift type devices. Those devices herein referred to as pad devices may also be referred to in the literature as flux leakage and eddy current tools. The phase shift devices may be referred to as electromagnetic phase shift devices. These devices can be run in all borehole fluids.[12-22]

The service company names for these services are listed below with an * indicating that the name is a mark of the company to which the tool is attributed.

The pad type devices are marketed under the following names:

Schlumberger: Pipe Analysis Tool (Log) **, PAT* (PAL*)
Atlas Wireline: Vertilog*, Digital Vertilog* (DVRT*)
Halliburton: Pipe Inspection Tool* (PIT*)

The electromagnetic phase shift type of tools are marketed under the following names:

Schlumberger: Electromagnetic Thickness Tool* (ETT*)
　　　　　　　Multifrequency Electromagnetic Thickness Tool*
　　　　　　　(METT*)
Atlas Wireline: Magnelog*, Digital Magnelog* (DMAG*)
Halliburton: Casing Inspection Tool* (CIT*)

The phase shift type of tool is also available through a number of other smaller service companies.

Pad Type Devices—Configuration and Operation

A schematic of the test section of a pad type device is shown on Figure 12.3.[10] Figure 12.3, part b, shows the tool test section for a PAT. There is an upper and lower array of sensors, each having six pads. The vertical spacing between the upper and lower arrays of pads is about 1 foot (.3 m) for the PAT, much less for the Vertilog and the PIT. The pads are sized to match the casing to be evaluated. The upper and lower pad arrays are staggered with respect to each other for 100% coverage of the pipe wall. Above and below the pad arrays are the magnetic pole pieces and the upper and lower centralizers (not shown). To the left, in Figure 12.3, part a, is the same test section of the tool with the pads and housing stripped away. Inside is a coil through which a DC current passes. That current generates magnetic flux lines in the core of the tool. These flux lines pass through the pole pieces to

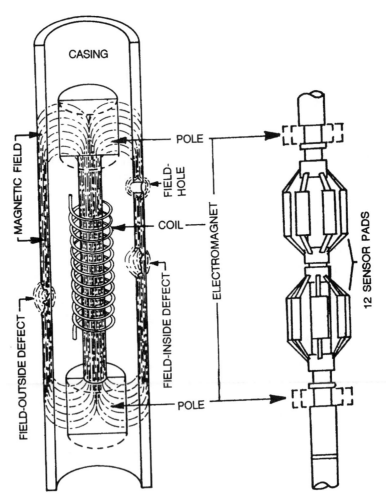

Figure 12.3 Electromagnetic flux leakage pad casing inspection tool (After Schlumberger, Ref. 14, and Western Atlas, Ref. 10)

and through the casing and back to the tool core. The purpose of the pole pieces is to enhance the magnetic coupling to the casing and they are sized for different casings. These flux lines pass through the casing parallel to the casing walls unless there is damage, at which point the flux lines are distorted as shown.

Each pad contains a series of coils to perform "flux leakage" and "eddy current" tests. In Figure 12.3, part a, when the flux lines encounter a defect in casing, they are distorted both to the inside and outside of the pipe wall regardless of whether the defect is on the inside or outside. As a result, when a pad slides past the defect, the coil(s) within the pad cut flux lines and an electrical current is generated. This current is the response which is presented on the log. Where no defect is present, the flux lines are parallel to the wall and the pad cuts no flux lines as it slides along the wall. This test is the "flux leakage" test. Since it detects defects anywhere within the wall, it is hereinafter referred to as the "total wall" test.

The "eddy current" test examines the inner wall surface, and is therefore also called the "inner wall" test. An alternating current is passed through a coil in the pad. This causes surface or eddy currents on the inside wall of the pipe in front of the AC coil. As long as the inner surface is smooth, the fields generated by the AC coil and the eddy currents balance. When a defect is encountered, these fields become unbalanced and an AC signal is detected in other coils of the pad. This inner wall signal is shown on the log.

Defects detected by these type of tools are qualitative at best. The shape of the defect as well as its size are relevant to the distortion of flux lines detected by the pad. For example, a rough defect with sharp changes in wall penetration would yield a much stronger signal than that same defect would after being sandblasted and smoothed, even though the depth of penetration of the defect has not changed.

Pad Type Devices—Log Presentation

The log presentation for a Vertilog is shown in Figure 12.4.[4] In this example, the FL-1 and FL-2 curves correspond to the flux leakage or total wall tests for the upper and lower array, respectively. The Discriminator is the eddy current test from both arrays, and is used to discriminate damage on the inside. It should be noted that if the inner wall test shows damage, it does not mean that the outer wall is damage free. The Average is just that. It is the average response of all of the pads. If it is large as at the collars, the defect has a substantial lateral extent. If it is small, the defect is confined to only one or two pads and is limited. In this example, the joint of pipe near the top of the log shows external corrosion, sometimes rather substantial, and two short intervals with internal corrosion. Lower downhole, scratchers and centralizers are noted, with a strong FL-1 and FL-2 response, but none on the inner surface discriminator test. Perforations show up quite strongly since there is both a hole and a burr which causes the pads to bounce around when sliding across perforations.

Some companies try to quantify the responses of pad type tools. Figure 12.5 shows a typical response classification.[4] This chart is for a Vertilog tool in 7 in. (17.8 cm), H-40, 20 lb/ft (24.87 kg/m) casing. The vertical axis goes from 0 to 50 chart divisions (maximum response). The curves show the tool response to machined test defects having certain wall penetration and located either on the inside or outside of the pipe. For example, if the FL response is 10 chart divisions, and the inner surface Discriminator reads zero, the defect is on the OD and the wall penetration is 50%, i.e., a class 3 defect. Similarly, if the Discriminator shows a defect, and no defect is assumed on the OD, then the defect is on the ID and the wall penetration of the pit is 30%, i.e., a class 2 defect.

In recent years, the pad devices have been upgraded to monitor the response of each pad individually. Both the total wall Flux Leakage and the inner surface Eddy Current tests are monitored. Such a presentation for the Digital Vertilog (DVRT) is shown on Figure 12.6.[22] The left 12 curves correspond to total wall FL1 to FL12, and the right-hand 12 curves correspond to the inner surface EC1 to EC12 measurements. Sometimes a color coded map of the pipe wall is shown, both for inner surface and total wall, to highlight defects. The log of Figure 12.6 shows the response for 3/8 in. (1 cm) drilled defects on both the inner and outer surfaces. This test shows that defects smaller than about 3/8 in. (1 cm) diameter and 25% wall penetration cannot readily be detected with this type of equipment.

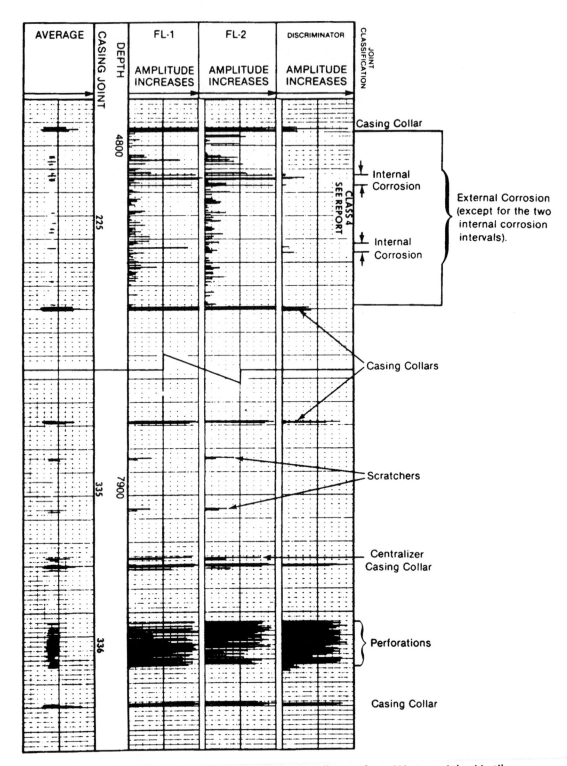

Figure 12.4 Response of inner surface and total wall tests for a Western Atlas Vertilog (Courtesy Western Atlas, Ref. 4)

Electromagnetic Phase Shift Devices—Tool Operation

This class of tool is used to measure the amount of metal remaining in a casing string. It is best suited for large scale corrosion, vertical splits, holes larger than about 2 in. (5 cm) diameter, and examining the outer of two concentric strings of casing. It is very suitable for time lapse logging to monitor metal loss from time to time over a well's lifetime.

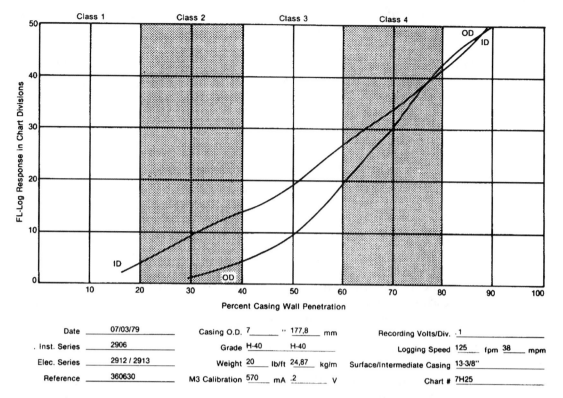

Figure 12.5 Vertilog response to various casing defects (Courtesy Western Atlas, Ref. 4)

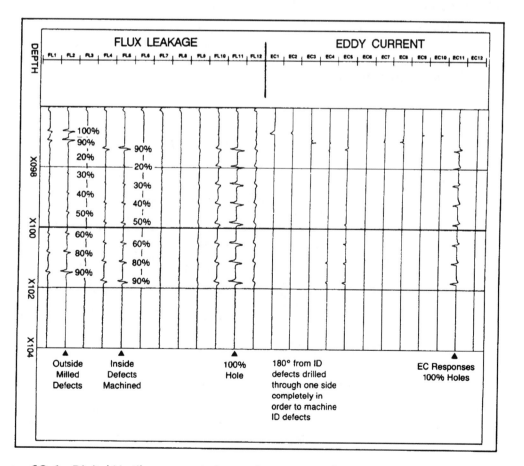

Figure 12.6 Digital Vertilog presentation and response to known defects (Courtesy Western Atlas, Ref. 22)

The basic tool is made up of two coils as shown in Figure 12.7.[14] A low frequency alternating current is put through the transmitter coil. This generates an AC field which interacts with the casing and induces an alternating current in the second coil. The phase shift of the AC signal in the receiver coil is a function of the electrical conductivity, magnetic permeability, and thickness of the casing, as well as the frequency of the transmitted AC field.

The equation relating these variables is

$$\Delta\phi = 2t\sqrt{\pi f \mu \sigma} \qquad \text{(Equation 12.1)}$$

where

$\Delta\phi$ = Phase Shift (Low Frequency), radians

t = Casing Wall Thickness, meters

f = Frequency, Hertz

μ = Magnetic Permeability, Henry/meter

σ = Conductivity, mho/meter

In older equipment, it is assumed that the magnetic permeability and conductivity of the casings are constant and equal. This is not the case, and as a result, phase shift tools have been developed which operate at a variety of frequencies to normalize variations in these parameters. Such new tools furthermore compute an average thickness. New tools also offer a number of transmitter to receiver spacings. For analysis in multiple strings of casings, where the greatest depth of investigation is required, the lower frequencies and longer spacings are best. These tools may be equipped with an electronic caliper or may be run with a multifinger caliper for assessment of the inside wall of the casing.

Electromagnetic Phase Shift Devices—Log Presentation

The basic presentation is a phase shift in degrees, with greater phase shift indicating more metal and lesser phase shift indicating loss of metal. Some equipment shows a series of curves for differing transmitter/receiver spacings and frequencies along with supporting quality control curves. Consult with your service company regarding the variety of presentations available.

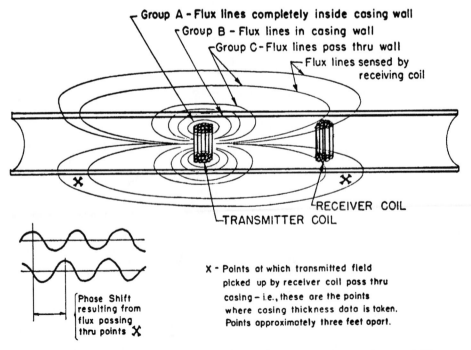

Figure 12.7 Schematic of electromagnetic phase shift casing inspection tool (Courtesy Schlumberger, Ref. 14)

It is interesting to compare the response of the phase shift tool to that of the pad type device. Figure 12.8 shows such a comparison for a hypothetical casing string. On the left is the pad tool log with the maximum total wall and inner surface response seen by any of the 12 pads presented. On the right is the phase shift response with more metal corresponding to greater phase shift. Notice that joint B only has occasional pits and therefore serves as a reference phase shift response for the particular size and weight of casing present in this string. A heavier joint of pipe would have a higher phase shift. A lighter joint would have a lesser phase shift. While most newer tools correct for variations in casing properties, some do not. If not, there often will appear to be a drift of phase shift along the casing joint. When looking at a number of joints, this drift creates a sawtooth pattern along the log, with each joint being one tooth of the saw.

Phase shift tools can examine the outer of two concentric strings. Across joint A on Figure 12.8 there is a concentric outer string, except for the interval where it is parted. The phase shift is larger over the intervals having concentric casings since there is more metal around the tool. It also shows a dramatic drop in phase shift back to the inner string value over the parted interval. Notice that the collars show for both the inner and outer string.

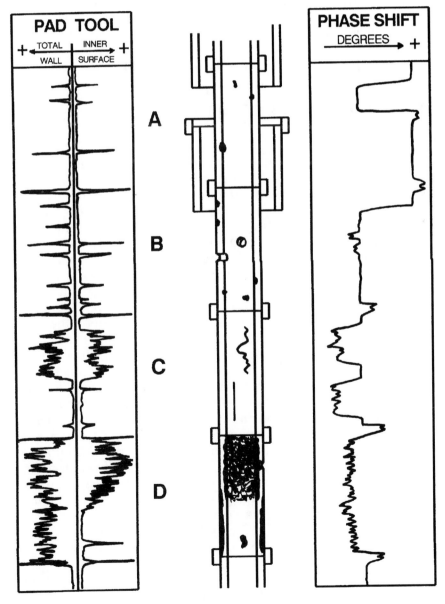

Figure 12.8 Response of flux leakage pad type and phase shift casing inspection tools to the same casing string

The pad device shows the collars only of the inner string. The increase in phase shift near the top of joint B may be mistaken for a heavier joint of pipe, except that the weight change appears to begin midway between and not at the collars. The inside and outside small pitting on joints A and B is detected by the pad tool but not detected by the phase shift tool. Damage on the outer string could be detected by the phase shift device provided it is severe enough and the phase shift tool is run with a pad tool to discriminate between damage on the inner or outer string.

Joint B has outer pits near the top, through holes midway, and inner surface pits on the lower section of the joint. Only the one larger hole (greater than 2 in. or 5 cm) shows as a reduction of phase. Note that point defects such as this, as well as collars, show up two times. This is due to the fact that the flux lines cross the casing wall at two points marked X in Figure 12.7. So, as the tool passes a defect, it is detected as a loss of metal first above then below the transmitter. Small pits and holes will generally not be detected by the phase shift tool. Perforations do not show up on a phase shift log while pad devices locate perforations easily. The pad tool detects the holes and inner wall pits on both the total wall and inner surface test. Outer surface pits are not detected on the inner wall test.

Splits, whether smooth or jagged, show up strongly on the phase shift log. Smooth splits, parallel to the casing axis, do not distort the magnetic flux lines, and hence only the ends of the split may be detected on the pad tool. The jagged split shows up quite well. Overall areas of corrosion show up on the phase shift log as a loss of metal. The pad tool reacts strongly to such corrosion as well. Not the hole seen on the lower inner surface of joint D. It is detected by the inner surface pad test.

ACOUSTIC CASING INSPECTION TOOLS

Log Presentations

The acoustic tools have been discussed in the previous chapter on ultrasonic pulse echo technology. These tools are either the fixed transduce CET/PET type (see Figure 11.1) or the rotating transducer variety (see Figure 11.17a). The transducer emits an acoustic pulse perpendicular to the casing wall. On the basis of the reflected wave, certain geometrical factors may be evaluated (see Figures 11.2 and 11.6).[23,24] Since the fluid acoustic velocity is measured independently by the tool, accurate distances from the transducer to the pipe wall may be calculated. The strength of the received signal is an indication of inner pipe surface roughness. The frequency of the received signal is a measure of the wall thickness. These tools may not be effective in highly attenuating heavy muds or with gas in the borehole. Consult with the service company prior to a job having such conditions.

The geometrical parameters of the CET are shown on the log of Figure 12.9.[23] On track A are the usual ovality, CCL, internal radius minimum and maximum, IRMN and IRMX, respectively. On track E are four sections cut across the casing diameter. Section E1, captioned ER1, IR1, and IR5, ER5, represents a section across the pipe through transducers 1 and 5. The IR and ER refer to the Internal Radius and External Radius, respectively. This cross section shows no damage. Section E2 is 45° over from E1 and shows damage to the second joint from the top on the inside. All other sections look to be in good shape. Track D shows the minimum and maximum thickness of the pipe. Track B shows a map of the internal radii based on the time of travel of the first reflection. A radius larger than nominal is shaded black. The damage to the second joint of pipe from the top is clearly evident. Track C shows a map of the inner surface roughness based on the strength of the first reflection. It is labeled "Internal Rugosity." This track shows the lower two joints of pipe to be clean on the inside with corrosion or roughness increasing uphole. The CET samples an area of about one in. (2.5 cm) diameter for each 45° sector of the casing. Since this is less than 100% coverage, such maps may tend to exaggerate the lateral extent of some defects or miss others.

Figure 12.10 shows the color coded maps that are available with special CET processing. This example, shown in black, white, and gray tones, shows the color tracks for metal thickness, internal radius, and first echo amplitude based on the Schlumberger Acoustic Corrosion Evaulation*, ACE* (* indicates mark of Schulmberger) processing. The color coding is

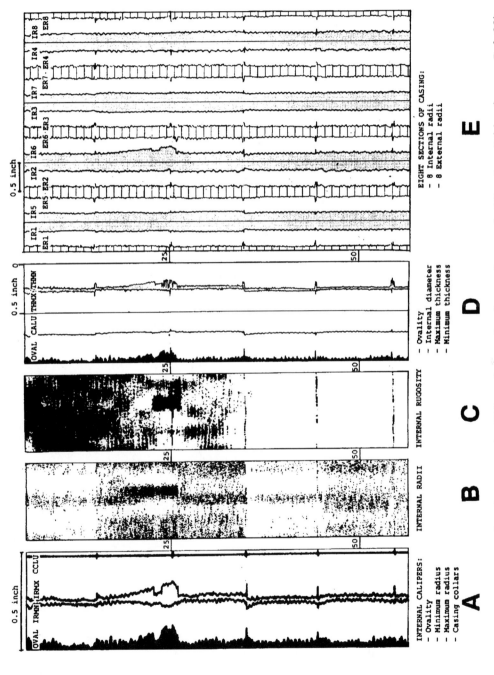

Figure 12.9 Results of an acoustic CET run in a casing inspection mode (Courtesy Schlumberger, Ref. 23)

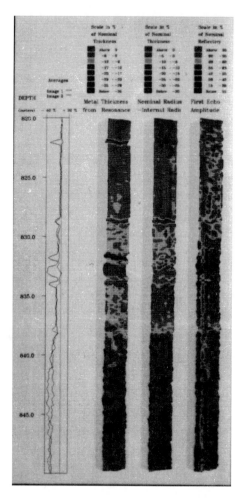

Figure 12.10 Schlumberger Acoustic Corrosion Evaluation (ACE) casing inspection processing from CET data (Courtesy Schlumberger)

shown at the top. The pipe section is reasonably good at the top but is progressively worse going downhole. Metal loss appears to be in the neighborhood of 30% and the damage is to the inside of the casing. The left-hand track shows a log curve of the inner radius and thickness.

Other special presentations are available for each of the measured parameters. These include color coded joint by joint histograms showing the range of values detected. Also included are "Pipe in Perspective" presentations in which each joint is examined either from the side or looking directly down the pipe. Such presentations which also show the orientation of the damage pattern can be very helpful in determining the mechanism causing the damage.

Tools with rotating transducers can provide 100% coverage of the casing wall. Figure 11.19 showed the USI presentation in which the same three acoustic parameters, first echo amplitude on track 3, internal radius on track 5, and wall thickness on track 7, are shown as color coded maps (black and white in this figure). Even fancier presentations are available such as the enhanced computer evaluation of a screen shown in Figure 12.11. This figure is from a Halliburton CAST* tool(* mark of Halliburton Energy Services). On the left is a map presentation of a screen over about a 24 ft (7 m) interval. This screen has three prominent areas of damage at XX12, XX17, and XX22. On the right is an artist's conception of the damage and its orientation looking at the casing from the outside. Of course, the outside representation, while dramatic, may not be accurate since the tool does not examine that region. The inside region is accurate.

Figure 12.11 Halliburton Energy Services CAST tool run for inspection of slotted liner (Reproduced with permission of Halliburton Co.)

BOREHOLE VIDEOS

Equipment and Presentation

In a sense, borehole videos are the ultimate tools to assess conditions downhole. The problems can be viewed directly without significant interpretation. They have, however, some very serious shortcomings. There are two basic types of operational tools available. One takes continuous movies viewed in real time at the surface. The other takes only snapshots, as frequently as 9 seconds or faster between photos. Without a doubt, the continuously run tool is more useful. However, it must be run on a special line, either a coaxial cable or a fiber optic cable. The snapshot tool can be run on a convention wireline. Coaxial cable units set up inside of a coil tubing unit are also available for horizontal borehole inspection logging.[25-27]

Another drawback is that the fluid in the well must be clear. Most videos run to date have been done in wells where the borehole was filled with filtered water or blown down with nitrogen to provide a clear fluid for logging. Recent reports indicate that wellbore fluids may be suitable if one can see through the produced water at least about 6 in. (15 cm). If oil and water are the downhole phases, there typically is much more water by volume than oil, and so oil entries can easily be seen. Gas entries cause the water to become murky.

Many novel applications for this equipment exist. Fishing jobs are a natural to see the fish and select the proper equipment to catch it. Oil and possibly gas entries can be identified. Serious damage and scale can be detected. Perforations can be inspected. For a frac job, the perforations taking the proppant can be identified. Needless to say, this equipment only looks inside the casing and outside damage will not be evident unless it has progressed through the casing. Furthermore, if quantification of metal loss or depth of penetration of pits is desired, these tools are of little use.

Figure 12.12 is a snapshot from a continuously run borehole video survey. In this case, an impression block showed the tool to be retrieved in the center of the well bore. However, after a number of failed attempts the well was blown down with nitrogen to provide a clear view for a video camera. The camera shows the heretofore unknown presence of the cable around the tool. With this knowledge, the cable and tool were easily retrieved.

The snapshot of Figure 12.13 shows a video camera with a light source hanging below the lens on two support arms. On the left, oil bubbles are observed entering the water-filled

Figure 12.12 Downhole video camera showing a stuck tool surrounded by cable (Courtesy BP Exploration [Alaska] and Western Atlas)

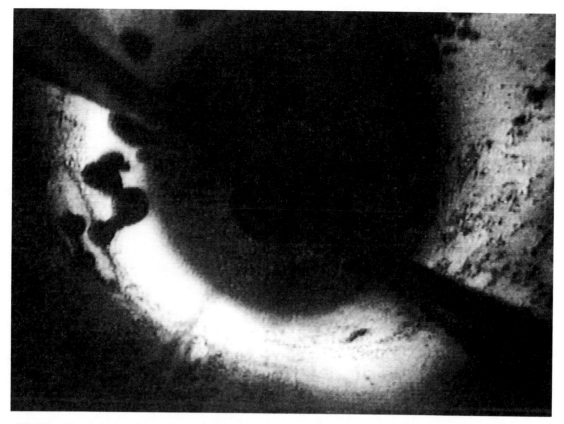

Figure 12.13 Downhole video showing oil entry (Courtesy BP Exploration and Western Atlas)

well bore and floating up to the high side of the pipe. Note also that the casing is coated with a light carbonate scale except for a small circular area around the perforation.

CASING POTENTIAL SURVEYS FOR CATHODIC PROTECTION
Tool and Log Presentation

Electrochemical corrosion is a primary cause of casing damage.[28-33] Electrical current flows in a casing string as a result of the potential differences along its length arising from changes in formation fluids or casing properties. Positive current flows out of casing at high potential areas called anodes. As a result of this current flow, metal is lost from the casing at such anodes, i.e., corrosion is taking place. Figure 12.14 shows current propagating from an anode to a cathode on the same casing string.[4]

The basic casing potential profile equipment is shown in the schematic of Figure 12.15, part a.[2] The tool is comprised of two sets of wheel contactors separated from each other by an insulated section of the sonde. These may be separated from each other by as much as 15 ft (4.6 m), but this distance may be adjusted with spacers at the surface. The voltage difference between the upper and lower contacts is measured with the tool stationary.

As shown in Figure 12.15, part b, the polarity of the current indicates the direction of the current flow. A negative slope indicates that current is leaving the casing (iron ions are leaving) and corrosion is occurring. In the figure, the intervals from about 1,400 to 2,300 and 2,900 to 3,300 are in the process of corroding.

In order to minimize electrochemical corrosion, wells may be cathodically protected. This procedure entails putting a current to the wellhead and driving the well to be a cathode with respect to the surroundings. A sacrificial anode is placed in the vicinity of the well. An example of the well's response to such current is shown in Figure 12.16.[4] The base voltage profile with the well in its natural state is shown in run 1. Corrosion is occurring from 1,500 to

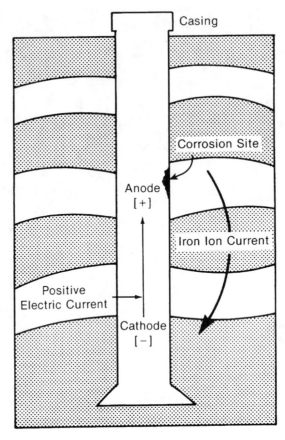

Figure 12.14 Electrochemical corrosion in casing (Courtesy Western Atlas, Ref. 4)

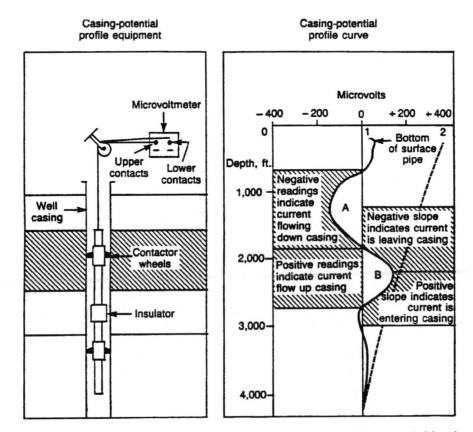

Figure 12.15 Casing potential profile equipment and log presentation (Courtesy Schlumberger, Ref. 2)

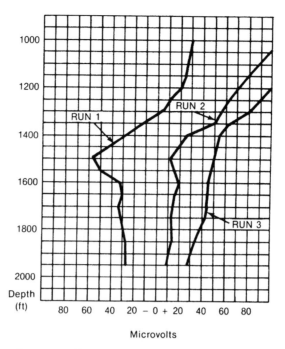

Figure 12.16   Casing voltage profile with varying currents applied to the casing (Courtesy Western Atlas, Ref. 4)

1,650 ft (457 to 503 m) and again below 1,700 ft. (518 m). Run 2 was made with 5 amps current. The negative slope remains over the interval from 1,500 to 1,600 ft (457 to 488 m). With 8 amps current, run 3 indicates that the entire casing string is a cathode and cathodically protected.

Cathodic protection does not necessarily mean that corrosion is stopped. Other corrosive mechanisms may still be active. Furthermore, cathodic protection on one well may have an adverse effect on other wells or pipelines in the area by making them anodes and subject to greater than anticipated corrosion.

Rate of Metal Loss

Other configurations exist with arrays of contacting electrodes to measure voltage and casing resistivity between electrodes. When both the casing voltage and resistance are known, the flowing current can be calculated using the equation

$$I = V/R$$ (Equation 12.2)

where

I = Current Flow, Amps

V = Casing Voltage, Microvolts

R = Casing Wall Resistance, Microohms

Suitable values of R are also available from service company tables. To compute the metal loss between two measure points 1 and 2, use the equation

$$M = ((I_1 - I_2) \times 20.13)/L$$ Equation 12.3

where

M = Metal Loss Rate ((lbs/yr)/ft)

L = distance between voltage measurements

For example, the natural state of the well of Figure 12.16 is measured with a tool having a 15 ft (4.47 m) space between the contact electrodes is shown as run 1. Supposing this to be 7 in., 26 lb/ft, J55 casing, the resistance is found to be 15.5 microohms/ft. This value may be measured or from service company tables. Since the contact points of the tool used to measure the voltage are spaced 15 ft (4.57 m), the resistance over this same interval is $15 \times 15.5 = 232.5$ microohms. To compute the metal loss rate for the interval 1,550 to 1,600, first read the voltages for these depths. These are 50 and 33 microvolts, respectively. The currents are then

$$I_{1550} = 50/232.5 = .215 \text{ amps}$$
$$I_{1600} = 33/232.5 = .142 \text{ amps}$$

The rate of metal loss is then

$$M = ((.215 - .142) \times 20.13)/(1600 - 1550) = .029 \text{ lbs/yr/ft}$$

Service companies may provide such metal loss information, either as a weight per year estimate or as a rate of penetration indication, thickness loss per year. The log of Figure 12.17 is a Schlumberger Corrosion Protection Evaluation Service* (CPET*) log (* mark of Schlumberger).[32] Corrosion rate is indicated on the left track in mm/yr.

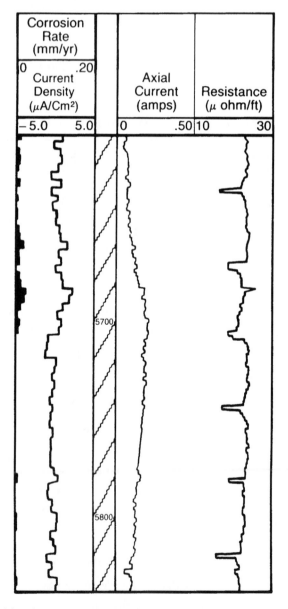

Figure 12.17 The Schlumberger Corrosion Protection Evaluation Service shows corrosion rate (Courtesy Schlumberger, Ref. 32)

POINTS TO REMEMBER

- Types of casing inspection surveys include mechanical calipers, electromagnetic, acoustic, video, and potential surveys.
- Mechanical calipers examine only the inside of the pipe.
- Mechanical calipers may show only the maximum, the max/min, or the response of each finger to damage.
- Electromagnetic tools evaluate the total wall of the pipe and can discriminate inside from outside damage.
- Acoustic tools can also evaluate the total wall and discriminate inside from outside damage.
- Except for the phase shift device, all tools need a prior bit and scraper run to clean the pipe inside.
- Pad and acoustic devices can see defects as small as 3/8 in. (1 cm) diameter and 25% wall penetration.
- Pad tools use a Flux leakage test for total wall evaluation and an eddy current test for inner surface evaluation.
- Pad tools do not respond to outer concentric strings of pipe.
- Phase shift tools cannot detect holes smaller than about 2 in. (5 cm) diameter.
- Phase shift devices measure the amount of metal remaining and are poor for detecting defects such as holes smaller than about 2 in. (5 cm) diameter.
- Phase shift tools easily detect splits in casing.
- Phase shift tools are affected by an outer concentric string.
- Casing inspection tools are larger diameter and not generally through tubing tools unless designed to examine the tubing.
- The mechanical and electromagnetic tools are insensitive to borehole fluids.
- Acoustic tools may have problems with mud weights greater than about 12 lb/gal (1.44 gm/cc) and oil based muds. Note that special heavy mud equipment is available.
- Video tools require a reasonably clear fluid.
- Casing potential profiles requires a non-conducting fluid, although equipment is available for conductive fluids.
- The fixed transducer acoustic tools do not cover the pipe 100%.
- Acoustic tools use a pulse-echo technique.
- Acoustic tools measure the first reflection amplitude, the inner wall radius, and the casing thickness.
- Acoustic tools cannot measure the full range of wall thickness, and may have problems if the reflected wave is weak.
- Video surveys require a special coaxial or fiber optic cable for a quality continuous real time view downhole.
- Casing potential surveys are used to locate where electrochemical corrosion is occurring in the casing string.
- Casing potential profiles are used for cathodic protection of casing.

References

1. Cryer, J., Dennis, B., Lewis, R., Palmer, K., and Wafta, M., "Logging Techniques for Casing Corrosion," *The Technical Review*, Vol. 35, No. 4, pp. 32–39, Schlumberger, Houston, Texas, 1987.
2. Morriss, S.L., Reinders, J.W., Lewis, R.G., and Carroll, J.F., "Completion Evaluation Using Well Logs," Schlumberger Educational Services, Houston, Texas, 1986.
3. Schlumberger, "Corrosion Evaluation," Document SMP–9110, Houston, Texas, 1989.
4. Atlas Wireline Services, "Casing Evaluation Services," Document 9426, Houston, Texas, 1985.
5. Welex (Now Halliburton Energy Services), "Casing Inspection," Document P–3012, Houston, Texas, 1985.
6. Schlumberger, "Corrosion Monitoring," Document M–090041, Montrouge, France, 1989.
7. Kading, H.W., "Computer Calioer, Fingerprints of the Hole, From Austin Chalk to Ellenburger," 18th Annual SPWLA Symposium, June, 1977.

8. NL McCullough (Now Atlas Wireline Services), "Multi-Finger Caliper Tool (MFC)," Brochure, Houston, Texas, 1985.
9. Kinley Corp., "Worldwide Petroleum Services Can Cut Your Production Costs," Houston, Texas, 1992.
10. Sondex, "Multi-Finger Caliper," London, UK, 1992.
11. Schlumberger, "Cased Hole Log Interpretation Principles/Applications," Document SMP–7025, Houston, Texas, 1989.
12. Cuthbert, J.F., and Johnson, W.M., "New Casing Inspection Log," 1975 AGA Operating Section Transmission Conference, Bal Harbour, Florida, May, 1975.
13. Bradshaw, J.M., "New Casing Log Defines Internal/External Corrosion," *World Oil*, pp. 53–55, September, 1976.
14. Smolen, J.J., "PAT Provisory Interpretation Guidelines," Document C–12013, Schlumberger, Houston, Texas, July, 1976.
15. Haire, J.N., and Heflin, J.D., "Vertilog—A Down-Hole Casing Inspection Service," Paper SPE 6513, Annual California Regional Meeting, Bakersfield, April, 1977.
16. Cuthbert, J.F. and Knepper, G.A., "Gas Storage Problems and Detection Methods," Paper SPE 8412, 54th Annual SPE Conference, Las Vegas, Nevada, September, 1979.
17. Smith, G.S., "The ETT-C, An Improved Corrosion Inspection Tool," 1980 AGA Operating Section Transmission Conference, Salt Lake City, Utah, May, 1980.
18. Smith, G.S., "Principles and Applications of a New In-Situ Method for Inspection of Well Casing," Paper SPE 9634, SPE Middle East Technical Conference, Manama, Bahrain, March, 1981.
19. Iliyan, I.S., Cotton, W.J., and Brown, G.A., "Test Results of a Corrosion Logging Technique Using Electromagnetic Thickness and Pipe Analysis Logging Tools," *JPT*, pp. 801–808, April, 1983.
20. McCann, C., "Simultaneous Mechanical Caliper—Electromagnetic Caliper Log," Southwestern Petroleum Short Course Association, Texas Tech University, Lubbock, Texas, 1987.
21. Wafta, M., "Downhole Casing Corrosion and Interpretation Techniques to Evaluate Corrosion in Multiple Casing Strings," Paper SPE 17931, SPE Middle East Technical Conference, Manama, Bahrain, March, 1989.
22. Atlas Wireline Services, "Digital Vertilog," Document 9640, Houston, Texas, 1991.
23. Dumont, A., Patin, J.B., and LeFloch, G., "A Single Tool for Corrosion and Cement Evaluation," Paper SPE 13140, 59th Annual Technical Conference, Houston, Texas, September, 1984.
24. Bettis, F.E., Crane, L.R., Schwanitz, B.J., and Cook, M.R., "Ultrasound Logging in Cased Boreholes Pipe Wear," Paper SPE 26318, 68th Annual SPE Conference, Houston, Texas, 1993.
25. Palmer, I.D., and Sparks, D.P., "Measurement of Induced Fractures by Downhole TV Camera in Black Warrior Basin Coalbeds," *JPT*, pp. 270–275, March, 1991.
26. BP Exploration (Alaska) Inc., "Video Survey of Well H–12," Anchorage, Alaska, September, 1991.
27. Halliburton and WesTech, "Downhole Video Services," Houston, Texas, 1994.
28. Dresser Atlas (Now Atlas Wireline Services), "Casing Evaluation Services," Document 9541, Houston, Texas, 1987.
29. Bradshaw, J.M., "Production Cost Reduction Through Casing Corrosion Monitoring," Paper SPE 7704, SPE Production Technology Symposium, Hobbs, New Mexico, October, 1978.
30. Pace, F.A., and Krupicki, S.E., "Reduction of Casing Leaks by Using Cathodic Protection—East Texas Field," *JPT*, pp. 1,437–1,442, July, 1982.
31. Gast, W.F., "Has Cathodic Protection been Effective in Controlling External Casing Corrosion for Sun Exploration & Production Co.? A 20 year Review Tells the Story," The International Corrosion Forum, NACE Paper 151, Boston, Massachusetts, March, 1985.
32. Monrose, H., and Boyer, S., "Casing Corrosion: Origin and Detection," *The Log Analyst*, pp. 507–519, November–December, 1992.
33. Schlumberger, "CPET—Corrosion Protection Evaluation Service," Document SMP–5089, Houston, Texas, 1988.

CHAPTER 13

FLUID MOVEMENT: TEMPERATURE SURVEYS

TEMPERATURE LOGGING OVERVIEW

Applications of Temperature Surveys

Temperature surveys are the mainstay of logging for fluid movement detection downhole. "Production logs," a term often used to categorize surveys run for this purpose, are almost always run with a temperature instrument on the tool string. While temperature logs may be tricky, if not downright difficult to interpret, they often provide flow information which cannot be detected by other means. For the most part, interpretations of temperature logs are qualitative in nature, although quantitative techniques are available.

Temperature surveys are run in producing wells to locate production sources downhole, assist in locating channels, and possibly discriminate gas from liquid entries. In injection wells, such surveys locate zones of injection and can highlight channeling behind pipe. Other applications include the evaluation of the height of an induced fracture, locate zones of acid placement, and detect the top of cement.[1,2] Certainly other applications exist.

Geothermal Gradient

Temperature in a well increases with depth. The actual rate of increase will depend upon the type of formation and its thermal conductivity. For purposes of production logging (PL), this temperature profile is assumed linear and called the "geothermal profile." If T_o is the mean surface temperature, the temperature at a depth downhole is given by

$$T = G \times Z + T_o \qquad \text{(Equation 13.1)}$$

where

G = Geothermal gradient, °F/100ft or °C/100m

Z = Depth, ft or m

T = Temperature downhole at depth Z, °F or °C

Geothermal gradients are typically in the range of .5 to 2.0 °F/100ft (.9 to 2.7 °C/100m). As a rule, the geothermal profile in PL is located by logging the rathole, the static region in the wellbore below the perforations. Extrapolation of this temperature gradient across the producing intervals is taken to indicate as the local formation temperature.

Types of Temperature Surveys

The most common temperature survey is the continuously run log of temperature in a well with depth. The term "temperature log" generally refers to a log of this type. Differential temperature surveys are measurements of the temperature gradient, dT/dZ, along the wellbore.

This log is measured by memorizing the temperature, say 10 feet (3 m) earlier, and recording the difference between that and the current reading. Such a rolling difference may be based on any vertical distance selected for dZ, typically between 1 and 10 feet (.3 to 3 m).[3]

Special techniques and tools are also used in temperature logging. Time lapse logs are commonly used and are very powerful techniques for locating injected fluids. While most temperature logs are measured with a sensor in the region of wellbore flow, wall contact devices have also been used. These offer excellent vertical resolution and have been used for detection of an injected steam floodfronts in monitor wells.

Running the Temperature Log

Modern temperature logs are usually recorded simultaneously with other PL sensors during every pass of the logging tool across the interval of interest. In PL, this means runs in both the up and down direction (required for the flowmeter). After successive runs, the temperature may become somewhat smeared regarding small details. As a result, the best temperature log is taken during the first run down into the hole. This should be run at a relatively slow logging speed of about 30 ft/min (10 m/min) or less. The full interval should be logged.

Main Causes of Well Temperature Deviating from Geothermal

There are at least three broad causes of well temperature varying from the geothermal profile. These follow with a short comment on each:

1. **Fluid movement in a single well**. When fluids are produced or injected in a single well, such fluid movement cannot be at equilibrium with the environment unless it moves infinitely slowly. Hence, such moving fluids will cause the wellbore temperature to deviate from the geothermal profile. This single well model is the main focus in the study of temperature surveys in production and injection wells.
2. **History and time lapse**. When the environment within a well is changed, such as by shutting in, new fluid production, or circulating fluids prior to a workover, the temperature log may be affected. While the effects may be small in regions where fluids are moving, the recent history is critically important when wells are shut in and fluids are not moving downhole. This occurs because the temperature of borehole fluids, by heat conduction, tries to approach the geothermal temperature of adjacent formations. This is the basis for very powerful techniques to locate injected fluids in a well.
3. **Production/injection in offset wells**. Breakthrough of fluids injected in neighboring wells can have a large effect on temperature logs. Even production of nearby wells may cause warmer fluids to be drawn from deeper or shallower within the reservoir. All of these factors cause the fluids to vary from the expected geothermal temperature. In all of the discussion to follow, this effect is assumed to be nonexistent unless specifically otherwise stated.

SINGLE WELL TEMPERATURE LOGS

Liquid Entry from a Single Point

The schematic of Figure 13.1 shows the temperature profile expected for a point liquid entry. It is assumed that the formation has a high permeability and the temperature of the liquid entering the wellbore is reservoir temperature. The liquid is assumed to be at a pressure much in excess of its bubble point, and hence no solution gas is present over the logged interval. When the temperature log is run in the rathole below the point of entry, the fluid is static and is in thermal equilibrium with the surrounding formation. As a result, the temperature recorded is geothermal temperature and this leg of the log is the basis for the placement of the geothermal gradient line.[1,2]

Above the point of entry, three possible logs are shown. Log A corresponds to a low flow rate. When the fluid enters, it begins moving uphole without a change of temperature. In

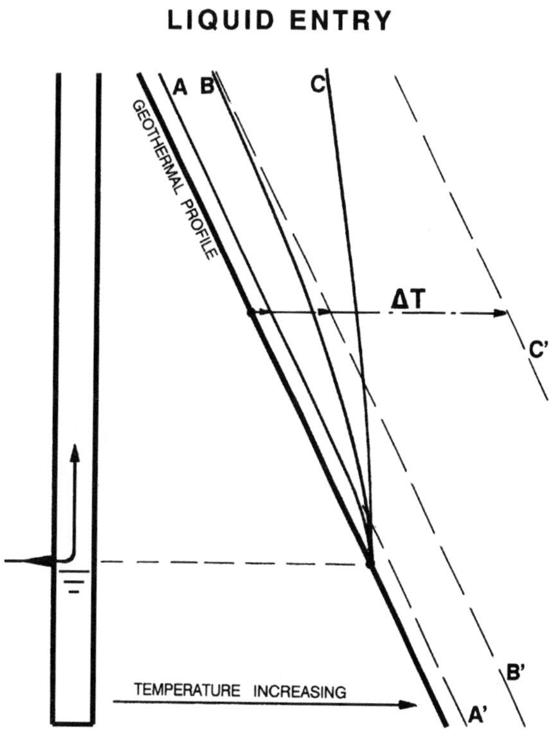

LIQUID ENTRY

Figure 13.1 Temperature survey for small, medium, or large liquid entry

a short distance, the fluid begins to cool since the surrounding geothermal temperature has decreased. After some distance, an equilibrium situation is reached, and the log asymptotically approaches the line A′. A′ is parallel to, but offset from the geothermal gradient. Log B corresponds to a higher flow rate, and asymptotically approaches line B′. Due to the higher flowrate witnessed in log B, the offset of B′ is larger than A′. Log C corresponds to the highest of the three flowrates shown, and asymptotically approaches the line C′.

The horizontal distance, ΔT, between the geothermal profile and the lines A′, B′, and C′ is an indication of the flowrate. This distance may be expressed as[4]

$$\Delta T \cong bM/G \qquad \text{(Equation 13.2)}$$

where

b = Coefficient depending on fluids and formation thermal conductivity

G = Geothermal gradient

M = Mass flow rate

Hence, the distance, ΔT, to the asymptote is proportional to the flow rate, other things being equal. At first glance, this may seem to be a possible quantitative technique. While plausible, the tubing and packer usually appear within a short distance above the perforations and such asymptotes are not usually apparent.

Sometimes a heating effect is observed at the point of entry, called Joule-Thompson heating.[5] This effect is discussed in the section on cooling and heating anomalies.

Single Point Gas Entry

When a gas enters the wellbore, there is a simultaneous pressure and temperature reduction. As a result, the gas entry is characterized by a cooling anomaly.[4,6] Figure 13.2 shows the temperature log response to three different gas flowrates having the same temperature drop but from formations of varying permeability. As with the liquid entry, the temperature log across the rathole is essentially on the geothermal profile. At the point of entry, each log exhibits the same temperature drop. If the flowrate is very low, as from a tight formation,

GAS ENTRY

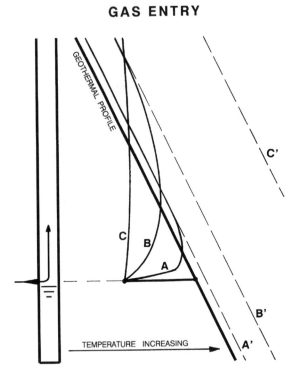

Figure 13.2 Temperature entry for small, medium, and large gas entry

the gas moves uphole slowly and has much time to heat up and approach equilibrium with the uphole formations. This low flowrate is denoted as log A and approaches asymptote A'. Log B shows an intermediate flowrate from a medium permeability formation, and approaches the asymptote B'. Log C shows a high flowrate from a high permeability formation and approaches line C'. Notice that in all cases, the gas heats up as it moves uphole until it reaches the geothermal temperature. Above that point, the gas begins to cool.

If the flow is varied from a single formation, the greater the drawdown, the greater the temperature drop. In this case, the temperature drop is indicative of the flowrate. Keep in mind for all of the above that only a single entry above the rathole is being considered.

Cooling and Heating Anomalies

While a gas entry appears as a cooling anomaly, all cooling anomalies are not gas entries. How can a liquid entry cause a cooling anomaly? Clearly, cooler liquid exists uphole from a point entry. If that liquid flows to the point entry through a channel behind pipe, the fluid entering the wellbore would be at a temperature lower than the geothermal profile. Figure 13.3 shows this example in the lower entry of the figure. Tracing the temperature of the liquid along its path shows the liquid entering the channel at point A and at the temperature of the formation from which it came. As the liquid moves downhole through the channel, it approaches line A' which is lower than and parallel to the geothermal profile. Assuming no big pressure drop at the entry point, the liquid enters the wellbore without a change of temperature at B. The liquid then moves uphole, eventually becoming parallel to the geothermal gradient. The net result is what appears to be a cooling anomaly on the temperature log at B, even though the entry is a liquid.

Are there other ways a cooling anomaly can be caused by a liquid entry? Consider the upper entry at C on Figure 13.3. Here a new liquid entry at the temperature of the geothermal profile mixes with the warmer flow from below. The result is that the new mixed flow is at a temperature between the two. This looks like a cooling anomaly on the log. This highlights a very important fact. Virtually every entry, once the wellbore fluid temperature is above the geothermal profile, causes a cooling effect regardless of that entry being a gas or liquid!

LIQUID ENTRY COOLING ANOMALIES

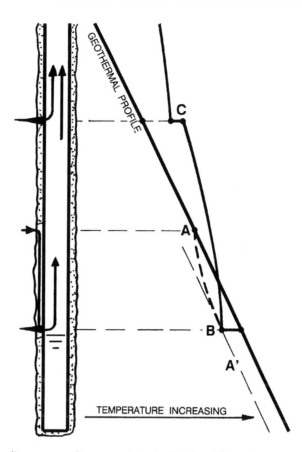

Figure 13.3   Cooling anomalies associated with liquid entries

Heating effects can occur on temperature logs. Such an effect can occur when a liquid is flowing across a large pressure drawdown. This is called Joule-Thompson heating and is sometimes attributed to friction heating as the liquid flows through the rock pore structure.[5,7] Joule-Thompson heating is limited to about 3°F/1,000 psi pressure drop (2.4°C/100 kg/cm²). This effect is most often observed when water enters at a low point in the well where there is no flow from below to mask it. It is not commonly observed with oil entries, probably due to the simultaneous cooling caused by solution gas emerging from the oil at high drawdowns.

Note that when the wellbore fluids are below the geothermal gradient, a gas entry may appear either as a heating or cooling anomaly depending on the temperature at which it enters the wellbore. A liquid entry may also appear as a heating anomaly under these conditions. Such a situation may occur when a gas enters at the bottom of the well. It also may occur if the rathole had recently been circulated with cool fluid and the well put back on production. The flowing part of the temperature survey would look like Figure 13.1 above the point of entry, but the rathole may not have fully returned to the geothermal profile, and hence a cooling anomaly might appear.

Injection in offset wells can also affect the temperature at a producing well, either as a cooling or heating anomaly depending on the temperature of the fluid at the point of breakthrough. Curiously, a channel from below will not generally show up as a heating anomaly. See the section following on channel detection.

Channel Detection

The temperature log is useful for detecting channels. Suppose that a channel exists and flows from a zone below the lowest entry point. This situation is illustrated in Figure 13.4.

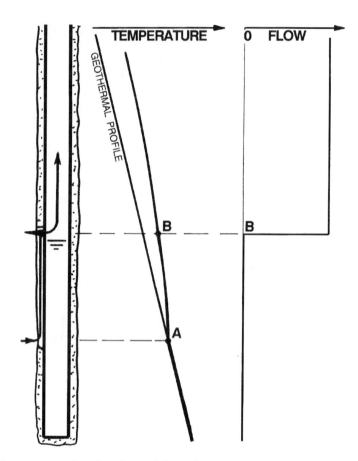

Figure 13.4 Temperature log for channel from below

The channeling fluid originates in a lower zone marked A and enters the wellbore at B. The wellbore fluid below the entry point B is static. These rathole fluids, being static, will be at an equilibrium temperature with respect to their surroundings, i.e., the adjacent formation or channeling flow. As a result, the temperature log shows an apparent entry at A, even though the flow enters the wellbore at B. A spinner flowmeter would confirm only the entry at B. This discrepancy between the two sensors occurs because under static flow conditions in the wellbore, the temperature log is affected not only by the wellbore, but the surroundings as well. The region of investigation for the spinner flowmeter is solely within the wellbore. The only way to rationalize these two conflicting readings is that a channel exists.

Suppose that the channel comes from above. Recall that Figure 13.3 shows a channel from above, with flow originating at point A and entering the wellbore at point B. The temperature log shows a cooling anomaly at point B and the spinner would show an entry at point B. Even so, two possibilities remain. The entry could be a gas or it could be a channel from above. A third sensor is required to characterize the entry as liquid or gas. If the entry is a liquid, a channel from above is indicated. If the entry is or contains gas, a channel may or may not exist.

The key to channel detection in producing wells is the use of combinations of sensors. The temperature, spinner flowmeter, and fluid identification devices are the traditional threesome used for production logging. Modern tool strings may include more sensors such as extra flowmeters, two fluid identification devices, tracers, and the like, but the techniques used are essentially unchanged from the basic PL threesome. The temperature responds to the in and near borehole environment while the other tools examine only the region inside the casing.

Other channel techniques exist, such as the pump in temperature survey (PITS). This is a time lapse technique wherein producing and shut in temperature surveys are compared. Cooler fluid from uphole invades the channel and provides a cooler signature when the well is shut in. Sometimes cooler fluids are injected to enhance the effect. Time lapse techniques are discussed in a later section.

Quantitative Evaluation of Temperature Logs

Referring to Figure 13.1, and as discussed in an earlier section, the distance between the geothermal profile and the asymptote is directly related to flow. So the flow indicated by log B is related to the flow of log C as

$$\frac{Q_B}{Q_C} \cong \frac{\Delta T_B}{\Delta T_C}$$

While useful conceptually, this technique is usually not practical since the tubing and packer intervene before the asymptote can be established.

Another method is that proposed by Romero-Juarez.[8,9,10] Assuming the flow has uniform thermal properties, consider a point, i, on a temperature log where the temperature is well away from the entry point and is smoothly moving toward its asymptote. Then, a parameter Z_i may be defined as

$$Z_i = \frac{T_G - T_w}{dT_w/dD}$$

(Equation 13.3)

where

$$T_G = \text{Geothermal temperature at depth i}$$

$$T_w = \text{Fluid (log) temperature at depth i}$$

$$dT_w/dD = \text{Slope of the temperature log at depth i}$$

The parameter Z can then be treated just like the ΔT associated with the asymptotes discussed earlier. See the following example.

Figure 13.5[10] shows a temperature log for two liquid entries, Q1 and Q2. To apply the Romero-Juarez method, first consider the total flow, Q1 + Q2, at point A. To compute Z_A, the numerator is $T_G - T_w = \Delta T_A = 1.1\,°F$. The slope is determined to be 0.0033 °F/ft. Hence

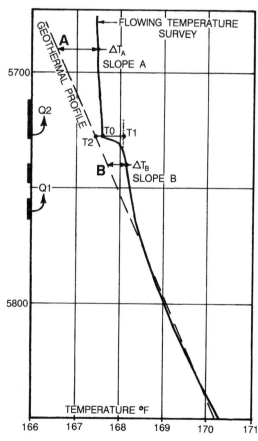

Figure 13.5 Quantitative evaluation of a temperature survey (After SPE, Ref. 10)

$Z_A = 333$, and corresponds to 100% of the flow. To determine the value of Q1, consider point B which is affected only by production Q1. The parameter Z_B is computed as equal to 50 (see Table 13.1).[10] As a result, the relative contributions are

$$Q1/(Q1 + Q2) = Z_B/Z_A = 50/330 = .15$$

and

$$Q2 = 1 - .15 = .85$$

The McKinley Mixing Method can also be applied to this example.[10] The entry Q2 enters the wellbore at the geothermal temperature T2. The flow from below, here attributed totally to entry Q1, has a temperature T1 when it meets the second entry, Q2. After mixing, the resultant mixture temperature T0 can be related to the two flows and their heat capacities, C_p, by the equation

$$Q_1 \rho_1 C_{P_1}(T_1 - T_0) = Q_2 \rho_2 C_{P_2}(T_0 - T_2) \qquad \text{(Equation 13.4)}$$

where the ρ term is the density and ρ_1 Q1 and ρ_2 Q2 are the mass flow rates associated with Q1 and Q2. The relationship between Q1 and Q2 is therefore given by

$$\frac{Q_2}{Q_1} = \frac{\rho_1 C_{P_1}(T_1 - T_0)}{\rho_1 C_{P_2}(T_0 - T_2)} \qquad \text{(Equation 13.5)}$$

and both Q1 and Q2 together equal 100% of the flow, hence

$$Q1 + Q2 = 1 \qquad \text{(Equation 13.6)}$$

Applying the McKinley method to the log of Figure 13.5, and assuming that the Q1 and Q2 phases are the same, then $\rho_1 C_{P_1} = \rho_2 C_{P_2}$, and the ratio $(T_1-T_0)/(T_0-T_2)$ is approximately equal to four. From equation 13.6, the flow contributions become Q1 = .20 and Q2 = .80, reasonably close to the values obtained from the Romero-Juarez technique.

TIME LAPSE TEMPERATURE LOGGING TECHNIQUES

Locating Zones of Injection

The technique of locating an injection zone by time lapse temperature logging is shown on Figure 13.6.[11] On the left of this schematic is shown an injection well with a mean surface temperature of 80°. Isotherms are shown increasing with depth in 5o increments. This well is an injection well with fluid being injected from the surface at a temperature of 80°. If the injected fluid is assumed to be water, the water is seen to heat up on its way downhole and causes distortions of the isotherms around the well. The zone of injection is located between the 105° and 110° isotherms. Below the injection interval, the isotherms return to their uniform geothermal temperature indicated by their horizontal shape. Notice that the water enters the zone of injection at a temperature substantially lower than the geothermal temperature.

If a temperature log is run during injection, the log will look like that shown on the right and labeled "injecting." The temperature increases with depth and abruptly returns toward the geothermal (static) profile below the zone of injection. When the well is shut in, the wellbore fluid is heated up by the adjacent formations and tries to return to the geothermal profile. The wellbore volume is small and this return to geothermal profile takes place relatively rapidly. Over the zone of injection, however, the volume of injected water is very large. This region of injection returns to the geothermal profile much more slowly. Temperature logs

TABLE 13.1 Romero-Juarez Temperature Log Interpretation, Example 1

Station	Depth (ft)	dT_w/dD (°F/ft)	$T_w - T_G$ (°F)	Z (ft)	Total Flow (%)
A	5,690	0.0033	1.1	333	100
B	5,740	0.008	0.4	50	15

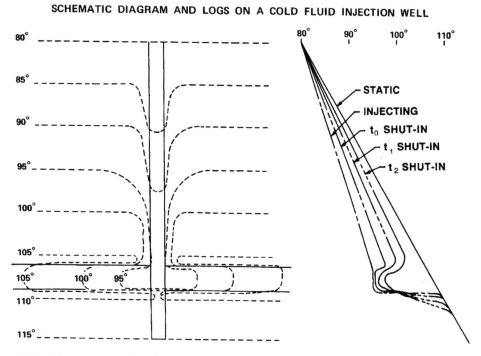

Figure 13.6 Technique using shut-in temperature surveys to locate zones of injection (After Halliburton Co., Ref. 11)

run after shut in will detect a "bump on the log" over the zones of injection due to this effect.[12-14] The shut-in runs are shown at times t_1, t_2, and t_3 after shut in.

This technique works regardless of how the water gets to the zone of injection. The water can exit the wellbore almost anywhere and channel to the zone of injection. The injected water can even get into the zone by way of a fracture deep in the formation. In any of these cases, the "bump" on the log occurs at the zone where the injected water ends up. While I have discussed the case of water injection, these principles will hold true regardless of the injected fluid.

Water Injection Guidelines

How long should the operator wait before running the shut-in temperature logs to locate zone of injection? One rule of thumb states that logging 24 hours after shut in would work out satisfactorily if the well's cumulative injection is less than about 100,000 barrels. If cumulative injection is greater than 100,000 but less than 1,000,000 barrels, then a 48 hour shut in would work well.[15] The reason for the longer period with greater injection is due to the cooling of the rock around the wellbore. The longer the injection, the more time available to cool the formation rock and the greater the mass of rock returning toward the geothermal profile after shut in.

A computer simulation was conducted by Amoco about 1970 to determine best conditions for shut-in temperature surveys in water injection wells.[14] This study showed that for shut-in temperature surveys to be effective at least 10°F (5.5 °C) temperature difference between the geothermal profile and the water as it enters the formation during injection was required. The "bump" is clearly visible on a 48 hour shut-in survey and marginally visible on a 6 hour shut-in survey. Obviously, the greater the temperature difference, the less time needed for good "bump" development. The study also pointed out other factors to consider, such as the effect of cumulative injection and time.

One important guideline is that the flow downhole must be stopped at shut in. If crossflow is taking place, the effect is to smear out the "bumps" over the interval of the cross flow and render the log useless over that interval.

Other Factors to Consider in Shut-In Temperature Surveys

Borehole and well configuration factors also affect the shut-in temperature log. The basic mechanism of heating or cooling of the borehole fluid during shut in is heat conduction to or from the adjacent formation. The following factors are relevant in this regard:

1. **Tubulars**. Casing, tubing and annular fluid, open hole, etc., all affect the rate of lateral heat transfer. As a result, other things being equal, wellbore fluids will approach the geothermal profile at differing rates depending on these factors.
2. **Washouts**. Washouts are actually pockets of water having a significantly larger volume of water than the gauge sections of the hole. These washed out regions will tend to look like zones of injection, especially in early logging runs.
3. **Hole conditions/Open hole caliper**. Even though the hole is cased and cemented, cement thickness can affect the lateral heat transfer. As a result, it is good practice to have an open hole caliper available should anomalies appear in unexpected places.

Oftentimes, there is a tendency to look only at the temperature logs for zones of injection. An open hole logging suite is helpful to confirm that expected zones of injection are, in fact, porous and permeable. Injection will not generally occur into a shale. As obvious as this might sound, this fact is frequently overlooked.

LOCATING FRACTURED AND ACIDIZED ZONES

Fracture Height Identification

The time lapse logging techniques of the previous section lend themselves to the determination of fracture height.[16,17] The basic technique is to compare a temperature run after fracturing operations to a base temperature log run prior to fracturing. The volume of the liquid in the fracture and the local leak-off around the wellbore return to geothermal temperature more slowly than the surroundings and the anomalous zone is associated with the fracture.

The logs of Figure 13.7 show a temperature survey used for the determination of the height of an induced fracture.[17] In this example, the static log is the initial temperature survey closest to the geothermal profile. The pre frac profile is a survey run after circulating cold water and before perforating. This was done to compare the thermal conductivities of various formations. The post frac profile is the temperature log run shortly after the fracturing operation. In this case, the cool anomaly shows the fractured interval. While this example shows a pre frac and static log, these may or may not be available for comparison with the post frac survey.

For best results with temperature surveys run for induced fracture height determination, run a base log before the fracturing operation. Be sure that there is adequate difference in temperature between the fluid injected and the geothermal temperature. Run the temperature logs as soon after the frac operation as possible, and continue for up to 12 hours after the job. Be sure the wells are shut-in when running the post fracture temperature surveys and that there is no backflow. Warm anomalies have occasionally been observed above the cooling anomaly associated with the fracture. Such anomalies are called "warm noses." They disappear after a few hours, but are believed to be part of the fracture height.

Detecting Acidized Zones

The time lapse method is again used for detecting acidized zones. Before acidizing, a pre acid temperature survey is run to get a base geothermal profile. If the well to be acidized is shallow and downhole temperatures are low, then the effect of the acid is to react with and heat the zone acidized. This log example is shown on Figure 13.8, part a.[15] The treated interval is interpreted to be the warm interval from where the log crosses the geothermal profile on top to the point below where the log begins to cool. Notice that the borehole above the acidized zone is cooler than the geothermal profile due to injection and passage of the cool acid from the surface. Temperature logs for location of acidized zones should be run as soon after acidizing as possible.

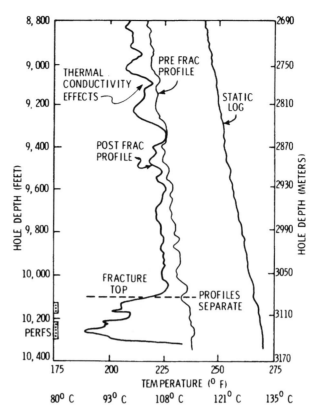

Figure 13.7 Temperature survey showing fractured zone (Copyright SPE, Ref. 17)

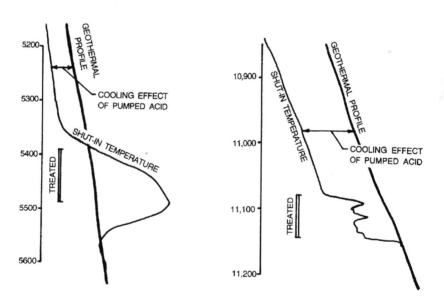

Figure 13.8 Temperature survey to detect zones of acid placement (After API, Ref. 15)

If the well is deep, the acid again reacts with the formation into which it enters. However, now the formation is so hot that even with the heat released from the reaction the spent acid is still colder than the formation. As a result, the treated zone is indicated as a zone warmer than the pumped down fluid but cooler than the geothermal temperature. This situation is shown on Figure 13.8, part b. If the acidized zone is overflushed, the flushing fluid now acts like typical injection fluid and creates a cooling anomaly which returns to geothermal temperature more slowly than the small volume in the wellbore.[15]

OTHER TEMPERATURE LOG APPLICATIONS

Monitoring Floodfronts

Temperature logs have been used to monitor flood fronts in nearby observation wells. Typically, the observation well is logged repeatedly at fixed time intervals. When the flood front reaches the observation well, the temperature log begins to deviate across the flooded zones. This practice is most commonly used for monitoring the progress of steam floods. Contact temperature devices, in which the sensors contact the pipe wall, appear to afford better vertical resolution than conventional sensors.

Location of Cement Tops

When cement begins to cure, it heats up. A temperature log run 6 to 36 hours after cement circulation can be used to locate the top of cement. It is best practice to log to at least 1,000 ft (300 m) above the point at which the cement top is expected. The example of Figure 13.9[15] shows a temperature survey on the right and a caliper on the left. A tentative geothermal profile based on the relatively low top and bottom readings is indicated. The effective cement top (good annular fill) is shown at 1,930 ft. The interval from 1,650 to 1,930 is slightly warmer than the assumed geothermal profile and is interpreted as highly contaminated cement. If cement top is defined as the very highest point of cement indication, then 1,650 ft would be the cement top in this example. Notice that the temperature survey varies with the caliper. This is a result of hole enlargements and larger volumes of curing cement, therefore higher temperatures.[15]

Location of Gas Zones in Air Drilled Wells

In some areas, very tight gas sands are difficult to detect with conventional logs. As a result, techniques have been developed where such intervals are drilled with air or nitrogen as the drilling fluid. With very low pressures in the wellbore, temperature logs are run. The tight gas zones show a large cooling anomaly with little flow rate. Two such tight gas zones are shown in Figure 13.10. To produce such zones, the well is cased, perforated, and the tight zones fractured.

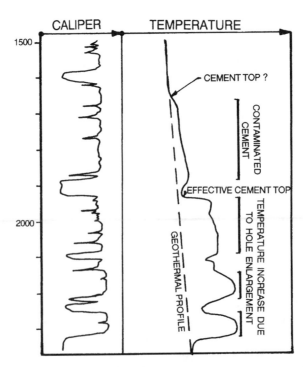

Figure 13.9 Temperature log response to curing cement (After API, Ref. 15)

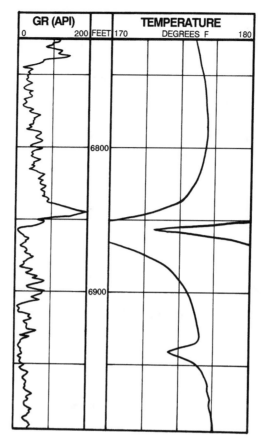

Figure 13.10 Gas entries into air drilled well, open hole (Courtesy Western Atlas)

Other Applications

Many other applications of temperature logs and specialized techniques are available. Procedures are commonly set up to locate tubing and packer leaks.[18] Papers have been written on how long it should take to return to geothermal temperature after circulating fluids downhole.[19–22] The rate of return of such fluid has also been used to predict the presence of channels in cement.[23] Differential temperature surveys have also had many applications, especially regarding zones of injection.[3,13] The differential is also very sensitive to changes of slope of the absolute temperature survey, and therefore useful to locate small entries whose only effect is to cause a small change of slope in the temperature survey.

RADIAL DIFFERENTIAL TEMPERATURE (RDT) LOGGING

Basics of Operation

The Radial Differential Temperature (RDT) tool measures changes in temperature laterally around the wellbore. The basic tool is a through tubing size, 1 11/16 in. (4.3 cm) or smaller, and has two retractable arms as shown on Figure 13.11a. If a channel exists behind pipe, as shown on Figure 13.11b, the channel will be hotter or cooler than the geothermal profile either because the channeling fluid originated at a different depth or the channel is a cooler gas. To locate channels, the RDT is stopped at a depth where its arms are extended to contact the wall. The arms rotate one revolution in about four minutes. If no channel exists, the difference between the sensors is constant. If a channel and temperature anomaly exists, the difference between Tw1 and Tw2 varies, repeating each 360° of rotation.[24, 25]

Figure 13.12 shows a conventional temperature log and an RDT.[24] The temperature shows a large cooling anomaly originating at the U sand. This is interpreted as gas channeling downward behind pipe from the U to the M and L sands. The RDT was set at 6,525,

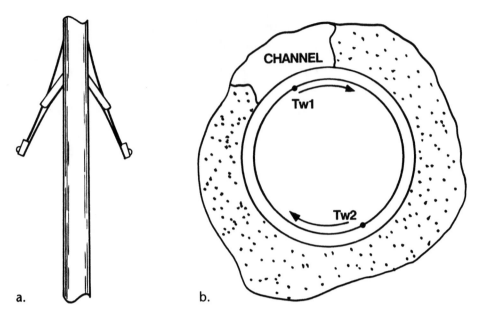

Figure 13.11 Radial Differential Temperature (RDT) measurement: a. Sonde configuration, and b. temperature sensor movement

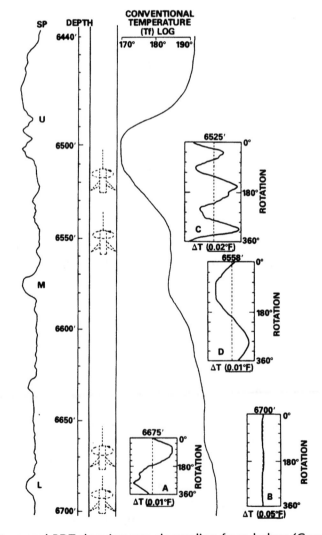

Figure 13.12 Temperature and RDT showing gas channeling from below (Copyright SPE, Ref. 24)

6,558, 6,675, and 6,700 feet. The test at 6,525 showed large temperature variations which did not repeat each revolution. This was interpreted as a large channel with little cement outside of pipe. The tests at 6,558 and 6,675 show the classic RDT response repeating each revolution and indicating a channel. The test at 6,700 shows a condition of no flow.

The RDT is often run with a perforating gun to orient the gun toward the channel, raise the tool, and shoot. This would enhance the likelihood of hitting the channel with the perforating gun. If the RDT tests are made both with the well flow and shut in, such tests can confirm that the wellbore is in communication with the channel, since the shut-in RDT will show no or reduced temperature difference.

The RDT is not applicable to deviated multiphase flowing wells, since the flow will segregate laterally (light phase to the high side of the pipe). The phases are moving at different velocities and hence the temperature profile will reflect the flowing phase and not a channel. This tool was originally developed by Exxon and Gearhart Wireline Services, now Halliburton Energy Services. The availability of this tool is limited to only a few geographic areas at this time.

LOG EXAMPLES

Producing Interval

Figure 13.13 shows an example of detecting produced fluids with a temperature survey.[4] In this example, the well started blowing oil and gas out of the annular space between the 7 in. casing and the surface pipe. This began about 20 hours after cementing. The temperature log shows a well defined geothermal profile below about 2,800 ft, and this interval is

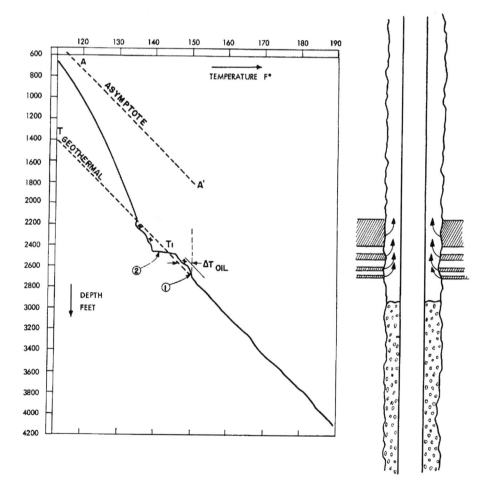

Figure 13.13 Temperature survey showing liquid and gas entries (Courtesy Schlumberger, Ref. 4)

certainly well cemented and exhibits no flow. A liquid entry is detected at point 1, above which the point "x" indicates another entry. This entry may be either a liquid or gas, since the temperature is above the geothermal profile. At point 2, a gas entry is observed to cause a decrease in flowing temperature to below the geothermal profile. The two entries marked "x" above entry 2 are likely gas entries since they cause an apparent cooling even though the flowing fluid temperature is below the geothermal profile. Above about 2,150 ft, the flow smoothly approaches the asymptote A', indicating no new entries or fluid losses.

Shut-In Injection Well

This example on Figure 13.14 shows a situation in which the desired injection interval is located at about 3,300 ft.[4] The spinner flowmeter shows that all of the injected water leaves the casing through at leak at 600 ft. The flowing (solid line) and shut in (short dashed line) are shown. The surface water temperature starts out at a rather warm 110 °F, then first cooling and heating as it is pumped downhole. As expected, the injected water is approaching an asymptote, A'. The transition from cooling to heating occurs when the injected water crosses the geothermal profile. After shut in, notice that the regions both above and below the crossover point have substantially returned toward the geothermal profile. The zone of injection is at 3,300 ft, and is indicated by the "bump" on the shut-in survey. It is clear in this case that it doesn't matter how the injected water gets there, the zone of injection can be identified by the "bump" on the log.

Pump In Temperature Survey (PITS)

A producing well had a higher than expected water cut. A pump in temperature survey (PITS), was run to check for a channel. This example is shown on Figure 13.15. Two sets of

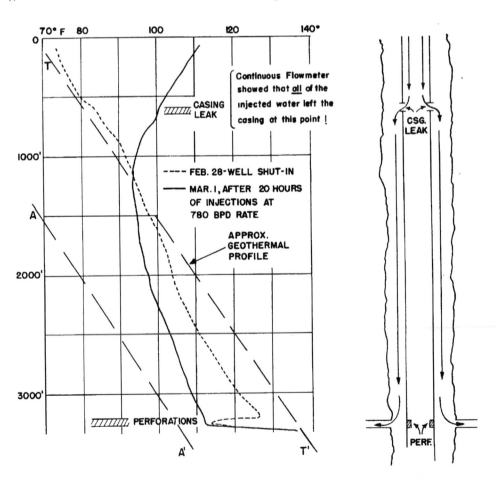

Figure 13.14   Shut-in injection survey to locate zone of injection (Courtesy Schlumberger, Ref. 4)

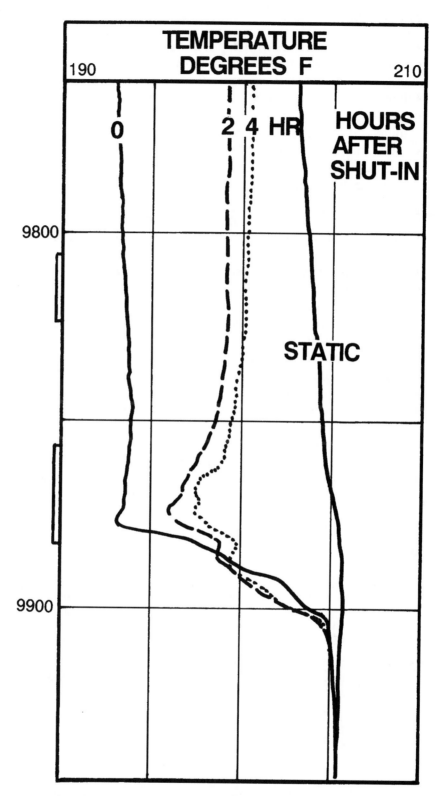

Figure 13.15 Pump In Temperature Survey (PITS)

perforations are indicated between 9,800 and 9,900 ft. A static base temperature log run is shown. A cool fluid was pumped in for a short time and temperature logs were run immediately following, 2 and 4 hours after shut in. These logs show a smooth return toward the geothermal profile above the lower perforations. However, a lingering cooling anomaly appears to extend about 20 feet below the lower set of perforations, indicating a channel down to the water leg of this zone.

POINTS TO REMEMBER

- Temperature increases with depth and is called the geothermal profile.
- The main causes of temperature character on a log are production, history, and offset well production/injection.
- Liquid enters the wellbore at the geothermal temperature.
- Liquids may enter the wellbore warmer than the geothermal temperature. This is the Joule-Thompson effect.
- Joule-Thompson heating is limited to about 3 °F/1,000 psi.
- The displacement of the liquid flow asymptote from the geothermal profile is proportional to flow rate, other things being equal.
- Gases cool when entering the wellbore.
- Once above the geothermal gradient, all entries are cooling anomalies.
- Combinations of tools, such as spinner, fluid identification, and temperature are necessary for channel detection.
- Quantitative techniques exist for production profiles.
- Shut-in surveys after injection are effective at locating zones of injection.
- For zones of injection, at least 10 °F (3 °C) difference between the geothermal temperature and the fluid temperature downhole is needed to locate such zones with a 48 hour shut-in temperature survey.
- Well bore fluids must be stationary during a shut-in log pass.
- Fracture height identification uses a shut-in log technique.
- Acid placement uses a shut-in log technique.
- For acid or fracture height identification, logging runs should be run soon after acidizing or fracturing.
- Temperature may be used to locate cement tops.
- The RDT measures variations of temperature laterally around the wellbore.
- The RDT is useful at detecting active channels in a vertical well.
- Pump-in temperature surveys are used to locate channels using a shut-in temperature log technique.

References

1. Schlumberger, "Production Log Interpretation," Document C-11811, Houston, Texas, 1973.
2. Atlas Wireline Services, "Interpretive Methods of Production Well Logs," 4th Edition, Document 9635, 1991.
3. Gearhart-Owen Industries (Now Halliburton Energy Serivces), "An Introduction to Single Element Differential Temperature Logging," Fort Worth, Texas, approximately 1982.
4. Schlumberger, "Interpretation of the Temperature Log and the Continuous Flowmeter," Paris, approximately 1967.
5. Steffensen, Roger J., and Smith, R.C., "The Importance of Joule-Thompson Heating (or Cooling) in Temperature Log Interpretation," Paper SPE 4636, 48th Annual Technical Conference, Las Vegas, Nevada, 1973.
6. Kunz, K.S., and Tixier, M.P., "Temperature Surveys in Gas Producing Wells," *Journal of Petroleum Technology*, July, 1955.
7. Maubeuge, F., Didek, M.P., Beardsell, M.B., Arquis, E., and Caltagirone, J.P., "Temperature Model for Flow in Porous Media and Wellbore," 35th Annual SPWLA Symposium, June, 1994.
8. Romero-Juarez, A., "A Note on the Theory of Temperature Logging," *SPE Journal*, pp. 375-377, December, 1969.
9. Curtis, M.R., and Witterholt, E.J., "Use of the Temperature Log for Determining Flow Rates in Producing Wells," Paper SPE 4637, 48th Annual SPE meeting, Las Vegas, Nevada, 1973.
10. Hill, A.D., "Production Logging-Theoretical and Interpretive Elements," SPE Henry L. Doherty Series Monograph Volume 14, SPE, Richardson, Texas, 1990.
11. Welex (Now Halliburton Energy Services), "Temperature Log Interpretation," Document CL-2002, approximately 1970.
12. Witterholt, E.J., and Tixier, M.P. "Temperature Logging in Injection Wells," Paper SPE 4022, 47th Annual Technical Conference, San Antonio, Texas, October, 1972.
13. Smith, C.S., and Steffensen, R.J., "Computer Study of Factors Affecting Temperature Profiles in Water Injection Wells," *JPT*, pp. 1,447-1,458, November, 1970.
14. Smith, C.S., and Steffensen, J.R., "Improved Interpretation Guidelines for Temperature Profiles in Water Injection Wells," Paper SPE 4649, 48th Annual Technical Conference, Las Vegas, Nevada, September, 1973.

15. Kading, H.W., and Hutchins, J.S., "Temperature Surveys; The Art of Interpretation," Paper No. 906-14-N, Spring Meeting of the Southwestern District, Division of Production, American Petroleum Institute, Lubbock, Texas, March, 1969.

16. Agnew, B.G., "Evaluation of Fracture Treatments with Temperature Surveys," Paper SPE 1287, 40th Annual Technical Conference, Denver, Colorado, October, 1965.

17. Dobkins, T.A., "Improved Methods to Determine Hydraulic Fracture Height," *JPT*, pp. 719-726, April, 1981.

18. Michel, C.M., "Methods of Detecting and Locating Tubing and Packer Leaks in the Western Operating Area of the Prudhoe Bay Field," Paper SPE 21727, Production Operations Symposium, Oklahoma City, Oklahoma, April, 1991.

19. Fertl, W.H., and Wichmann, "Static Formation Temperature Determined From Well Logs," Dresser Atlas Technical Memorandum Vol. 7, No. 3, Dresser Atlas (Now Atlas Wireline), Houston, Texas, April, 1976.

20. Wooley, G.R., "Computing Downhole Temperatures in Circulation, Injection, and Production Wells," *JPT*, pp. 1,509-1,522, September, 1980.

21. Beirute, R.M., "A Circulating and Shut-in Well-Temperature- Profile Simulator," *JPT*, pp. 1,140-1,146, September, 1991.

22. Kutasov, I.M., "Tables Simplify Determining Temperature Around a Shut-In Well," *Oil & Gas Journal*, pp. 85-87, July 26, 1993.

23. Barnette, J.C., Lanuke, E.W., and Carlson, N.R., "Exponential Coefficient Plots for Identifying Cement Channels from Temperature Logs," 25th SPWLA Annual Logging Symposium, June, 1984.

24. Cooke, C.E., Jr., "Radial Differential Temperature (RDT) Logging—A New Tool for Detecting and Treating Flow Behind Casing," Paper SPE 7558, 53rd Annual Technical Conference, Houston, Texas, October, 1978.

25. GO Wireline Services, "Radial Differential Temperature," Document WS-267, Fort Worth, Texas, 1978.

CHAPTER 14

FLUID MOVEMENT: NOISE LOGGING

NOISE AND FLUID MOVEMENT DOWNHOLE

Primary Applications of Noise Logs

Noise logs, sometimes called sound surveys, utilize microphones to detect fluid movement downhole. Such fluid movement is characterized by both flow and pressure drop. If these factors are adequately large, turbulence is generated and the flow can be detected even if it is outside of casing. Therefore, noise logs are useful for detection of channels behind pipe. They can also detect entries into the wellbore and distinguish such entries from the wellbore flow. Single phase flow can be discriminated from bubbly two phase flow. Even sand entries can be discriminated.[1,2]

Combinability of Noise with Other Surveys

Noise logs are most often run in combination with the temperature survey, and are not usually combinable with most industry production logging combination tool packages. The noise and temperature combination can, however, be a very powerful tool for the qualitative and sometimes quantitative evaluation of the downhole flow profile in a producing well. One obvious asset of this combination is the fact that both the temperature and noise logging sensors can detect events occurring both inside and outside of the casing string.

Types of Noise Surveys

There are basically two types of noise surveys available. Far and away, the most common is the stationary noise log. This measurement is made with a microphone and records both the amplitude and frequency spectrum of the sound at various stations downhole. The noise log is not run continuously since tool and cable road noise, arising from their scratching the casing wall, would dominate the signal trying to be measured. If run in combination with a temperature survey, the temperature log is run continuously before performing the noise log measurements.

Continuous sound surveys have also been available. These focus on frequencies high enough to not be affected by road noise. The information they supply is limited to only amplitude. The main application of such tools has been to locate gas entries and leaks in casing.

Running a Sound Survey

Noise logs are very sensitive, and in shallow wells even surface equipment must be shut down where possible to avoid sound from non-downhole sources. Such measurements are stationary, and so it is prudent to wait for about a minute after stopping the tool to stabilize tool movement before recording the sound signal. Tools are run without centralizers, and so they are laying against the casing when making a measurement. If possible, a shut-in or "dead well" sound survey should be run for comparison. Earphones or speakers are usually available to listen to the well.

The stationary positions are initially run at a large spacing, say from about 10 to 50 feet (3 to 15 meters), depending on the flowrate and length of interval to be evaluated. The initial continuously run temperature log, if available, is useful to locate positions downhole which may be of interest. After the well has been initially scanned, the operator can then focus on intervals where noisy events are concentrated, performing measurements at more frequent intervals as little as one foot (30 cm) apart.

NOISE AMPLITUDE AND SPECTRUM

Amplitude

Noise amplitude is an indicator of the depth of the flow anomaly.[3–6] Figure 14.1 shows a channel behind pipe with a flow moving from the zone at A to that at C.[4] Noise is associated with both flow and pressure drop. In this schematic, sharp pressure drops are observed at the entry A, a restriction at B, a pressure drop into a low pressure zone at C. The overall sound amplitude is shown at the right. This schematic teaches two things, that the amplitude locates the depth of the noisy event, and that the noisy event is not necessarily an entry, but may be a restriction in the channel.

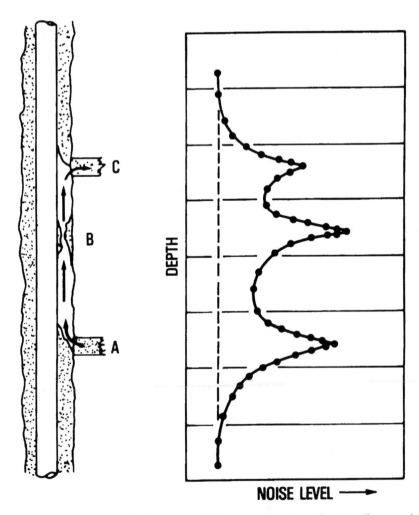

Figure 14.1 Noise amplitude increases adjacent to locations having flow and pressure drop (Copyright SPE, Ref. 4)

Noise Spectrum, Single-Phase Flow

Studies by Dr. R.M. McKinley at the Exxon Production Research Center showed that single and multiphase flow can be differentiated by their noise spectra.[3] Figure 14.2 shows both the single phase frequency spectrum for gas and liquid.[7] The solid curve is the sound spectrum for a 70 barrels per day (BPD) (11.1 m³/d) water expanding across a 90 psi (6.3 kg/cm²) differential into a water filled channel. The dashed curve is the spectrum caused by 3.8 Mcf/day (108 m³/d) gas expanding across a 10 psi (.7 kg/cm²) pressure differential into a gas filled channel. Both are characterized by the greatest area under the curve (energy) in the 1,000 to 2,000 hertz range. While single phase flow spectra may differ from these under different flow conditions, the single phase flow spectra is characterized by energy in the higher frequency ranges above about 1,000 hertz.

Noise Spectrum, Two-Phase Flow

The spectrum of Figure 14.3 is caused by .3 Mcf/d (8.5 m3/d) gas expanding into a water filled channel.[7] In this case, the bulk of the noise energy is concentrated in the region less than 600 hertz. It is this low frequency energy, less than 600 hertz, that characterizes two phase flow. This rule applies to both bubbling and slugging flows.

NOISE LOG PRESENTATION AND RESPONSE

Log Presentation

The noise log presentation is intended to show a simplified noise spectrum to highlight events downhole. Figures 14.2 and 14.3 contain bold lines at the 200, 600, 1,000, and 2,000 hertz levels. The log is recorded at these levels with high-pass filters, i.e., filters which eliminate sound below these frequencies and allow measurement of energy above. Obviously, the 200 hz cutoff energy is greatest, followed by the 600 hz, 1,000 hz, and 2,000 hz levels.

The noise logs for both the single phase and multi phase flows are shown on Figure 14.4.[7] Noise logs are recorded on a logarithmic scale. The single phase noise concentration is in the 1,000 to 2,000 hertz band or greater, and hence the energy levels associated with the lower frequency cutoffs are all nearly equal and tightly grouped together at the point of entry (highest amplitude). This single phase noise source is shown at the depth of maximum

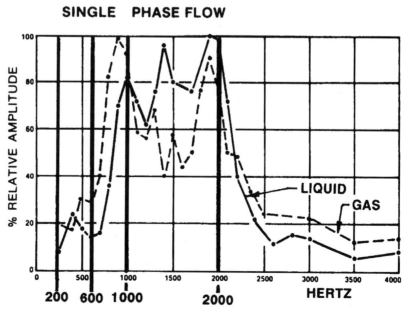

Figure 14.2 Noise spectrum for single phase flow (After Western Atlas, Ref. 7)

amplitude in Figure 14.4, part a. Notice that the noise level dies off as data is taken at points further away from the depth of the noise source. For two phase flow, as shown on Figure 14.4, part b, the greatest area is in the low end of the spectrum, and hence the separation of the frequency cutoff curves is spaced apart, a clearly different signature than the single phase entry. Notice also that the die away occurs downhole from the entry point, but that the amplitude stays high above the entry. This is a result of bubbles in the flow. These bubbles are the source of most of the noise.

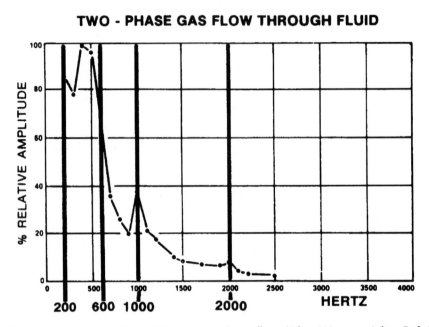

Figure 14.3 Noise spectrum for bubbling two phase flow (After Western Atlas, Ref. 7)

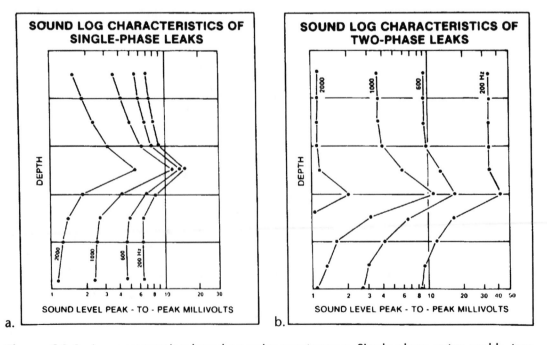

Figure 14.4 Log presentation based on noise spectrum: a. Single phase entry, and b. two phase flow (Courtesy Western Atlas, Ref. 7)

Log Response in Lab Tests

Lab tests of the noise log response were performed by Dr. McKinley. The two following tests are very enlightening.

Figure 14.5 shows a lab fixture on the right with tubing, partially filled with water, and an air filled annulus.[8] Air is flowed into this air filled annulus through a restriction at a rate of about 5 Mcf/d (142 m³/d). The log response clearly shows the entry at the point of maximum noise level. The data points are bunched together, indicating a single phase event. Above the entry point, the noise energy dies away. When the tool is pulled above the liquid/gas contact in the tubing, the signal is greatly reduced. Note also that signal reduction similar to this may occur when the tool is lowered to a point below the casing shoe. Both the liquid and the casing are carriers of the acoustic signal.

In the test of Figure 14.6, the same fixture is used, except that the annulus is now filled with water.[8] The air is flowed into the annulus at a rate of about 1 Mcf/d (28 m³/d). Logging from the bottom, the amplitude increases to the entry point. At the entry point and above, the frequency cutoff data is spread out, indicating a bubbling two phase flow. This noise level stays high since the noise source is the bubbling flow outside of the tubing. The annulus depth is so shallow in this test fixture that bubble volume as well as amplitude are increasing as the bubbles move upward. Above the gas/liquid contact in the tubing, there is again an acoustic decoupling.

Noise Die Away from Source

Noise die away from a noise source depends on its frequency content. Figure 14.7 shows the rate at which noise dies away as a function of distance from the source in a water filled pipe.[1] This chart shows the die away rates for two different size casings. Notice that the high frequency cut dies away much more rapidly than the lower frequencies. For example, at 100 ft from the source and in 8 5/8 inch casing, only 10% of the 2,000 hz signal remains compared to over 40% of the 200 hz signal. The attenuation rates for gas in the pipe are about twice that of the case of water. This means that the amplitude will be half what would be the case in water, other things being equal.

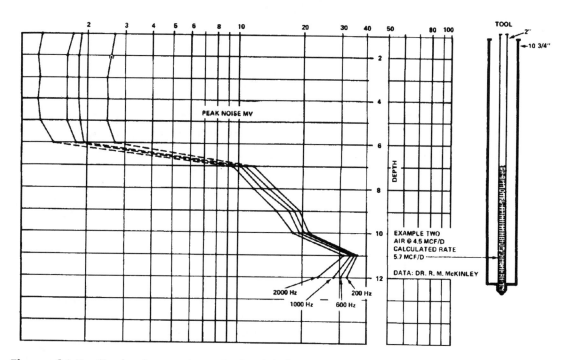

Figure 14.5 Single phase entry noise log lab fixture and test (Courtesy SPWLA, Ref. 8)

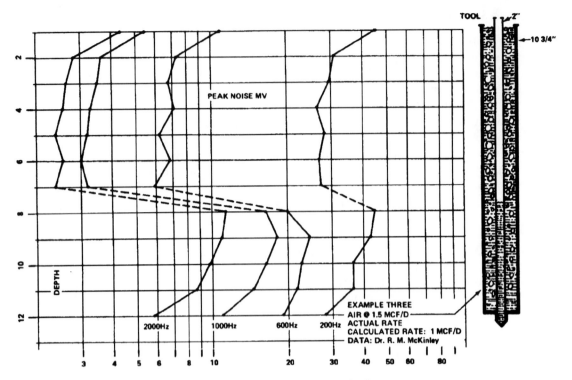

Figure 14.6 Bubbling flow noise log test (Courtesy SPWLA, Ref. 8)

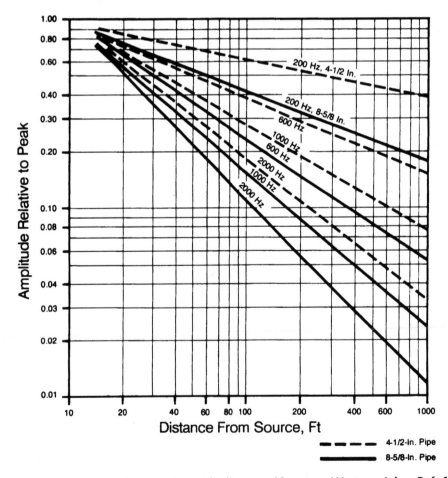

Figure 14.7 Attenuation of noise frequencies with distance (Courtesy Western Atlas, Ref. 1)

QUANTITATIVE CONSIDERATIONS IN NOISE LOGGING

Exxon Production Research Co. (EPRC) Standard Sonde

Provided that a noise sonde is calibrated to the standard EPRC sonde, a quantitative analysis can be made. Very few noise measurements are made with such a calibrated sonde. However, the equations presented in this section may still prove useful to compare various points in a producing wellbore. Furthermore, the equation and sonde may be calibrated together against a known level of production in the well.

The raw log can be corrected to an Exxon Production Research Co. (EPRC) standard sonde for quantitative work.[3,8] Generally, the noise level, N, must be normalized for four different factors as follows:

$$N^* = N \times F_L \times F_T \times F_M \times F_G$$

where

N^* = Normalized Noise Level

F_L = Line factor—depends on type and length of cable

F_T = Gain Factor—depends upon the service company tool

F_M = Meter factor—depends on millivolt scale used

F_G = Geometry factor—depends on casing and fluids present

The line factor and tool gain factor are usually unique to a service company and they may have appropriate values for these. The meter factor relates to the measurement scale. The standard is peak to peak millivolts. If only peak millivolts is recorded, then the reading must be multiplied by 2 to be comparable to peak to peak millivolts. If RMS millivolts is recorded, then the factor becomes 2.83.

The well geometry factor, F_G, is a correction due to the tubing, casing, and fluids therein. If the detector is in a water filled tubing string and the noise source is outside that string, the correction is unity. (See Table 14.1 for well geometry correction factors.[3])

Single Phase Entry

The leak rate of a single phase through a restriction or perforation can relate to the power that it dissipates, or at least that part which is converted into fluid turbulence.[3,8] This power can be expressed as the product of flow rate times pressure drop,

$$\text{Power} = \text{delP} \times q$$

TABLE 14.1

Type of Well	Fluid Content	Multiply Noise Amplitude by
A. Tubingless completion	Liquid in string	1
	Gas in string	3.5
B. Tubing string in casing	Liquid in tubing, liquid in annulus	4
	Gas in tubing, liquid in annulus, or vice versa	10
	Gas in both tubing and annulus	30
C. Leak into same string as detector	Liquid in string	0.06
	Gas in string	0.20

Studies at EPRC have shown that this power term may be related to noise log measurements as shown on Figure 14.8 and according to the following equation:[3]

$$delP \times q = 5 \times (N^*_{1000} - 6)$$

where

$delP$ = Pressure drop, psi

q = Flow rate, ft^3/day at downhole conditions

N^*_{1000} = Peak to peak noise greater than 1,000 hz

The above equation applies to both liquids and gasses.

Two-Phase Gas-Liquid Entries

This section applies to gas entries into a liquid. The peak to peak reading in the 200 – 600 hertz band was correlated with gas flowrate, at downstream conditions.[3,8] This correlation is shown on Figure 14.9.[3] The straight line portion is described by the equation:

$$q = (del^* - 10)/20$$

where

q = Flow rate, Mcf/day at downhole conditions

$del^* = N^*_{200} - N^*_{600}$, peak to peak mv in the 200 – 600 hertz band

Single-Phase Flow Past Sonde in Pipe

This equation applies only away from noise sources and must only reflect those noises generated as the flow passes the sonde. This situation is called "free flow." This technique, while marginally quantitative as are the others, may be quite useful to compare flowrates

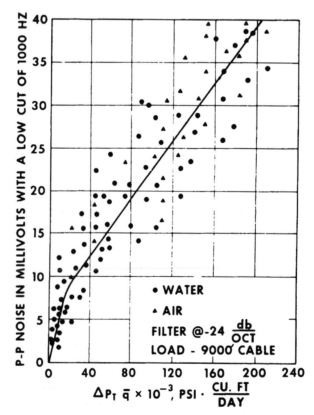

Figure 14.8 Empirical noise correlation for single phase entry (Copyright SPE, Ref. 3)

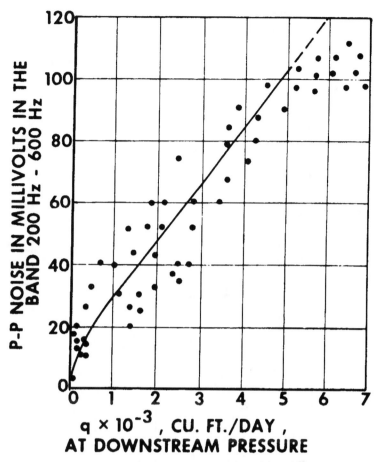

Figure 14.9 Empirical noise correlation for gas bubbling in liquid (Copyright SPE, Ref. 3)

between widely separated entries or at various points in the casing string. The equation describing this flow is:[4]

$$q = \left\{ \frac{A_s^2 \times N_{600}^*}{4 \times 10^{-6} \times \rho} \right\}^{\frac{1}{3}}$$

where

q = Flowrate at downhole conditions, Mcf/day

ρ = Fluid density downhole, lb/ft³

A_s = .785 × (ID²$_{pipe}$ − D²$_{tool}$), ft²

N_{600}^* = Peak to peak mv, 600 hz and greater

OTHER NOISE LOGGING TECHNIQUES

Continuous Noise Logs

Continuous noise logs have been run primarily for detection of gas leaks in gas storage wells. These tools focus on high frequency noise, greater than about 15,000 hertz. At these frequencies, road noise attenuates for the most part before reaching the sonde. The only two versions of this tool known to the author are the "Audio Log*" of Schlumberger and the "Sibilation Survey*" of Birdwell (Now Western Atlas).[9,10] The * indicates a mark of the company indicated. Tools such as these offer no insight into the frequency spectrum of the noise.

An example of this type of log is shown in Figure 14.10.[9] This example shows temperature and continuous noise (sibilation) surveys. Notice that the sibilation survey shows at least six noise peaks and possibly more. Each such peak corresponds to an entry. The

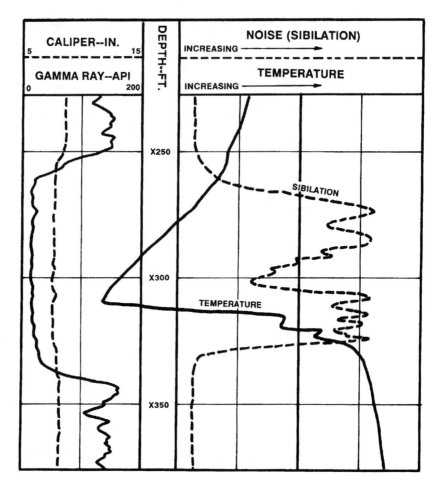

Figure 14.10 Continuous Sibilation survey (noise log) and temperature log (Courtesy Western Atlas, Ref. 9)

temperature survey, which is below the geothermal profile, shows only three cooling anomalies. While the other entries may be evident as changes in the slope of the temperature log, it would be difficult to confidently raise that suspicion without the continuous noise log.

Active Listening

The "active listening" technique is currently under study and is directed to detecting active channeling of fluid behind casing.[11] The technique entails directing a very short pulsed acoustic beam perpendicular to the flow and accurately recording its reflection. After a short interval of time, a second such acoustic signal is pulsed and recorded. If no flow is present, both reflections will be identical. If, however, flow exists, and contains scattering particles or bubbles, the reflected beams will be different. By varying the time between pulses, a velocity of the channeling flow may be established. It is anticipated that current pulse-echo cement evaluation and casing inspection technology can be used as the basis for a tool of this type.

NOISE LOG EXAMPLES

Noise Log Detects Gas Entries

In Figure 14.11 the well is perforated over two intervals and is producing about 3 MMcf/d, although much more production was expected.[12] A noise log was run to determine the relative contributions of the two sets of perforations. It is apparent that the perforations at B are producing much more than those at A. The well was reperforated at A and production

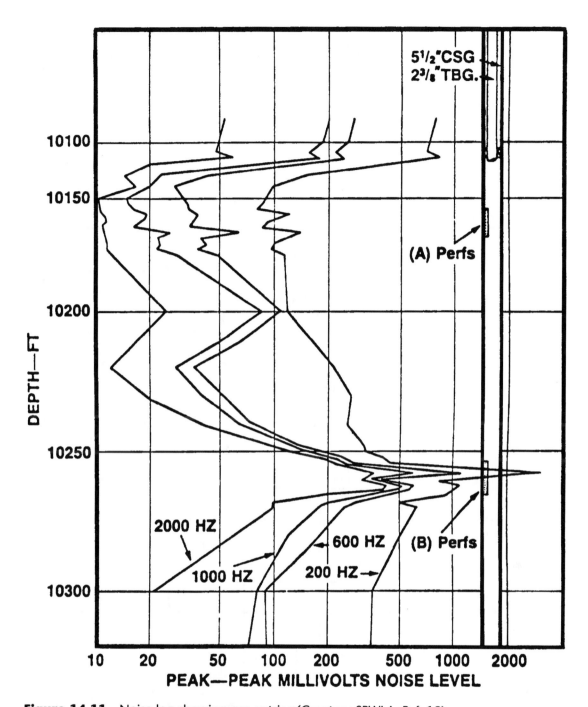

Figure 14.11 Noise log showing gas entries (Courtesy SPWLA, Ref. 12)

was increased to 12 MMcf/d. Notice that at the entry the noise levels are sharply peaked, somewhat tight, and die away smoothly both above and below the source. This indicates a single phase entry. The spread among the curves at A is due to the incomplete die away, especially of the lower frequencies. The noise level increase significantly when the tool is pulled into tubing, since the velocity of the gas is now quite large.

Gas-Liquid Interface

The Log of Figure 14.12 shows a spread in the frequency cuts below about X020 feet.[7] This indicates a bubbling flow up to the liquid-gas interface at X020. Above this point, there is a strong acoustic decoupling. The frequency cut clearly shows single phase flow above the interface, and the perforations at X850, X875, and X955 appear to be active gas entries.

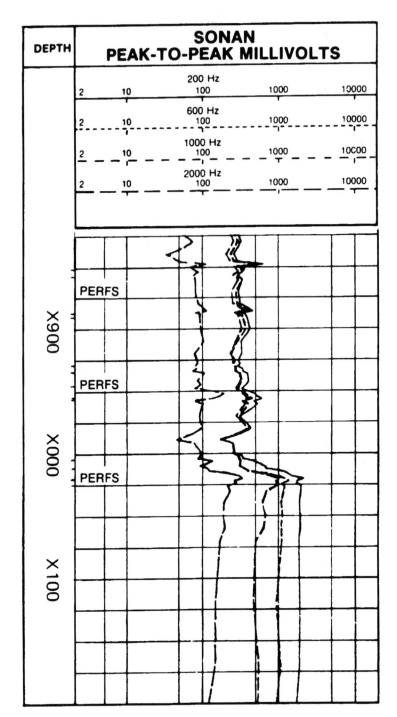

Figure 14.12 Noise log showing gas bubbling up to a gas-liquid interface (Courtesy Western Atlas, Ref. 7)

Channel Detection

A noise log was run to confirm a channel in the well shown on Figure 14.13.[13] This well is a dual completion and was producing oil through zones A and B, plus some water. A temperature indicated a channel from the water sand at C to the oil producing zones at A and B. The noise log was run to confirm this interpretation. The noise log sonde was run in the long string and the channel is outside of casing. The log was first run while the well was shut in and stabilizing. The solid and dashed curves with peaks of about .7 millivolts show noise at A and B as wellbore fluids invade these zones during shut in. When the well is flowed, the channel is clearly evident by the noise peaks now at A, B, and C.

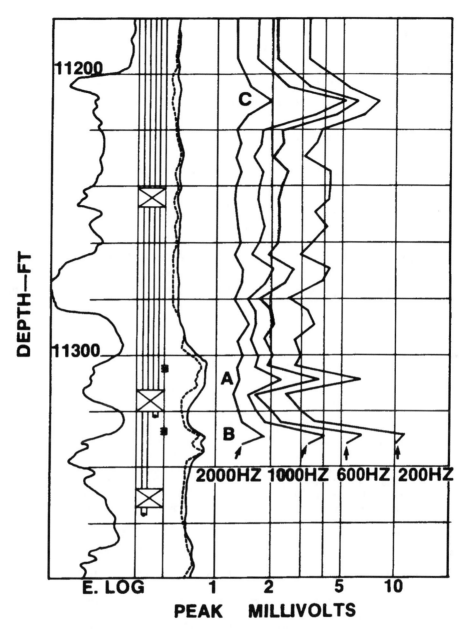

Figure 14.13 Noise log detects channels behind casing (Copyright SPE, Ref. 13)

POINTS TO REMEMBER

- Noise logs "hear" fluid movement downhole, even though the noise originates inside or outside of the casing.
- Noise logs can discriminate single phase from bubbling flow.
- Noise logs are most often combined with a temperature log.
- Noise surveys use stationary and not continuous readings.
- Most tools cannot be run continuously due to road noise.
- Sound amplitude indicates the depth of the anomaly.
- A noise anomaly must have flow and pressure drop, and hence it can be an entry or a restriction to flow.
- Noise spectrum can discriminate single phase from two phase entries.
- The log represents a series of cuts by high-pass filters at frequencies of 200, 600, 1,000, and 2,000 hz.
- Surface equipment may have to be shut down, especially in shallow wells.

- When taking a measurement, it is prudent to wait about one minute to let the tool settle down at each station.
- It is good practice to record a "dead well" response.
- Single phase entry noise is concentrated above the 1000 hz level.
- Two phase entries are characterized by noise in the 200 to 600 hz range.
- An acoustic decoupling and reduction in amplitude takes place above a gas/liquid interface.
- An acoustic decoupling and reduction in amplitude takes place below the casing shoe in open hole.
- Quantitative techniques are available, but require a calibrated tool—usually not available.
- Continuous noise surveys are available, but measure amplitude only at high frequencies and offer no spectral information.
- Active listening is a technique in which acoustic signals are pulsed perpendicular to a channel.
- In active listening, the reflected signal will change between acoustic pulses, indicating flow.
- Active listening requires particles or bubbles in the flow to be detected.

References

1. Atlas Wireline Services, Interpretive Methods for Production Well Logs, Fourth Edition, Document 9635, Houston, Texas, 1991.
2. Hill, A.D., "Production Logging—Theoretical and Interpretive Elements," SPE Monograph Volume 14, Henry L. Doherty Series, SPE, Richardson, Texas, 1990.
3. McKinley, R.M., Bower, F.M., and Rumble, R.C., "The Structure and Interpretation of Noise from Flow Behind Cemented Casing," *JPT*, pp. 329-338, March, 1973.
4. McKinley, R.M., and Bower, F.M., "Specialized Applications of Noise Logging," *JPT*, pp. 1,387-1,395, November, 1979.
5. Koerner, H.B., and Carroll, J.C., "Use of the Noise Log as a Downhole Diagnostic Tool," Paper SPE 7774, SPE Middle East Oil Technical Conference, Manama, Bahrain, March, 1979.
6. Pennebaker, E.S., and Woody, R.T., "The Temperature-Sound Log and Borehole Channel Scans for Problem Wells," Paper SPE 6782, 52nd Annual SPE Technical Conference, Denver, Colorado, October, 1977.
7. Bohn, F.O., "Technical Memorandum—An Introduction to Sonan Logging," Document 9385, Dresser Atlas (Now Atlas Wireline), Houston, Texas, February, 1979.
8. Britt, E.L., "Theory and Applications of the Borehole Audio Tracer Survey," 17th Annual SPWLA Symposium, Denver, Colorado, June, 1976.
9. Birdwell Company (Now Atlas Wireline Co.), "Gas Entry Detection Through Wellbore Sibilation Logging," Document BC77070A15, Tulsa, Oklahoma, 1977.
10. Schlumberger Well Service, "Schlumberger's Audio Log for better Answers in Air or Liquid," Advertising Brochure, Houston, Texas, about 1980.
11. Rambow, F.H.K., "Active Listening: An Alternative Method for Detecting Flow and Measuring Flow Velocity behind Casing," *The Log Analyst*, pp. 645-653, November-December, 1991.
12. Robinson, W.S., "Recent Application of the Noise Log," 17th Annual SPWLA Symposium, June, 1976.
13. Robinson, W.S., "Field Results from the Noise-Logging Technique," *JPT*, pp. 1,370-1,376, November, 1976.

CHAPTER 15

FLUID MOVEMENT: RADIOACTIVE TRACER LOGGING

INTRODUCTION TO RADIOACTIVE TRACER LOGGING

Tracer Logging Overview

Radioactive (RA) tracer techniques are usually used to monitor fluid movement downhole by detecting radioactive tracer materials released into in the flowing stream by the tool. These techniques are effective and quantitative, especially in single phase flow. Due to the inherent danger and fear of radioactive materials, tracer techniques are most often used in injection rather than producing wells. Radioactive tracers are also used when it is necessary to detect the placement of tagged material downhole, such as proppant used in a hydraulic fracture, gravel, and the like. Due to the lack of local availability of tracer materials and the logistics and regulations necessary to bring such materials into a country, tracer use in not common in most areas outside of North America.

Applications of Tracer Logging Techniques

The basic tracer techniques are used for detection of either fluid movement or locating RA tagged materials downhole.[1,2] The applications for fluid movement include:

1. Measure the injection (production) profile quantitatively.
2. Locate channels behind casing while injecting onto a well.

The applications for tagging materials include:

1. Cement top or squeeze placement by tagging cement with RA materials.
2. Gravel pack quality by tagging gravel with RA materials.
3. Detect RA tagged acid.
4. Measure fracture or propped height with tagged frac fluid or proppant.
5. Locate RA tags placed downhole with shaped charge or bullet perforators.

Sometimes a more complex operation may use a number of different RA tracers and spectral gamma ray (GR) detector tools. These same measurements may also be done with tracers having widely differing half lives, logging immediately following the operation and again at a later time after one of the tagging isotopes has died away. Such applications include:

1. A fracture is accomplished in two stages and each stage is separately tagged with different RA tracers.
2. The frac height, detected with tagged frac fluid, is discriminated from propped height where the proppant is tagged with another RA tracer.

Tools capable of detecting the direction from which gamma rays are coming have recently been developed for measurement of fracture orientation, i.e., does the fracture run North-South, East-West, etc. Later parts of this section discuss the hardware necessary for each of the applications alluded to above.

RADIOACTIVE TRACER TOOLS AND ISOTOPES

Basic RA Tool Configuration

Figure 15.1 shows a common RA tracer tool configuration for use in an injection well.[3] The casing collar locator (CCL) is used for depth control. The ejector portion of the tool is comprised of a chamber containing a RA tagged fluid compatible with the fluid being injected into the well and an ejector port. When activated at command from the surface, the tool ejects a small cloud or slug of RA fluid into the injection stream. This cloud then passes the GR detectors where it is detected both in terms of position and time.

Many variations of this tool configuration exist. The larger service companies use a 1 11/16 in. (4.29 cm) diameter tool, although such tools are available in sizes as small as 1 in. (2.5 cm) in diameter or less. Certain tools are available with dual ejectors, one for, say, a water–based RA fluid and another for an oil–based RA fluid. Port configurations vary from a single to numerous ports at the same depth through which tracers are ejected. Duration of time of ejection may be controlled from the surface. Tools may have as few as one GR detector located below the ejector for an injection well and above for a production well, to as many as four or more GR detectors placed at various positions above and below the ejector.

Special RA Tools

Under certain circumstances, such as the injection of polymers into a formation, the well fluids are non-Newtonian. In this case, it has been observed that RA tracer responses were quite erratic, primarily due to lack of mixing of the tracer solution and the injected polymer.[4-7] Furthermore, ejection from the side of the tool provided haphazard placement of

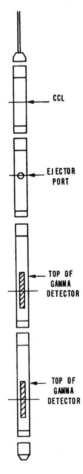

CCL

EJECTOR PORT

TOP OF GAMMA DETECTOR

TOP OF GAMMA DETECTOR

Figure 15.1   Radioactive tracer tool with gamma ray detectors positioned for an injection well (Courtesy Schlumberger, Ref. 3)

the tracer across the velocity profile of the polymer. As a result of this study, Western Atlas developed a tool called the "Swing-Arm Tracer*" (*mark of Western Atlas). The novelty of this tool is that when an ejection is made, an arm swings out to eject consistently into the middle of the highest velocity part of the flowing stream. This consistent positioning of the tracer assures an excellent quality of data , even though the tracer and the polymer do not mix well. This tool is suitable for any wellbore fluids, provided that its density and the tracer density are nearly equal.

Western Atlas also developed a tool to measure injection rates lower than about 200 bbl/day (32 m³/D).[7] This tool consists of an inverted collapsing basket assembly mounted on a centralizer cage. The very slow flows are collected by and diverted through the basket. After such diversion, the flow is forced through a smaller diameter section at a higher velocity, where meaningful measurements may be made. RA tracer and GR detectors internal to the tool are used to measure this higher velocity flow.

RA Detectors

The detectors used on most tracer surveys are either Geiger Mueller or Scintillation detectors. Of the two, the scintillation detector is far more efficient at counting gamma rays. The chart of Figure 15.2 shows the response for both types to an Iodine 131 gamma ray source placed 2 to 16 inches (5 to 41 cm) from the detector and separated by cement.[8,9] Comparing the 1 11/16 in. (4.3 cm) size tools, the Geiger Mueller detector loses its sensitivity in this test after about 6 inches (15 cm), whereas the scintillation detector can discriminate the signal from background out to about 12 inches (30 cm).

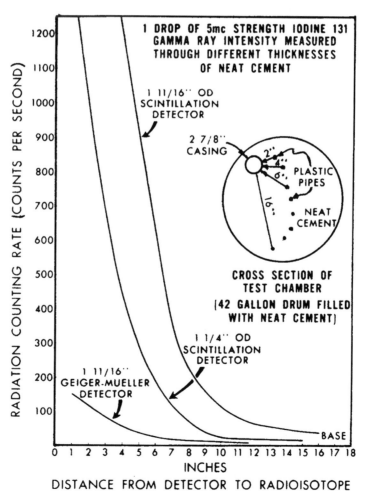

Figure 15.2 Test showing relative gamma ray counting efficiencies of various gamma ray detectors (Copyright SPE, Ref. 8)

Both types of tools have been used on a single tool string to detect channels. Since the scintillation detector has a greater depth of investigation, channels would present a strong signal on the scintillation detector log while the Geiger Mueller detector log would be very weak. If flow is inside the casing, both signals are high. The major service companies primarily use scintillation detectors today.

Radioactive Tracer Isotopes in Use

Table 15.1 lists common isotopes used in tracer work.[10] When monitoring fluid movement, typically water injection profiles, the most common isotope is iodine 131 in the form of NaI dissolved in water. It has a convenient half life of 8.04 days, which means that it can be shipped to the wellsite and retain enough strength to be usable even with the usual

TABLE 15.1

Tracer	Isotope	Half-life (days)	Energy (keV)	Intensity
Gold-198	^{198}Au	2.70	412	0.96
			676	0.01
Xenon-133	^{133}Xe	5.25	81	0.36
Iodine-131	^{131}I	8.04	284	0.06
			364	0.81
			637	0.07
			723	0.02
Rubidium-86	^{86}Rb	18.7	1,077	0.09
Chromium-51	^{51}Cr	27.7	320	0.10
Iron-59	^{59}Fe	44.6	1,099	0.57
			1,292	0.43
Antimoy-124	^{124}Sb	60.2	606	1.05
			720	0.15
			1,353	0.05
			1,691	0.49
			2,091	0.06
Strontium-85	^{85}Sr	64.8	514	0.99
Cobalt-58	^{58}Co	70.8	511	0.15
			811	0.99
Iridium-192	^{192}Ir	74.0	311	1.42
			468	0.48
			603	0.18
Scandium-46	^{46}Sc	83.8	889	1.00
			1,121	1.00
Zinc-65	^{65}Zn	244.0	511	0.015
			1,116	0.51
Silver-110	^{110}Au	250.0	666	1.32
			773	0.34
			885	0.73
			937	0.34
			1,384	0.24
			1,502	0.18
Cobalt-57	^{57}Co	271.0	122	0.86
			136	0.11
Cobalt-60	^{60}Co	1,925.0	1,173	1.00
			1,332	1.00
Krypton-85	^{85}Kr	3,098.0	514	0.004

delays in operations. Tracers are also used for monitoring steam injection. Krypton-85 is often used to follow the steam since it will remain in the gas phase. For tagging materials, the most common isotope is iridium 192. It has a half life of 74 days and is commonly used with proppants, gravels, and cements.

Table 15.1 lists many more elements. In a later section of this chapter, the use of multiple tracers in the same well is discussed. Spectral gamma ray tools are used to discriminate the isotopes. Obviously, it is critical that the isotopes used have very different gamma ray energy level peaks for proper discrimination.

Most service companies are avoiding the handling of radioactive materials unless absolutely necessary. As a result, in the use of multiple tracer programs, the procurement and handling of the tracers is usually done by companies specially qualified to handle such materials.

Factors Affecting Strength of GR Signal

The distance and material between the tracer and detector have significant effects on the intensity of the radiation at the detector. In general, the intensity is related to the distance from a point source by the inverse square law, i.e., by doubling the distance from the radiation source to the detector, the intensity is reduced to one fourth its original value. The following equation describes this phenomenon.

$$\frac{I_2}{I_1} = \frac{d_1^2}{d_2^2}$$

(Equation 15.1)

where

$$d = \text{Distance from source}$$

$$I_1 = \text{Intensity at } d_1$$

$$I_2 = \text{Intensity at } d_2$$

If the tracer is more distributed than a point source and close to the detector, then the rate of drop-off with distance may be somewhat less than indicated above.

The material between the source and detector causes the intensity at the detector to be reduced. The material is shielding the detector from the RA source. In general, this shielding effect is measured in materials by their characteristic Half Value Layers (HVL). The HVL is the thickness of a material required to reduce the intensity of the source to one half its original value. Table 15.2 indicates the approximate HVL values for various materials commonly encountered downhole and for a range of gamma ray energies. The higher gamma ray energies have more penetrating power.

How easy is it to detect tracers in a channel outside of pipe? If iodine 131 is the tracer, most of its energy is concentrated in the .364 mev level. At this energy level, Table 15.2 indicates that the intensity of the iodine drops to about one half of its original intensity due to a casing wall of about .35 inches (.9 cm). If the channel is outside of a 1 inch (2.5 cm) layer of cement, the intensity would again drop by about one half. If this is the only effect, the intensity is reduced to about one fourth of its original value due to the casing and cement.

TABLE 15.2 Half Value Thickness in Inches

Gamma Ray Energy (MEV)	Water	Concrete	Dense Limestone or Sandstone	Steel
0.2	2.0	1.10	0.83	0.27
0.5	2.91	1.46	1.22	0.43
1.0	4.00	1.96	1.80	0.60
2.0	5.59	2.99	2.45	0.89
3.0	7.00	3.59	3.00	1.13

CALCULATING FLOW RATES WITH TRACERS

Basic Technique

In the sections which follow, many tracer techniques to evaluate fluid movement are discussed. Most can be quantitative. The techniques essentially detect the tracer cloud position and its changes with time. Given such distance and time information, a velocity and volumetric flowrate can be calculated along the wellbore. The following equation can be used in all such instances.

$$Q = (83.93 \times F \times L \times (D^2 - d^2))/\Delta t \qquad \text{(Equation 15.2)}$$

where

$$Q = \text{Flow Rate, B/D}$$

$$F = \text{Flow profile correction factor}$$

$$L = \text{Distance between detections, feet}$$

$$D = \text{Flowing borehole diameter, inches}$$

$$d = \text{Tool diameter, inches}$$

$$\Delta t = \text{Time between detections, seconds}$$

This equation is further discussed with each tracer technique as required.

Flow Profile Correction Factor

The flow profile correction factor in equation 15.2 is necessary since the tracer tends to concentrate in the highest velocity part of the flow stream. As a result, the computed velocity is too large and must be reduced. The key to this correction is the Reynolds Number of the flow. For Reynolds Numbers less than about 2,000, the flow is laminar, while if greater than about 4,000, the flow is turbulent. The Reynolds number is defined as:

$$N_{Re} = \frac{\rho \overline{V} D}{\mu} \qquad \text{(Equation 15.3)}$$

where

$$\rho = \text{Density of the fluid}$$

$$V = \text{Average velocity of the fluid}$$

$$D = \text{Flowing borehole diameter}$$

$$\mu = \text{Viscosity}$$

and Vbar may be expressed as

$$\overline{V} = \frac{Q}{A} = \frac{4Q}{\pi D^2} \qquad \text{(Equation 15.4)}$$

where

$$A = \text{Cross sectional area of the flow}$$

$$Q = \text{Flow rate}$$

The chart of Figure 15.3 may be used to evaluate the Reynolds Number.[3] It is interesting to note that in a casing with an internal diameter of about 5 inches (12.7 cm), water (one centipoise viscosity) is fully turbulent at flow rates just above 200 B/D (32 m³/D). This would indicate that in most injection wells which are rather high in rate, the flow is mostly turbulent.

Figure 15.4 shows a correction factor as a function of Reynolds Number.[12] This chart is for a centralized tool with flow in the annulus between the tool and casing. If the tool is not centralized, the correction is approximately that on the chart plus .1.

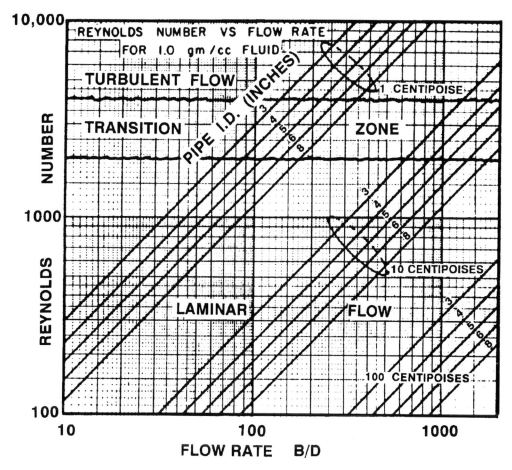

Figure 15.3 Reynolds number vs. flow rate for 1.0 gm/cc fluid (Courtesy Schlumberger, Ref. 3)

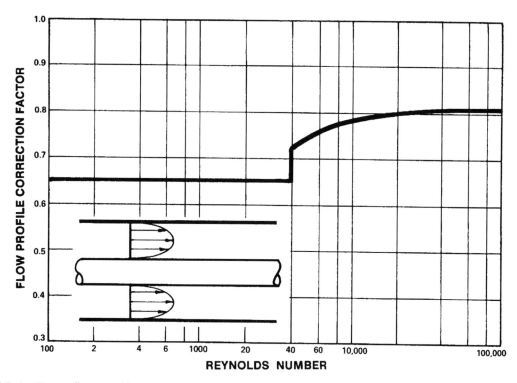

Figure 15.4 Tracer flow profile correction factor as a function of Reynolds number (Copyright SPE, Ref. 12)

While charts for correction factors may be useful, it is recommended that an empirical factor be established by matching the surface injection rate to that downhole above all points of fluid loss. This factor should be reasonably close to that indicated by the charts. For most computations, using this empirical factor without change across the whole interval to be evaluated is adequate.

Effects of Area Change of Injected Stream

When the flow stream undergoes changes in area, as from tubing into casing, the injected fluid jets into the larger pipe and tracer measurements in that region yield incorrect results. The chart of Figure 15.5 indicates the distance required, in feet, from an area increase to minimize its effect on the tracer measurement.[12] This chart relates the diameter change ratio, D_{large}/D_{small}, to the distance required. For example, if the ratio of casing ID to tubing ID is three, then measurements should be taken no closer than nine feet (2.75 m) from the bottom of tubing.

RA TRACER TECHNIQUES FOR FLUID MOVEMENT

Sequential Passes to Follow RA Tracer Cloud

This technique is typically used to initially scan the well to get an idea of the flowing conditions downhole. It is good practice to run a base gamma ray log before ejecting any RA

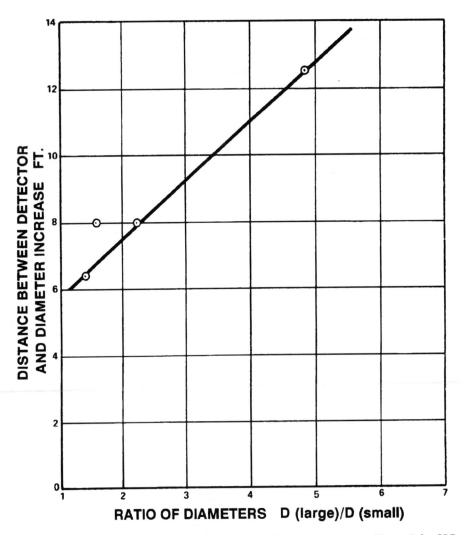

Figure 15.5 Distance required between detector and diameter increase (Copyright SPE, Ref. 12)

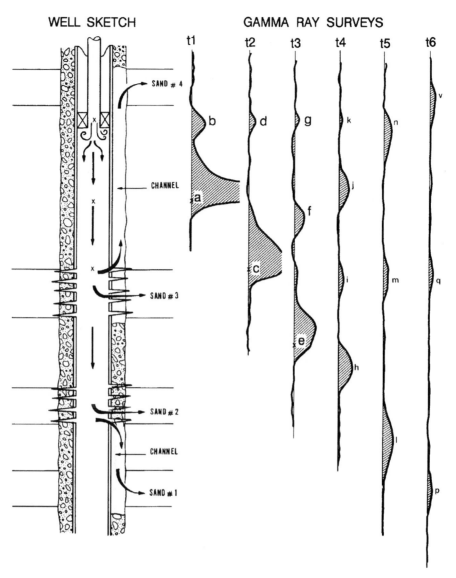

WELL SKETCH   GAMMA RAY SURVEYS

Figure 15.6 Sequential passes technique to follow radioactive cloud (Courtesy Schlumberger, Ref. 1)

tracer downhole. This technique defines the zones of injection and locates possible channels. A schematic of this technique is shown on Figure 15.6.[1]

The well sketch shows an injection well with flow going out through the two sets of perforations, and the traveling up and down through channels to nearby sands. Initially, at time t_o, a slug of RA tracer is ejected at the bottom of the tubing. The tool is then lowered into the well and logged up at time t_1. A detection of the RA slug is noted at a and another at b. The slug at a is the main RA cloud in the flowing injection stream, while the slug at b is a remnant trapped in the boundary layer at the bottom of tubing. The tool is again lowered and logged up through the RA cloud at time t_2. The main RA slug is now at c and the remnant remains at d. Note that a velocity can be measured since the slug has moved from a to c in the time between runs. The tool is again lowered and logged up at time t_3. The main RA slug has become two slugs at f and e. This splitting of the RA tracer cloud suggests that the injection stream has split up, with some portion proceeding up above the perforations (slug f) and some proceeding down below the perforations (slug e). Slug f suggests a channel above the perforations.

To confirm the channel above the upper set of perforations, a "standing channel check" would be performed. First, place the tool (ejector and GR detectors) above the upper set of perforations. Second, eject a small cloud of RA tracer into the injection stream, keeping the tool stationary. Third, monitor the small RA cloud. It should be detected going down past the GR detector and again a short time later as a portion of that RA cloud passes the GR detector going up in the channel. Similarly, if a down channel is indicated by tracer moving down below the lowest set of perforations, a small slug must be ejected in the rathole to check if flow is occurring inside of the casing below the perforations.

The procedure continues until the slug is too strung out and not readily detectable. At that point, the tool is moved to another position and a second RA tracer slug is ejected into the flowing injection stream. Usually, this technique will begin its tests at the bottom of the well, i.e., above the lowest set of perforations. This would ensure against RA tracer from a previous shot interfering with current tests.

Velocity Shot Method

The velocity shot method is probably the most accurate of the injection well profiling techniques. A schematic of this technique is shown on Figure 15.7. In this figure, a fluid, say water, is being injected into the two sets of perforations indicated. Sequential passes have been performed and confirm that no channels or leaks in the casing are present. The

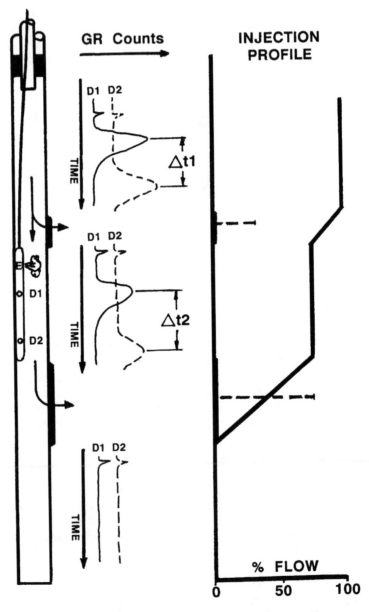

Figure 15.7 Velocity shot technique for evaluation of injection profile

tool is stopped and remains stationary at a depth of interest. The schematic shows a test being made for the velocity at a station between the perforations.

With the tool stationary, the recorder is put on time drive, i.e., the logs are recorded as a function of time and not depth.[13-15] An ejection of a small amount of RA tracer into the injection stream is made. This slug moves past the upper D_1 and lower D_2 detectors. The time is measured on the time drive log shown to the right. The spike is an indication of the time of ejection. The increase in GR counts seen by the detectors is due to the RA slug passing. The time of travel between these detectors is indicated as Δt. From the tool configuration, which should be noted on the log heading, the distance between the detectors is measured.

Using the equation 15.2, a velocity can easily be computed at each tool test position. Notice that in a cased well the numerator term in the equation is constant, except for the factor F. If F is also assumed constant since the flow is mostly turbulent, then the flowrate at each station i is determined only by the $delt_i$ measured. If Q_1 is the surface injection rate, then the flowrate at station i is given by

$$Q_i = Q_1 \times \left(\frac{\Delta t_1}{\Delta t_i}\right)$$

(Equation 15.5)

The result of a velocity shot test is an injection profile as shown at the right of the figure. An alternative bar chart presentation is also shown.

The measurement of time is shown as a peak to peak measurement. While this may be most accurate, the RA slugs often get strung out, and sharp peaks cannot be readily detected. Another technique is illustrated on Figure 15.8. A reference point is located at the intersection of the background and a line drawn through the rising part of the GR curve. Even when the GR curve is spread out, this reference point can usually be detected even though the peaks cannot. The chart corrections will not agree with the use of this reference point, since the delt measured now is a function of slug velocity and diffusion. However, if the correction factor is evaluated downhole, or if the numerator of equation 15.2 is assumed constant, then the answers are substantially identical.

Area Method and Related Techniques

When a GR detector is passed through a slug of RA tracer material dispersed as a cloud in the borehole, the counts detected are proportional to the amount of RA tracer present. Some error may be introduced if the velocity of the slug changes and hence the exposure time to the RA tracer is slightly longer, or the diameter of the borehole changes and some RA material is further away from the detector. However, if tool speed is the same for each

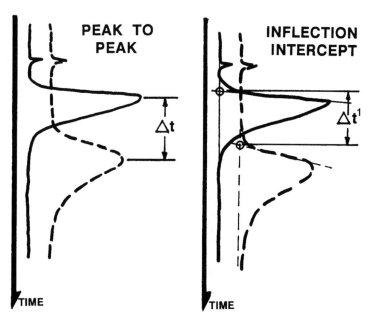

Figure 15.8 Techniques to measure time between detections of tracer cloud

pass and much larger than the velocities downhole, and the well is cased and of constant size, then the statement is approximately true. Even if conditions are far from ideal, this technique can provide useful, if only qualitative, information.[16-18]

Figure 15.9 shows an injection well with four sets of perforations: A, B, C, and D. Below tubing and above the top set of perforations, a large slug of RA tracer is ejected into the flowing injection stream. As with the sequential pass technique, the tool is lowered and pulled through the RA cloud. This is noted as pass no. 1. The tool is again dropped down and logged up through the slug, and this procedure is repeated until the slug disappears.

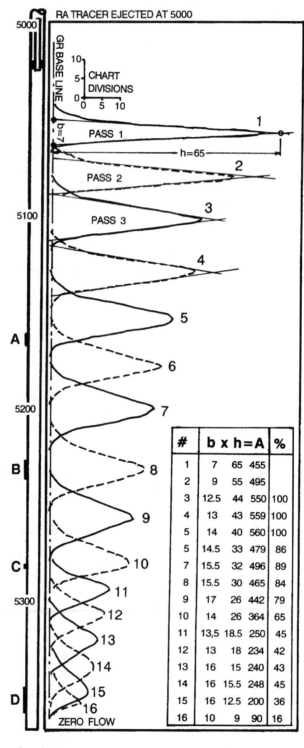

#	b	x h	= A	%
1	7	65	455	
2	9	55	495	
3	12.5	44	550	100
4	13	43	559	100
5	14	40	560	100
5	14.5	33	479	86
7	15.5	32	496	89
8	15.5	30	465	84
9	17	26	442	79
10	14	26	364	65
11	13,5	18.5	250	45
12	13	18	234	42
13	16	15	240	43
14	16	15.5	248	45
15	16	12.5	200	36
16	10	9	90	16

Figure 15.9 Area under the gamma ray count curve technique (After Production Logging Services, Inc.)

In the figure, there were 16 passes through the RA tracer slug. It is also clear that the area under the GR curve is declining downhole. If the tracer detections are approximated with a triangle, their areas can be easily measured. The areas of GR passes 1, 2, 3, and 4 should all be the same since no material is lost from the RA cloud. Note that the peaks decline and the bases are increasing, tending to keep the areas constant. In this case, passes 1 and 2 are slightly different in area than 3 and 4, due primarily to the RA tracer taking some time to mix into the injection stream. As a result only passes 3 and 4, or their average area, represent 100% of the flow. The flow remaining at any other pass is expressed as a ratio of the areas, i.e.,

Fraction remaining at pass i =   (Equation 15.6)

(Area under pass i)/(Area under 100% flow)

The table within the figure shows that after passing the perforations at A, the cloud drops from 100% of the flow down to between .86 to .89 of the initial area. Therefore, about 12.5% of the injected flow was pumped into the perforations at A. Similarly, between B and C only about 79% remains, indicating a loss of $.875 - .79 = .085$, or 8.5% injected into perforations B. Continuing in this vein, the chart and the injection profile is completed as shown. When the area is adjacent to a set of perforations, it is not representative of either the flow above or below the perforation, but is someplace in between.

This technique has been used successfully to detect and quantify flows outside of tubing. It is also used for open hole injection profiles when hole conditions are poor. Note that there is a technique called the "Self Method" which used the sum, rather than the product, of the base and height of the GR triangular areas of each pass.[18] This technique is otherwise the same as that described above.

A closely related technique is to eject two small slugs a short distance apart. The separation of the slugs initially corresponds to 100% of the injected flow stream at the initial test station. When some portion of the injected flow stream exits the pipe, the spacing between the two slugs becomes smaller.[19,20] If l_o is the initial distance between the slugs corresponding to 100% of the flow, then the fraction remaining at test position i is:

$$Q_i = Q_o \times (l_i/l_o)$$ (Equation 15.7)

This technique loses resolution in slow flows when the slugs come close together and the slugs become spread out.

Other Velocity Measuring Techniques

Velocity computations can be made with the recording of Figure 15.9 if the times of the passing of the peaks is recorded. Suppose that during pass no. 1 the peak is detected at time t1, and that during pass no. 2 the peak is detected at time t2, and so on. Then, equation 15.2 can be used with delt equal to the time between detections and the distance, L, equal to the distance between detections. Since the tool only passes through the slug during the moments of detection, the tool diameter in the flowing stream is effectively zero. The correction factor should be adjusted to match surface injection rates above the highest points of fluid exit.

A closely related technique is variously called the drop shot method or Ford Method. The idea is to eject a slug of RA tracer into the injection stream, lower the tool to a known depth, put the recorder on time drive, and wait for passage of the RA slug. After the slug passes, drop the tool again to another known depth and similarly wait. Continue the process until the slug disappears or is otherwise unusable. Computationally, this technique is identical to that of the preceding paragraphs, except that now the recording of the slug passage is done as a function of time with the tool stationary, and the tool depths are manually recorded.

Tracer Distribution Technique

In this technique, a large slug of RA tracer is released into the injection stream, typically at the surface, and the RA tracer is monitored by a number of logging passes as it is pumped into the formation. As the tracer is pumped into the formation, it is assumed that the RA tracer plates out at the formation face and that the hottest spots have taken the most fluid. The schematic of Figure 15.10 illustrates this technique.

If the RA tracer is pumped into the formation, this technique will not work. To assure that the RA tracer plates out at the formation face, service companies have used flakes or beads tagged with the RA tracer. These flakes or beads are deposited on the formation face much like mud cake during drilling operations, but are too dispersed to act as a diverting agent. These flakes or beads will remain for a few hours before dissolving away. Logs run during these few hours would be indicative of the injection distribution.

Conditioning Survey Displacement

This technique may be used in naturally producing wells and wells on beam pump. Care must be taken to properly handle the RA tracer which is brought to the surface in the produced fluid. An example of this technique is shown in Figure 15.11.[21] Velocity shot tests showed entries of 125 and 225 B/D from the lower and upper perforations, respectively. The well was then conditioned by continuously and uniformly releasing RA tracer across the logged interval while moving the tool downhole. The first up run shows an otherwise uniform distribution of RA tracer in the well except for reduced GR counts at the upper perforations. This is due to an entry of fresh formation fluid into the well. A second and third pass show drops in GR counts at both the lower and upper perforations due to dilution from produced fluids. The length of the diluted interval, with consideration that fluids may be coming from below, provides an indication of the rate of fluid entry at each set of perforations.

Interface Detection

This uncommon technique may be used to evaluate injection profiles when hole conditions are poor and rugose. In this method, tubing is run to bottom. If injection is maintained both in the tubing and annulus, the two flows converge to a stagnation plane outside of tubing. Above this stagnation plane, the annulus injection is going into the formation. Below the plane, the tubing injection is going into the formation. By tagging the annular flow with RA tracer, the stagnation plane can be detected by a GR detector run in the tubing. If the total flow is maintained at, say, 500 B/D, then a rough injection profile can be obtained by varying the tubing and annular flows as shown in Figure 15.12.[3, 22]

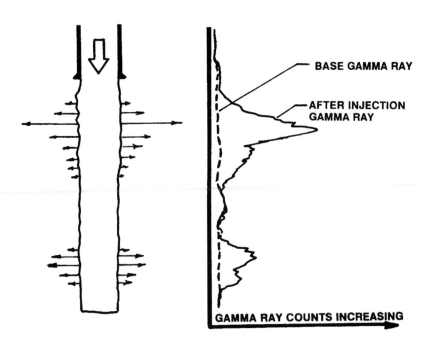

Figure 15.10 Tracer distribution technique

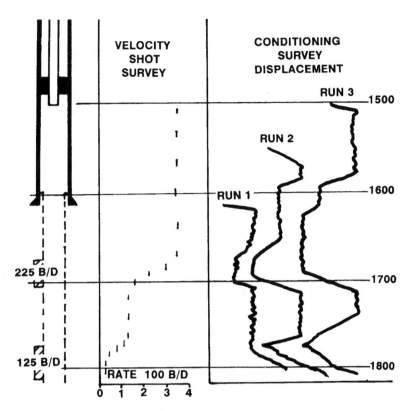

Figure 15.11 Conditioning survey and displacement technique (Copyright SPE, Ref. 21)

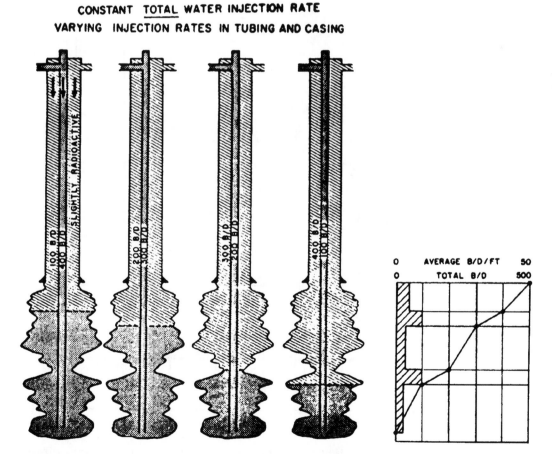

Figure 15.12 Interface detection technique (Courtesy Schlumberger, Ref. 3)

In this figure, the first run has 100 B/D annular injection and 400 B/D tubing injection. In the second run, the mix has changed to 200 B/D annular injection and 300 B/D injection through tubing. Continuing this process, each movement of the interface in subsequent runs represents an injection of 100 B/D. An injection profile is shown at the right.

TAGGING MATERIALS DOWNHOLE, MULTIPLE TRACERS, ROTASCAN

Fracture Height Evaluation—Single Tracer

Tagging of proppant with iridium to detect the propped height of a hydraulic fracture has been common practice for many years. RA tracer and temperature have frequently been run together to detect such fractures. Typically, a base GR is taken before the hydraulic fracture operation. During the operation, proppant tagged with RA tracer is pumped down to hold the fracture open for easy passage of produced fluids. Figure 15.13 shows a set of base and after frac temperature and RA tracer gamma ray detector logs.[23] The region shaded black shows an excessive RA count rate due to the proppant in the nearby fracture wings, and is therefore interpreted as propped fracture. Notice the accumulation of RA tagged proppant has accumulated at the bottom of the well. This is not part of the fracture.

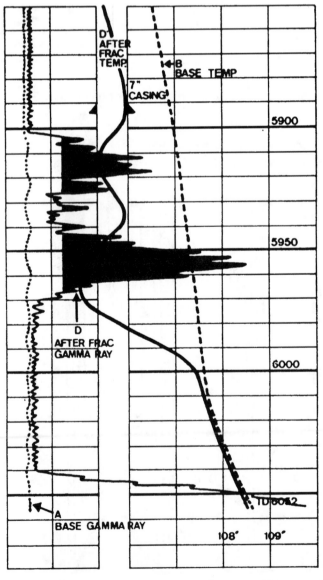

Figure 15.13 Fracture height evaluation using a single tracer (Courtesy Southwestern Petroleum Short Course, Texas Tech University, Ref. 23)

Multiple Tracers

Multiple tracers are well suited to detecting the effectiveness of different phases of certain downhole operations since each phase is tagged with a separate RA tracer material.[24-28] Figure 15.14 shows the relative intensity of differing RA tracer materials.[24] This is a graphical presentation of selected isotopes listed on Table 15.1. Isotopes with well separated and high peaks can be most easily discriminated with a spectral gamma ray tool such as those described in chapter 4. The 1 11/16 in. (4.3 cm) through tubing size tools are most usually used for these applications. From this figure, the best selection would probably be scandium-46 and one of the other three, probably iridium-192 due to its high intensity and low GR energy.

Fracture Height Evaluation—Multiple Tracers

In the example of Figure 15.15, a hydraulic fracture was induced using fluid tagged with scandium-46 and proppant with iridium-192.[10, 29] After the fracing operation, the well was logged with Halliburton's 3 5/8 in. (9.21 cm) Compensated Spectral Natural Gamma* (CSNG*) tool. The spectral data was processed using the techniques of their Tracer Scan* processing (* denotes a mark of Halliburton Energy Services).

In the middle track are the scandium and iridium GR records calibrated in API unit count rates. These show that the frac height runs from about x880 to x956 and that the fracture is propped along its full height. By measuring the degree of downscattering of GR energies due to passage through the formation and wellbore fluids, the Tracer Scan technique is also able to determine a relative distance to the GR source.[30] The scandium and iridium curves on the right are measurements of relative distance to the GR sources. Notice that the iridium signal above and below the perforations is inside the casing (near relative distance scale) while the signal across the fractured interval originates well outside the casing.

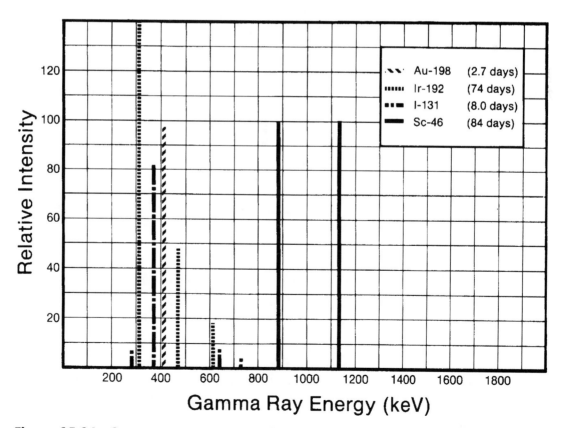

Figure 15.14 Gamma ray energy spectra for various tracers (Reproduced with permission of Halliburton Co. and SPWLA, Ref. 24)

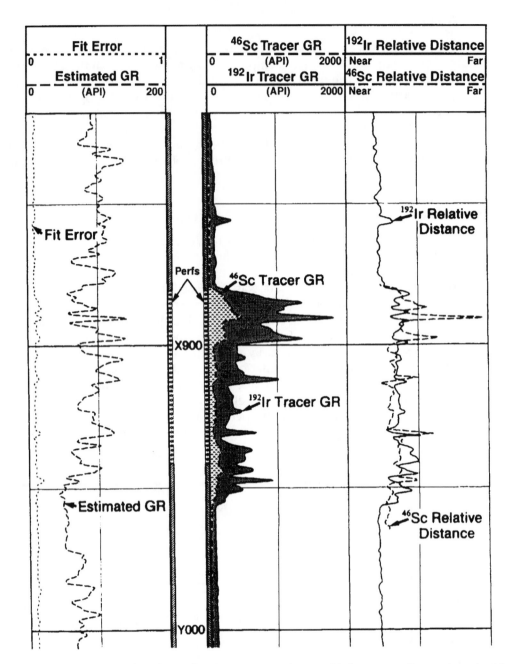

Figure 15.15 Fracture height evaluation survey using multiple tracers (Reproduced with permission of Halliburton Co. and SPWLA, Ref. 10)

RotaScan* (*Mark of Halliburton Energy Services)

In some instances, it is desirable to determine the orientation of RA tracer material around the wellbore. A most obvious application is to evaluate the orientation of RA tagged induced fractures in a formation. Halliburton Energy Services offers a directional gamma ray device called the "RotaScan*" tool.[31, 32] (* denotes a mark of Halliburton Co.) It consists of a Sodium Iodide scintillation type crystal detector which is shielded and collimated as shown in Figure 15.16.[31] The detector rotates at a rate of 20 rev/min. At a suggested logging speed of 5 ft/min (1.5 m/min), the RotaScan samples one complete rotation for every 3 inches (.1 m) depth interval. The RotaScan has no spectral capability.

Halliburton performed a number of tests to simulate the RotaScan response to RA tagged fracture planes at various orientations around the wellbore. The results are shown in a

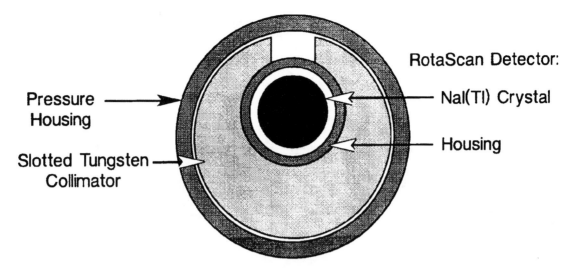

Figure 15.16 Halliburton Energy Services RotaScan tool cross section (Reproduced with permission of Halliburton Co. and SPWLA, Ref. 31)

polar plot presentation on Figure 15.17.[31] The fracture is simulated with two RA tagged plates, with the plate on the right being somewhat stronger than the one on the left. Fractures passing through the center of the wellbore, labeled A, leave a "footprint" shaped signal. A similar response is noted for case B where the fracture cuts the casing 1 3/8 in. (3.5 cm) off center. When the fracture is tangent to or away from the casing, the shape changes to a more circular pattern reaching a maximum in the direction of the fracture. Polar plots are available for field logs as well.

RotaScan Log Example

Another RotaScan presentation is shown on Figure 15.18.[31] In this well, two sets of perforations are shown. The upper set was perforated in an East-West orientation, the same as the known fracture orientation for the area. The lower set was fractured in a North-South orientation. A mini frac was performed and the wellbore cleaned up to minimize tracer contamination. On the left track, the maximum GR intensity is recorded along with the direction of that intensity as shown by the tadpoles.

The track on the right shows the detected GR intensity as a function of direction for each rotation. The intensity is shaded black above a certain threshold value. The direction of maximum GR signal is southerly at the lower perforations while the upper set shows a much smaller peak toward the east and perhaps toward the west. The upper set was oriented with the expected fracture orientation. The large difference between the two intensities is probably due to the ease with which the frac fluid moves away from the wellbore when the perforations are aligned with the fracture, while a significantly larger amount of tracer fluid remained in the borehole vicinity in the other case.

TRACER LOG EXAMPLES

Sequential Passes Show a Channel

The example of Figure 15.19 shows a water injection well in which the injection zone is supposed to be in the open hole below the casing shoe.[33] Tracer was ejected into the injection stream, and a number of up passes were made to monitor the movement of the large RA slug. Run #1 begins at time 0:00, run #2 begins at time 1:17, run #3 at time 2:25, and so on. Runs # 1–4 show downward movement with a reduction of height and increase of base dimensions, and approximately constant area. Run #5 and #6 show a rapid decrease in amplitude and area, as well as a second bump above the casing shoe. Run #6 has been

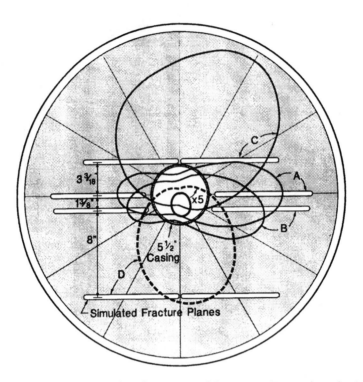

Figure 15.17 RotaScan response to simulated RA tagged fractures (Reproduced with permission of Halliburton Co. and SPWLA, Ref. 31)

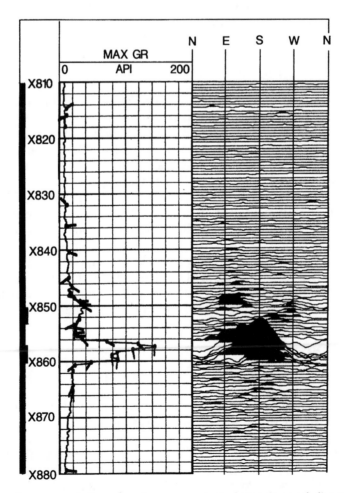

Figure 15.18 RotaScan log presentation showing gamma ray intensity and direction (Reproduced with permission of Halliburton Co. and SPWLA, Ref. 31)

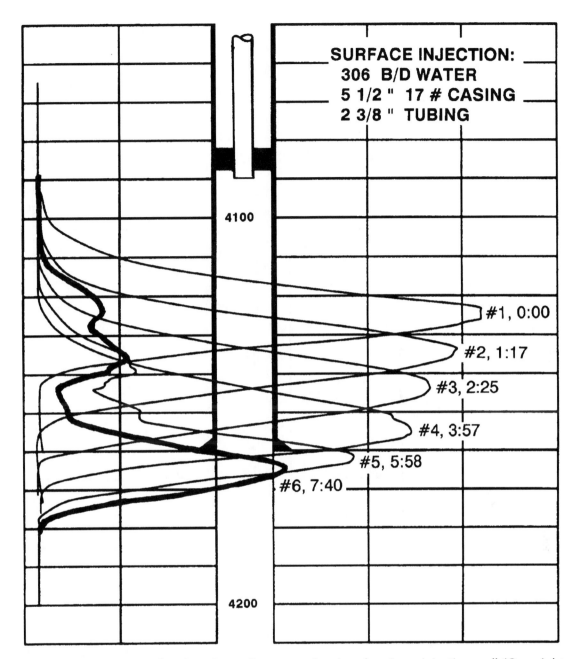

Figure 15.19 Tracer showing channeling around casing shoe in an injection well (Copyright SPE, Ref. 33)

drawn heavy to aid in seeing this second bump which has moved up to about 4,110 feet behind casing. Notice that run #6 began only 7 minutes and 40 seconds into the program. This slug was monitored for over an hour, and showed the slug moving to a place at least 150 ft uphole from the casing shoe.

RA Tagging of a Cement Top

Figure 15.20 shows a base GR before cementing and another GR run afterwards.[2] The radioactive cement top is readily detected in the example. Two RA hot spots are indicated below the RA tracer associated with the cement top. These are adjacent to shales and likely are due to an accumulation in a washed out interval. Notice that the shale zone of the upper shale zone is washed out somewhat before circulating cement as indicated by the caliper.

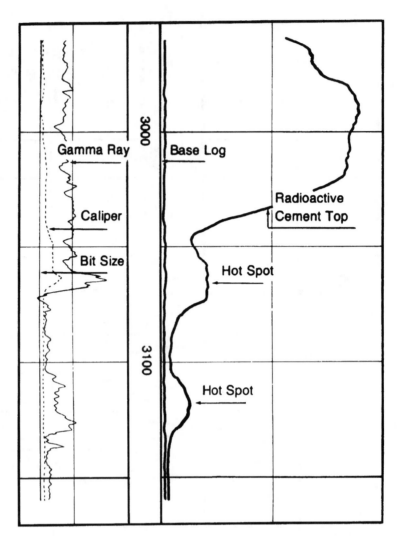

Figure 15.20 Detection of cement top tagged with radioactive cement (Courtesy Western Atlas, Ref. 2)

Multiple Tracers with Differing Half Lives

Tracer techniques are often used to monitor placement of proppant in fracturing operations. In this example, shown on Figure 15.21, the lead in proppant is tagged with iridium-192, having a half life of 74 days, and the tail in proppant is tagged with a short lived iodine-131 with a half life of eight days. A base log is shown dashed. The heavy solid line is the GR survey run shortly after the frac job. The gamma ray counts over background (base log) are due to both iodine and iridium and show the total propped frac height. A month later, the iodine-131 has died away, leaving the iridium remaining. The excess counts over background now represent the lead in proppant. The difference between the two in terms of height is the tail in proppant.

Tracers in Limited Entry Steam Injection

Limited entry steam injection is a technique to control the injection profile by placing only a handful of perforations in a well and pumping steam with a large enough pressure differential to maintain sonic velocity through each. Tracers are used only to confirm that the perforations are, in fact, taking steam. In Figure 15.22 the tracer shows that the seven perforations all show RA spikes, confirming injection into each.[34] While this example is done with RA tagged water, krypton is also used with steam. A temperature log in a nearby observation well confirms this injection profile.

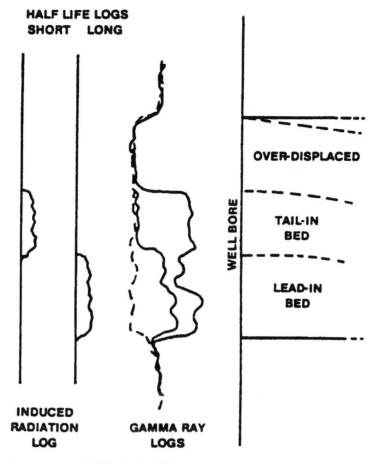

Figure 15.21 Test using tracers of differing half lives to discriminate lead-in from tail-in fracture proppant (Courtesy Schlumberger)

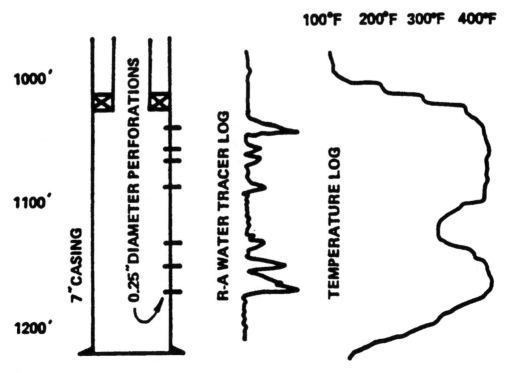

Figure 15.22 Detection of active perforations in limited entry steam injection well (Copyright SPE, Ref. 34)

POINTS TO REMEMBER

- Tracer surveys are best suited to single phase injection wells, and the use in producing wells is not common.
- The tool configuration is set up of the nature of the flow (injection or production) expected.
- Spacers may be placed between detectors for optimum velocity measurement.
- Scintillation detectors have a greater depth of investigation than Geiger Mueller detectors.
- Radioactive materials require careful handling and their use must be consistent with applicable laws.
- The well should be stable (72 hours flowing is rule of thumb).
- A tool sketch showing the tool diameter along with ejector and detector spacings should be on the log heading.
- Verify spacing by measurement at the surface.
- Fluid RA tracer material must be soluble or neutrally buoyant relative to well fluids.
- The tracer tool is best run centralized.
- The ejection time should not change from shot to shot.
- Caliper surveys are required for open hole flow profiling.
- Caliper logs may be required in cased hole if scale or other buildup is present in the wellbore.
- A base GR survey should be run before any tracer has been ejected into the well.
- An injection well should be initially scanned by ejecting a slug of tracer above all exits and making drag (up) runs at intervals of 2 minutes or less, depending on the velocities.
- Perform velocity shots after scanning the well.
- Velocity shots are done from the bottom up to prevent earlier shots from interfering.
- Channel checks should be done to verify channels and check for packer leaks.
- Accurate time records indicating the start of each run are helpful in reconstructing the job and in interpretation.
- If flow is detected below the lowest perforations, be sure to test if the flow is inside or outside of the pipe.
- In multiple tracer tests, be sure to select isotopes whose GR energies are widely spaced for more certain detection.
- For a velocity profile, the correction factor should be computed above the top set of perforations where the flow should equal that injected at the surface.
- The correction factor may be greater than one in a producing interval having two-phase flow.
- Check that data is not affected by a hole diameter increase.
- The intensity of the radioactivity at the detector is a function of the concentration of RA material being detected; the type of detector; and the distance, type, and thickness of material between the detector and tracer.
- The RotaScan tool can discriminate the directional orientation of GR sources downhole.
- RotaScan is most commonly used for fracture orientation.

References

1. Schlumberger, "Production Log Interpretation," Document C-11811, Houston, Texas, 1973.
2. Atlas Wireline Services, "Interpretive Methods for Production Logs," Document 9635, 4th edition, Houston, Texas, 1991.
3. Schlumberger, "Production Log Interpretation," Schlumberger Limited, Houston, Texas, 1970.
4. Roesner, R.E., LeBlanc, A.J., Strassner, J.E., and Bragg, J.R., "Visual Flow Loop Investigation of Nuclear Flolog Performance in Non-Newtonian Fluids," 23rd Annual SPWLA Symposium, July, 1982.
5. Roesner, R.E., Sloan, M.L., and Turney, R.A., "New Logging Instruments for Polymer and Water Injection Wells," 24th Annual SPWLA Symposium, June, 1983.
6. Knight, B.L., and Davarzani, M.J., "Injection Well Logging Using Viscous EOR Fluids," Paper SPE 13143, 59th Annual SPE Conference, Houston, Texas, September, 1984.

7. Atlas Wireline Services, "The Nuclear Flowmeter and Swing-Arm Tracer Injector," VHS NTSC Format Movie, Atlas Communication Services, Houston, Texas, 1984.
8. Kelldorf, W.F.N., "Radioactive Tracer Surveying—A Comprehensive Report," *JPT*, pp. 661-669, June, 1970.
9. Wylie, B.W., and Newman, G., "Trac-III: Annulus Production Logging," SPE Trans-Pecos Production Study Group, February, 1988.
10. Gadeken, L.L., Gartner, M.L., Sharbak, D.E., and Wyatt, D.F., "The Interpretation of Radioactive-Tracer Logs Using Gamma-Ray Spectroscopy Measurements," *The Log Analyst*, pp. 24-34, January-February, 1991.
11. Atlas Wireline Services, "Radiological Procedures," Document 9107, Houston, Texas, 1981.
12. Bearden, W.G., Cocanower, R.D., Currens, D., and Dillingham, M., "Interpretation of Injectivity Profiles in Irregular Boreholes," *JPT*, pp. 1,089-1,097, September, 1970.
13. Walker, T., Sherwood, J., Sumner, C., and Marshall, R., "The Fluid Travel Log," Paper SPE-701, 38th Annual SPE Conference, New Orleans, Louisiana, October, 1963.
14. Welex (Now Halliburton Energy Services), "Technical Bulletin Utility of the Fluid Travel Log," Document Number CL-2003, Houston, Texas, July, 1968.
15. Ellenberger, C.W., and Aseltine, R.J., "Selective Acid Stimulation to Improve Vertical Efficiency in Injection Wells," *JPT*, pp. 25-29, January, 1977.
16. Wiley, R., and Cocanower, R.D., "A Quantitative Technique for Determining Injectivity Profiles Using Radioactive Tracers," Paper SPE 5513, 50th Annual SPE Meeting, Dallas, Texas, September, 1975.
17. G.J. Lichtenberger, "A Primer on Radioactive Tracer Injection Profiling," Proc., 28th Annual Southwest Pet. Short Course, Texas Tech U., pp. 251-263, Lubbock, Texas, April, 1981.
18. Self, C., and Dillingham, M., "A New Fluid Flow Analysis Technique for Determining Bore Hole Conditions," Paper SPE 1752, SPE Symposium on Mechanical Engineering Aspects of Drilling and Production, Fort Worth, Texas, March, 1967.
19. Hill, A.D., and Solares, J.R., "Improved Analysis Methods for Radioactive Tracer-Injector Profile Logging," Paper SPE 12140, 58th Annual SPE Conference, San Francisco, California, October, 1983.
20. Anthony, J.L., and Hill, A.D., "An Extended Analysis Method for Two-Pulse Tracer Logging," SPE Production Engineering, pp. 117-124, March, 1986.
21. Stratton, R., Chase, R., and Schaller, H., "Case Histories of Production Logging," Paper SPE 2335, 39th Annual SPE Conference, Bakersfield, California, November, 1968.
22. Peterson, J.E., "Use of the Interface Survey in Irregular Bore Holes," Paper SPE 4021, 47th Annual SPE Conference, San Antonio, Texas, October, 1972.
23. Moon, K.E., "An Improved Radioactive Tagging System for Stimulation Evaluation," Presented at the Southwestern Petroleum Short Course, Texas Tech U., Lubbock, Texas, April, 1978.
24. Gadeken, and Smith, H.D. Jr., "Tracerscan, A Spectroscopy Technique for Determining the Distribution of Multiple Radioactive Tracers in Downhole Operations," Paper ZZ, 27th Annual SPWLA Symposium, Houston, Texas, June, 1986.
25. Gadeken, L.L., Smith, H.D., and Seifert, D.J., "Calibration and Analysis of Borehole and Formation Sensitivities for Gamma Ray Spectroscopy Measurements with Multiple Radioactive Tracers," Paper V, 28th Annual SPWLA Symposium, London, England, June, 1987.
26. Lopus, T.A., Seifert, D.J., and Schein, G.W., "Production Improvement Through Identification of Conductive Natural Fractures Utilizing Multiple Radioactive Isotope Technology," Paper SPE 16192, SPE Production Operations Symposium, Oklahoma City, Oklahoma, March, 1987.
27. Pemper, R.P., Flecker, M.J., McWhirter, V.C., and Oliver, D.W., "PRISM—Hydraulic Fracture Evaluation with Multiple Radioactive Tracers," *Geophysics*, Vol. 53, No. 10, October, 1988.
28. Abernathy, S.E., Woods, S.E., and Taylor, J.L. III, "Radioactive Tracers in Oil and Gas Production: Practical Considerations in the 1990's," Paper SPE 27236, 2nd International Conference on Health, Safety, & Environment in Oil & Gas Exploration and Production, Jakarta, Indonesia, January, 1994.
29. Gadeken, L.L., Gartner, and Sharbak, D.E., "Improved Evaluation Techniques for Multiple Radioactive Tracer Applications," 12th International Logging Symposium of SAID, Paris, France, October, 1989.
30. Gadeken, L.L., and Smith, H.D. Jr., "A Relative Distance Indicator from Gamma Ray Spectroscopy Measurements with Radioactive Tracers," Paper SPE 17962, 6th SPE Middle East Oil Technical Conference, Manama, Bahrain, March, 1989.
31. Gadeken, L.L., Ginzel, W.J., Sharbak, D.E., Shorck, K.A., Sitka, M.A., and Taylor, J.L. III "The Determination of Fracture Orientation Using a Directional Gamma Ray Tool," 32nd Annual SPWLA Symposium, Midland, Texas, June, 1991.
32. Simpson, G.A., and Gadeken, L.L., "Interpretation of Directional Gamma Ray Logging Data for Hydraulic Fracture Orientation," Paper SPE 25851, SPE Rocky Mountain Regional/Low Permeability Reservoirs Symposium, Denver, Colorado, April, 1993.
33. Hill, A.D., "Production Logging—Theoretical and Interpretive Elements," SPE Monograph Volume 14, Henry L. Doherty Series, Richardson, Texas, 1990.
34. Small, G.P., "Steam-Injection Profile Control Using Limited-Entry Perforations," Paper SPE 13607, SPE California Regional Meeting, Bakersfield, California, March, 1985.

CHAPTER 16

FLUID MOVEMENT: OXYGEN ACTIVATION AND OTHER PULSED NEUTRON APPLICATIONS

OVERVIEW

This section deals with the use of pulsed neutron capture and carbon/oxygen (C/O) devices applied to detect and measure parameters associated with fluid flow downhole. The main application is related to oxygen activation and used to detect the movement of water. However, many new applications have developed in recent years oriented toward detecting the types of borehole fluids present or locating channels. These applications offer a whole new technology of fluid movement measurements done with one tool and having the potential to work in horizontal logging environments. This section will not deal with the basics of the pulsed neutron capture (PNC) tool as it is discussed in an earlier chapter.

WATER MOVEMENT DETECTION USING OXYGEN ACTIVATION

Principle of Water Detection Using Oxygen Activation

It was observed that the background curves of a PNC logging run would occasionally respond oddly in a producing well. This was attributed to the oxygen activation effect.[1, 2] The background curves of a PNC survey may be affected by upward water flow in or around the borehole. When the neutron burst occurs, the oxygen associated with the up flowing water is activated to an unstable nitrogen isotope having a half life of 7.35 seconds. When the nitrogen isotope returns to its native oxygen, gamma rays are emitted which may be detected by the near or far background count rate measurement. Notice that the times under consideration after the neutron burst are long after the inelastic or capture gamma rays have ceased. Figure 16.1 shows a PNC tool in a wellbore with a flow profile containing water moving past the tool. The activated oxygen is shown as a bell shaped curve whose area is decaying as the water flows up past the detectors. A number of factors affect what is seen by the detectors.[3]

1. The water flow rate must be upward relative to the tool, i.e., the speed of the water must be greater than the logging speed to be detected.
2. The amount of water affects the size of the signal.
3. Relative velocity and decay rate affect the size of the signal. The maximum signal occurs at the detector when the velocity relative to the tool is:

$$V_{opt} = 5.657 \times L \qquad \text{(Equation 16.1)}$$

where L is the source to detector spacing. L is typically 1 and 2 feet for the near and far detectors, respectively. Too fast or too slow of a water movement relative to the tool may result in a weak or undetectable background signal.

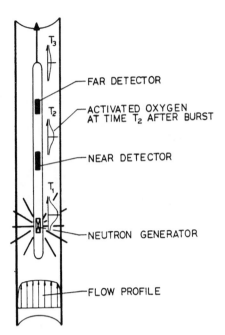

Figure 16.1 Oxygen activated water causes increased counts on the near and far detectors (Courtesy SPWLA)

N-F Background Response to Activated Water

When water is encountered moving past the tool, the effect is detected as excess counts on the near and far background count rates. The log of Figure 16.2 shows a Halliburton thermal multigate decay (TMD) over about a 300 ft (100 m) interval with three sets of perforations in the lower half of the interval.[3] On this log, the near (SS-BKG) and far (LS-BKG) background count rates increase just above the lowest and middle sets of perforations. This clearly shows the oxygen activation effect resulting from increased water flow associated with those two entry points.

The distance from the source of the water entry may be evaluated easily since the source to near detector and inter-detector spacings are about the same (about 1 ft [30.5 cm]). If these spacings are equal, the distance up the well bore that the water traveled between the near and far detections is equal to the distance from the entry point to the near detection. A small correction of about 1 ft (30.5 cm) should be added to compensate for filtering delays in the tool. [2, 3]

For example, in the log of Figure 16.2, the near background (SS-BKG) first detects the activated oxygen at point A, at a depth of Y66 feet. The far background (LS-BKG) detects the activated oxygen at Y63 feet. Assuming that the source to near and N-F spacings are equal, then the water entry is actually located at the depth Y66 ft + 3 ft + 1 in. (filtering correction) = Y70 ft.

The displacement distance, DD, the distance above the entry at which it is detected by a background gamma ray detector, may be related to the source-detector spacing, L (ft), the tool velocity, V_t (ft/min), and the water velocity, V_w (ft/min), by the following equation:

$$DD = L + \frac{L + V_t}{V_w - V_t} \qquad \text{(Equation 16.2)}$$

If in the metric system, use meters for distance and meters/min for velocities in the above equation. If displacement distance is known, then the equation above may be rewritten to solve for the water velocity.

$$V_w = \frac{DD \times V_T}{DD - L}$$

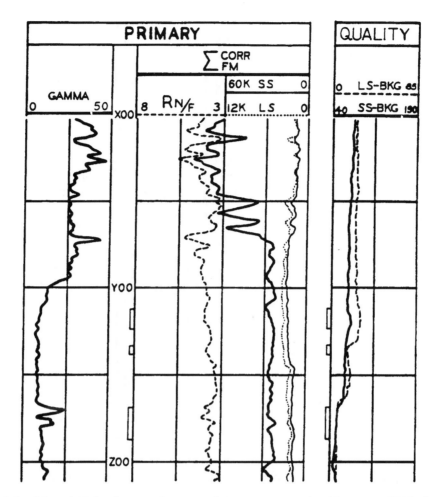

Figure 16.2 SS and LS background counts detect water entries (Courtesy SPWLA)

This technique will only work when the tool is logged in the up direction. If logged in the down direction, activation of the formation would cause excessive gamma rays and mask the water detection.

When two phases are flowing in the wellbore, the bubbles of the light phase may carry or sweep some of the heavy phase with it. If there is no or little net production of the heavy phase, it will "fall back" so that there may be concurrent flows in both directions of heavy phase fluid, even though there is no or little net heavy phase flow. If the heavy phase is water, a water up flow may be detected due to this sweeping effect caused by the lighter phase bubble flow. Consequently, water up flow indications may provide false indications under some conditions. See the chapter on fluid identification for more information on the "fall back" phenomenon.

Stationary Oxygen Activation Measurements

At present, there are two tools available for the standing oxygen activation measurement. These are as follows, with the * indicating that the designated names are trademarks of the companies indicated:

Western Atlas: Hydrolog*

Schlumberger: Water Flow Log * (WFL*)

These tools were developed at the prodding of the United States Environmental Protection Agency to detect upward moving channels in hazardous waste wells which might contaminate fresh water sands. The configuration for which these tools were designed is shown on Figure 16.3. Fluid injected into this well seeps upward through a channel adjacent to the

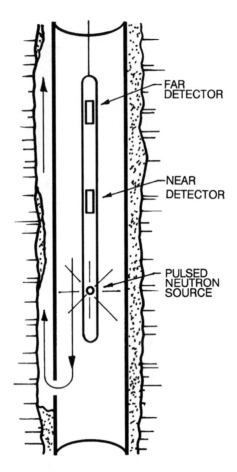

Figure 16.3 Environment for which stationary oxygen activation tools are designed

pipe. The tools, which are stationary during a test, activate the upward moving water and detect the excess counts on the near and far detectors. These tools can also detect flow in the wellbore. If the channel is downward, these tools may also be used for its detection, but must be brought out of the hole and inverted to place the detectors below the neutron source.

These oxygen activation measurements are being merged with conventional production logging tools such as spinners, fluid identification devices, and the like. It is expected that these enhance interpretation and assist in the evaluation of complex three-phase flows.[4, 5]

The Atlas Hydrolog tool[6] is a special tool with the near (SS) and far (LS) detectors optimally spaced for best detection of water movement. The Hydrolog pulses continually for 28 milliseconds after which it is turned off for 8 ms, then the cycle is repeated every 36 ms. When the pulsing is turned off, the first 1 ms will be dominated by capture events. The count rate during the following 7 ms stabilizes at a value consistent with the formation activation background counts plus those counts caused by activated water flowing past the detector. By comparison of these measured near (SS) and far (LS) count rates with those which one would get if no flow was present, then a water velocity can be calculated by the following equation:

$$\text{Velocity} = \frac{0.097215 \times (\text{Detector Spacing})}{\ln\left[\dfrac{N_{\text{counts}} - N_{\text{background}}}{F_{\text{counts}} - F_{\text{background}}}\right]} \times 60 \qquad \text{(Equation 16.3)}$$

where

$$\text{Velocity} = \text{Water velocity, ft/min}$$

$$0.097215 = \text{Natural decay constant of } N^{16} \text{ isotope}$$

$$\ln = \text{Natural logarithm}$$

Counts = Accumulated number of gamma ray counts for each detector over a fixed measurement time at the tested condition

Background = Counts as above, except under non-flowing conditions

60 = Conversion from ft/sec to ft/min

A typical logging presentation for the Hydrolog is shown on Figure 16.4. The company and well name, location, and other details are shown, along with the test information. In this example, the Oxygen SS and Oxygen LS correspond to the near and far flowing counts, and the BKG SS and BKG LS are the near and far background. This test was for 8,433 cycles at 36 ms/cycle, which means that this test took slightly more than 5.0 minutes to complete. The velocity indicated is 18.4 ft/min.

The lowest reasonably certain velocity measurable with this technique is about 3 ft/min (1 m/min). Stationary reading typically take about 5 minutes or so per station. The main problem with this technique is that the background counts measurement must usually be taken at a point different from that of the test. This is so because simply shutting in an injection well does not assure no flow at the tested interval. This is mainly a problem at low volume flow rates.

The Schlumberger approach to water flow measurement is quite different and much more like conventional tracer techniques.[7, 8] Furthermore, the Schlumberger Water Flow Log is a Dual Burst TDT with a modified pulse sequence, so that the conventional TDT information is also available while downhole. Unlike a conventional TDT logging run however, the WFL is run centralized. The tool configuration is the same as that shown in Figure 16.3. With the Schlumberger system, the N, F, and GR are used for water movement detection. The N is rarely used in practice and so another gamma ray sonde may be placed midway in the Schlumberger tool string. If this is done, the water velocity is measured by detectors spaced at about 2, 6, and 20 feet (0.6, 1.8, and 6.1 m), which affords better midrange resolution.

Company Name	EPA
Well Name	LEAK TEST WELL #1
Field Name	EAST CENTRAL UNIT
County Name	PONTOTOC
State Name	OKLAHOMA
Service Name	HYDROLOG
Bkg. File Name	EPA5.DAT
Disk File Name	EPA13.DAT
Tool Position	DNO1
Real Time	301.0
Depth	300.0
Station Number	14
Spectrum Number	1
Comment	FLOW DOWN 2-3/8 APX. 4 GAL/MIN

OXYGEN SS (cts)	BKG SS (cts)	FLOW IND. SS (cts)
17.917 ± 0.576	8.403 ± 0.395	9.515

OXYGEN LS (cts)	BKG LS (cts)	FLOW IND. LS (cts)
6.522 ± 0.348	0.242 ± 0.067	6.280

VELOCITY (ft/min)	LODR	ISS (cts)	ILS (cts)	GR (cts)	BGR (cts)
18.430 ± 4.107	28.83	3683.0	125.0	22.8	

# CYCLES	SYNCS/ CYCLE	# BKG GATES	BKG WIDTH (µs)	SPACING (ft)	SS LLD	LS LLD
8433	28	16	400.0	1.31	240	240

Figure 16.4 Data presentation for Western Atlas Hydrolog (Courtesy Western Atlas)

The operation of the Schlumberger system is illustrated on Figure 16.5. The neutron generator is turned on for either 2 or 10 seconds, then turned off. If no water flow is present, the count rate decays as shown, reaching background after about one minute. If water flow is present, then the count rate decays as before, until the activated water moves adjacent to the detector. When that occurs, excess counts are observed. After the cloud of activated water passes, the counts return to the background decay curve. With this system, the velocity is simply the distance between the source and the detector divided by the time from the middle of the burst period to the centroid of the bump associated with the excess counts.

This approach offers an interesting added piece of information. Figure 16.5 shows three bumps, all arriving at the detector at the same time. Obviously, each flow condition associated with a bump has the same velocity. The area under the bump is dependent on the volume of activated water molecules, and therefore the area under the bump can be used to determine volumetric flow rate.

The data are recorded on three detectors, typically the near (N), far (F), and gamma ray (GR). Only one will be typically optimized to provide good data. While each burst and decay sequence takes about one minute, the data collected may be highly statistical, and therefore the burst and decay sequence will typically be repeated up to about 15 times (about 15 minutes). Figure 16.6[9] shows a log test using the F detector which measure a flow velocity of 4.8 ft/min and a water flow rate of 850.5 b/d. The burst duration is indicated as 10.0 sec. and the data shown was summed over 15 bursts and decay cycles. The curve at the bottom of the chart is the excess counts above background, i.e., the smoothed difference between the observed counts and the background curve.

This technique has a minimum velocity detectable of about 3 ft/min (1 m/min). Station readings typically take up to 15 minutes. Theoretically, the measured bulk flow rate can be accurate for flow either in the wellbore or in a channel if the flowing environment is modeled properly. Modeling of the variety of downhole environments has not been done at this time.

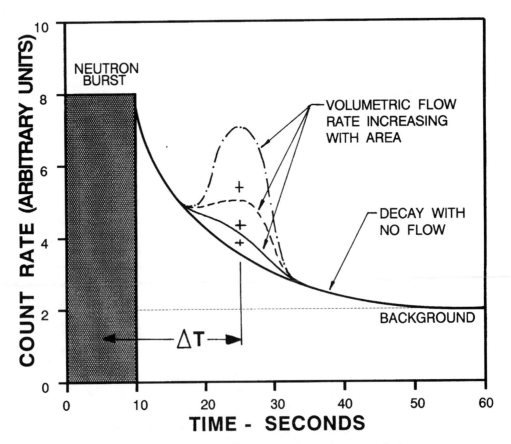

Figure 16.5 Schlumberger Water Flow Log (WFL) measurement technique

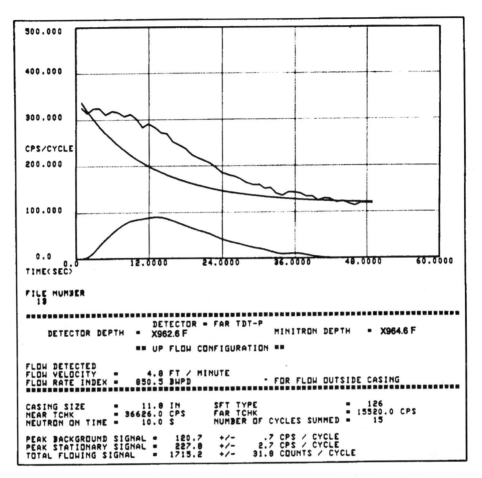

Figure 16.6 Example WFL velocity and flow rate measurement from the Far detector (Copyright SPE, Ref. 9)

The Halliburton TMD-L includes the stationary oxygen activation capability and utilizes a technique similar to the Atlas Hydrolog. The TMD-L also offers a measurement of downscatter from the high energy 6.13 and 7.12 mev oxygen activation peaks. As with the multiple tracers, an estimate of whether the observed flow is inside or outside of the casing may be made. This measurement, if it proves credible, may be important in evaluating low flow rates and discriminating channel flow from convection in the wellbore (See next section).

Effect of Convection on Stationary Oxygen Activation Velocity Measurement

Natural convection currents in a wellbore occur at temperature gradients well below the geothermal gradients commonly seen in wells. As a result, the detection of water movement, where the wellbore is filled with water, may be suspect. Such instability is much worse in deviated wells since there is a temperature gradient both along and across the wellbore.[9]

When slow flows are detected, i.e., in the range of 2-4 ft/min (.6 to 1.2 m/min), such flows may be due to convection in a standing water column. Velocities associated with convection currents also appear to increase with the local temperature gradient. If such slow water movement is observed, the interval should be reexamined for flow in the opposite direction. Detection of slow flows in both directions would indicate convection and no net flow across the interval. The assessment of whether the flow is inside or outside the well bore may also be helpful to resolve such an issue.

Convection currents have also been observed due to the logging tool itself. In one case, where the logging tool had been at a zone in the wellbore 14 minutes earlier which was 30 °F (16.7 °C) cooler, flow was indicated for 15 minutes after which such indications ceased. This effect was presumably due to convection caused by the tool until the tool finally reached a temperature close to ambient.

The above cases of convection arise in a water filled wellbore. If hydrocarbon bubbles are moving through the standing water, those bubbles entrain water and carry water uphole. A simultaneous water fallback occurs and there is no net water flow. Such entrained water may, in some cases, be detected by the oxygen activation tool. This effect is similar to thermal convection, except that the convection that is occurring is due to the passage of the bubbles.

Applications for Stationary Oxygen Activation Velocities

The stationary oxygen activation measurement can be used to measure water movement either inside casing or in a channel. The direction of the flow can be either up or down, but both cannot be done in the same trip in the hole. An application shown on Figure 16.7 uses the WFL to measure velocity of water moving up an annulus when water is injected through an injection mandrel at the bottom of the interval.[10, 11] The water is injected through tubing and a plug was set in the tubing tail. The water injection rate is 2,500 barrels per day (397.5 m³/day), and the flow area behind the tubing varied somewhat since the tubing was 3 1/2 inch (8.9 cm) along with 4.5 inch (11.4 cm) blast joints. Figure 16.7, part a, shows the neutron source stop position. Tests were made every 3 ft (.91 m). The line drawn upward from the stop point represents the distance between the source and the detector which was used to measure the velocity (either the F at 2 ft [.6 m] or the GR at 15 ft [4.6 m]). This line representation gives a sense of the vertical resolution of the velocity measurement. On Figure 16.7, part b, is shown the interpreted result, both a cumulative injection or an injection on a per foot basis.

IDENTIFICATION OF BOREHOLE FLUIDS

Phase Holdup from Borehole Capture Cross Section (Sigma)

When the produced water has a high capture cross section (sigma), it may be discriminated from oil or other hydrocarbon production by using the borehole capture cross section

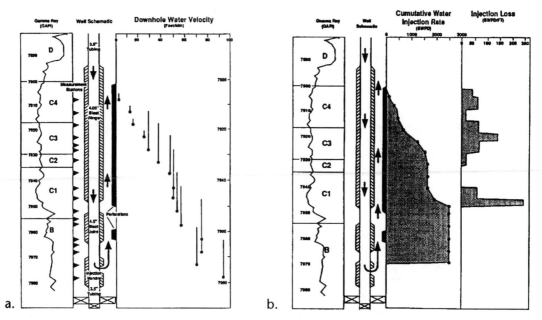

Figure 16.7 Example of measurement of water injection profile in the annulus with WFL in the tubing (Courtesy Schlumberger, Ref. 10)

curve of the PNC log.[12] The example following, shown on Figure 16.8, is a Western Atlas PDK-100 run open hole in a Mid East well.[12] The casing shoe is located at XX324. This example shows the gamma ray (GR) and formation sigma curves, the near (BKS) and far (BKL) background curves, and the borehole sigma indicator, RBOR. Notice that both the GR and RBOR are shown both flowing and shut in. Sigma was also available shut in and flowing, but both were nearly identical and only one sigma curve is shown. The BKS and BKL curves shown were recorded flowing and their shut-in counterparts are not shown. RBOR is only a relative indication of borehole sigma and not a measure of it.

In this example, the shut-in RBOR shows two distinct and reasonably constant readings. Below XX395 the reading is about 16 while above this point it is about 12.5. This is the result of a standing column of fluid which has segregated with water below XX395 and oil above. XX395 is a water oil contact in the wellbore. When flowing, RBOR reads the water value of 16 up to about XX390, above which RBOR drifts toward the oil line until going into casing. The change is a result of oil production beginning at XX390 and producing uniformly across the remainder of the interval. The relative reading of RBOR may be taken

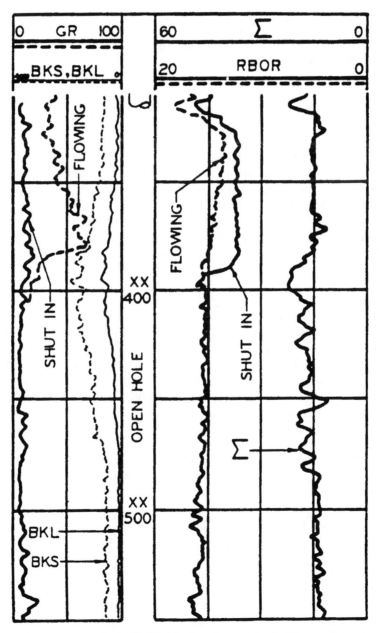

Figure 16.8 Example PDK-100 with RBOR showing water holdup and background counts showing water flow (Courtesy Western Atlas, Copyright SPE, Ref. 12)

as an indication of the fraction by volume of oil or water in the wellbore. Assuming a linear relation between the fraction by volume or holdup and the relative borehole sigma reading, then holdup may be calculated according to the equations:

$$Y_w = \frac{\Sigma_{BH} - \Sigma_{BHO}}{\Sigma_{BHW} - \Sigma_{BHO}}$$

(Equation 16.4)

and

$$Y_w + Y_o = 1 \quad \text{(2 Phase Flow)}$$

(Equation 16.5)

where

Y_o = Oil holdup (fraction by volume)

Y_w = Water holdup (fraction by volume)

Σ_{BH} = Borehole sigma indication

Σ_{BHO} = Borehole sigma indication in oil interval (shut-in)

Σ_{BHW} = Borehole sigma indication in water interval (shut-in)

For example, just below the casing shoe at XX340 where RBOR reads about 13.75, the borehole water holdup is

$$Y_w = (13.75 - 12.5)/(16.0 - 12.5) = .36$$

and the borehole oil holdup is

$$Y_o = 1.0 - .36 = .64$$

This example shows another interesting effect. The flowing background curves, BKS and BKL, increase beginning at about XX480 up to about XX380. This indicates that water is flowing in the lower part of the logged interval and it comes in at a depth somewhat below XX480. The water flow apparently increases until the effects of the oil entry cause a reduction of RBOR above XX380. The excess counts on the flowing GR are also a result of activated oxygen associated with the water flow, but the counts are not detected until much later by the GR curve since the GR sensor is many feet above the neutron source.

Phase Holdup from Carbon/Oxygen Ratio

The Schlumberger Reservoir Saturation Tool (RST) series is designed to measure formation carbon/oxygen properties even under conditions of oil flowing through an otherwise water filled wellbore. The tool measures a C/O ratio with both the near and far detectors. The combination of responses is uniquely related to formation oil saturation and wellbore oil and water holdup for a given formation and borehole environment. This effect has been discussed earlier in the chapter on Carbon/Oxygen logging. The possible area in which near and far C/O count data may fall is defined by the quadrilaterals for the RST tools shown on Figure 7.12. Such areas are unique for casing sizes, borehole sizes, and formation types. When the near and far data is plotted on such a chart, its relative position between the uppermost and lowermost boundaries defines the formation water saturation. Its relative position between the leftmost and rightmost boundaries is a measure of the oil and water holdup.[13]

For example, Figure 16.9 shows such a chart with three points designated A, B, and C.[13] If the near and far C/O data plots at point A, this indicates that the formation is wet and water is filling the borehole, $Y_w = 1.0$ and $Y_o = 0.0$. At point B, the formation has an oil saturation of about .6 and the wellbore oil holdup is about .35. At point C, the formation is again an oil zone with an oil saturation of .6 while the borehole is filled with oil, $Y_o = 1.0$ and $Y_w = 0.0$.

The RST is destined to replace the TDT-P, and therefore, this single tool potentially offers a wide variety of flow evaluation measurements. Furthermore, it has the ability to evaluate the formation in terms of sigma or C/O to further confirm that what appears as a water entry is, in fact, coming from a water bearing zone.

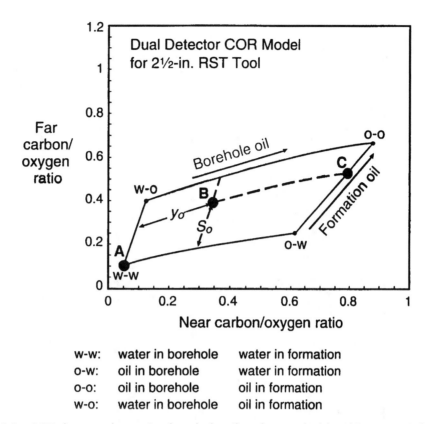

w-w: water in borehole   water in formation
o-w: oil in borehole   water in formation
o-o: oil in borehole   oil in formation
w-o: water in borehole   oil in formation

Figure 16.9   RST chart to determine borehole oil and water holdup (Courtesy Schlumberger, Ref. 13)

Ratio of Inelastic Near to Far Count Rate for Gas Holdup

Gamma Rays detected during and within a few microseconds after the neutron burst are often associated with inelastic collisions. The Atlas PDK-100 takes a ratio of such inelastic counts observed in the near detector divided by the counts of the far detector. This ratio is called RIN. When a transition is made from a water to an oil column in the wellbore, RIN is observed to become smaller. This is due to the relatively greater wellbore fluid sensitivity of the near detector. Hence, when the oxygen and chlorine associated with water are no longer present, the reduction of gamma ray counts is relatively greater on the near detector. As a result, the RIN is reduced.

When a gas is present, the effect on RIN is much more pronounced. This is due to the much greater ability of neutrons to propagate through the gas environment. As a result, relatively more neutrons are detected at the far detector and the value of RIN is greatly reduced.

This effect has been used to detect changes in gas holdup in a horizontal wellbore.[14] It was not used to quantify such holdup. The general procedure is to run a flowing and static RIN curve and overlay these. If the flowing RIN is significantly less than the RIN under static conditions, gas entry is indicated. A gas entry is shown in Figure 16.10 at about XX821.[14] This well produces 735 BOPD, 8.01 MMscf/D, and 220 BWPD.

Characterizing Three-Phase Flow with PNC Tools

Work has been done to combine the aforementioned approaches to evaluate the holdup of three phases: gas, oil, and water.[15] This work was flow loop confirmed. The technique is to use the borehole sigma, borehole C/O, and N/F count rate ratio. Note that the N/F used is not the inelastic, but the ratio of N/F in the capture mode. Recall that the N and F capture count rates separate significantly when in a gas environment and act in a very much similar manner to the N/F inelastic counts discussed in the previous section.

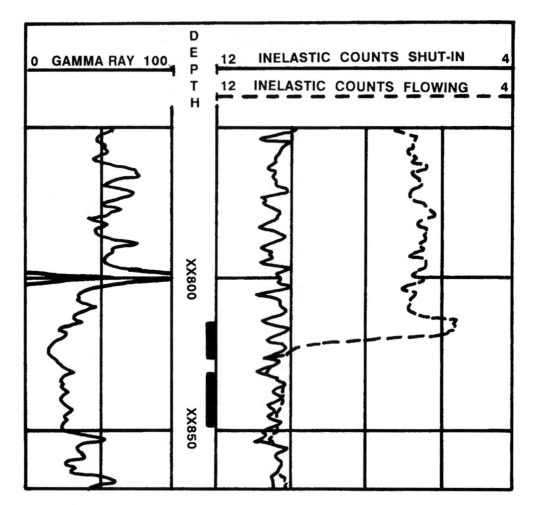

Figure 16.10 Inelastic count rate shows gas entry (Courtesy Western Atlas, Copyright SPE, Ref. 14)

The study concluded that these three parameters can be used to characterize the volume fractions of gas, oil, and water with an accuracy of 5 to 10%, depending on hole size. These measurements appeared to maintain their integrity independently of wellbore deviation angle and flow regime. Preproduction base logs are helpful and desirable. Obviously, the produced water must have significant salinity to obtain meaningful borehole sigma information. Field tests are currently being conducted on this approach.

SPECIAL PNC APPLICATIONS FOR FLUID MOVEMENT

Channel Detection Using Boron Logs

Channel detection using temperature or noise logs is often ambiguous. In certain areas, radioactive tracers cannot be used either due to safety, environmental, or political reasons. As a result, a technique based on the high capture cross section of boron has been developed[16] in Alaska to locate channels (see chapter 5 for discussion of capture cross section).

The technique uses a solution of a borax compound. These may include borax, boric acid, or sodium tetraborate pentahydrate dissolved in water. The latter, whose chemical formula is $Na_2B_4O_7$, offers the greatest increase in capture cross section per pound. The mix rate used in Alaska is 7 pounds per barrel (20 kg/m³) of seawater. The technique is a log-inject-log technique where before and after boron injection runs are compared. Saturated brine or other less common salts may be used instead of a boron solution. Boron compounds are commonly used due to high capture cross section, ready availability, and relatively low cost.

A suggested operational procedure as used in Alaska is as follows:

1. Run a baseline PNC log in a liquid filled hole.
2. With logging tool above perforations and below the tailpipe, pump in about 1 to 6 bbls of boron solution per foot (.5-3.0 m³/m) of perforations. Pump at a high rate and pressure to assure flow into channels, but stay below fracture gradient.
3. Begin PNC logging passes as soon as boron solution reaches perforations.
4. Displace the fluid into the formation and make a final PNC logging pass after such displacement to asses changes in formation sigma due to residual boron.

The example of Figure 16.11 shows the PNC base and a during/after boron injection PNC logging run.[16] The difference between the two PNC runs is shaded and is due to the increased formation capture cross section resulting from the boron solution. This comparison clearly shows that boron solution has channeled up about 50 ft (15 m) above the perforations and 80 ft (24 m) below.

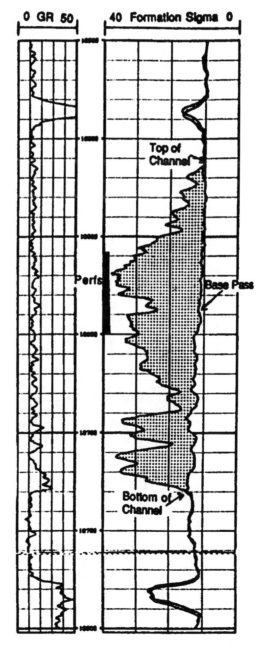

Figure 16.11 Example of a PNC log to detect injected boron solution filling a channel (Copyright SPE, Ref. 16)

Detection of Acidized Zones Using a PNC Log

HCl acid treatment causes two effects which help detect the placement of the acid with PNC logs. The first is that the acidizing will tend to increase the porosity, thereby leaving more room for higher capture cross section formation fluids. The second is that the acid residue will tend to remain in the formation for some time, and such residue contains chlorides, which are detected with a PNC tool. Figure 16.12 shows a before and after acidizing run of a TDT.[13] The separation is the increased apparent formation sigma response due to acidizing.

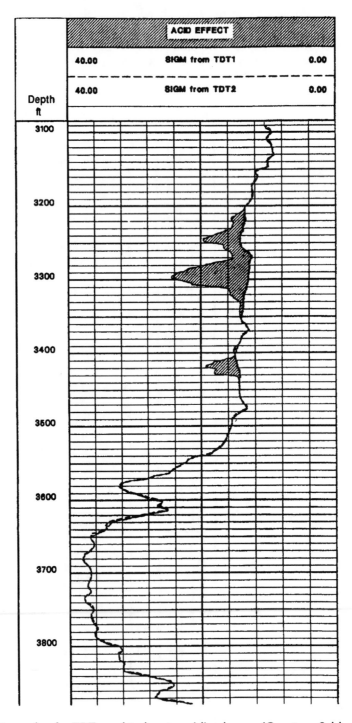

Figure 16.12   Example of a TDT used to locate acidized zone (Courtesy Schlumberger, Ref. 13)

POINTS TO REMEMBER

- High energy neutrons emitted by a PNC tool activates oxygen, creating an isotope with a half life of 7.35 seconds.
- When returning to native oxygen, a gamma ray is emitted.
- An increase in N and F background counts may be caused by activated oxygen (water) moving up past the neutron source.
- Activation logging can detect water movement either in or out of the casing.
- Stationary oxygen activation logging can detect either up or down flows, but cannot do both direction on a single trip in the hole.
- Minimum flow velocity is about 3 ft/min (1 m/min).
- The Schlumberger WFL can give indications of volumetric flow rate as well as velocity, and can make a formation sigma measurement.
- Stationary oxygen activation logs take from 5 to 15 minutes at each measure point.
- Holdup of a phase in the wellbore means its fraction by volume.
- If water salinity is high, borehole sigma measurements may provide water and oil holdup information.
- The C/O ratio can be used to measure oil and water holdups.
- The ratio of inelastic counts, a ratio such as RIN, can indicate gas entries.
- The boron log may be used to detect channels when temperature or noise logs may be ambiguous, or where radioactive tracers cannot be used.
- The PNC log is effective at locating acidized zones.

References

1. Arnold, D.M., and Paap, H.J., "Quantitative Monitoring of Water Flow Behind and in Wellbore Casing," *Journal of Petroleum Technology*, pp. 121-130, January, 1979.
2. deRosset, W.H.M., "Examples of Detection of Water Flow by Oxygen Activation," 27th Annual SPWLA Symposium, June, 1986.
3. Buchanan, J.C., Clearman, D.K., Heidbrink, L.J., and Smith, H.D. Jr., "Applications of TMD Pulsed Neutron Logs in Unusual Downhole Logging Environments," 25th Annual SPWLA Symposium, New Orleans, Louisiana, June, 1984.
4. Muttoni, A., and Cannatelli, D., "Integration of Production Log and Oxygen Activation Techniques for Diagnosing Water Production Problems," *JPT*, pp. 890-898, October, 1994.
5. Boyle, K., "Shell Canada Experience with Oxygen Activation in Three Phase Flow," 33rd Annual SPWLA Symposium, June, 1992.
6. Hill, F.L., Barnette, J.C., Koenn, L.D., and Chace, D.M., "New Instrumentation and Interpretive Methods for Identifying Shielded Waterflow Using Pulsed Neutron Technology," CWLS Formation Evaluation Symposium, Calgary, Canada, September, 1989.
7. McKeon, D.C., Scot, H.D., Oleson, J.R., Patton, G.L., and Mitchell, R.J., "Improved Method for Determining Water Flow Behind Casing," Paper SPE 20586, 65th Annual SPE Technical Conference, New Orleans, Louisiana, September, 1990.
8. McKeon, D.C., Scott, H.D., and Patton, G.L. "Interpretation of Oxygen Activation Logs for Detecting Water Flow in Producing and Injection Wells," 32nd Annual SPWLA Symposium, June, 1991.
9. Wydrinski, R., and Katahara, K.W., "Effects of Thermal Convection on Oxygen Activation Logs," Paper SPE 24739, 67th Annual SPE Technical Conference, Washington, D.C., October, 1992.
10. Scott, H.D., Pearson, C.M., Renke, S.M., McKeon, D.C., and Meisenhelder, J.P., "Applications of Oxygen Activation for Injection and Production Profiling in the Kuparak River Field," Paper SPE 22130, International Arctic Technology Conference, Anchorage, Alaska, May, 1991.
11. Schnorr, D.R., "Logs Determine Water Flow Behind Pipe in Alaska," *Oil & Gas Journal*, pp. 77-81, November 8, 1993.
12. Randall, R.R., Gray, T., Craik, G., and Hopkins, E.C., "A New Digital Multiscale Pulsed Neutron Logging System," 60th Annual SPE Technical Conference, Las Vegas, Nevada, September, 1985.
13. Schlumberger, "RST—Reservoir Saturation Tool," Paper SMP-9250, Schlumberger, Houston, Texas, June, 1993.
14. Barnette, J.C., Copoulos, A.E., and Biswas, P.B., "Acquiring Production Logging Data With Pulsed Neutron Logs From Highly Deviated or Non-Conventional Production Wells With Multiphase Flow in Prudhoe Bay, Alaska," Paper SPE 24089, SPE Western Regional Meeting, Bakersfield, California, March 30 - April 1, 1992.
15. Peeters, M., Oliver, D.W., and Wright, G.A., "Pulsed Neutron Tools Applied To Three-Phase Production-Logging in Horizontal Wells," 35th Annual SPWLA Symposium, June, 1994.
16. Blount, C.G., Copoulos, A.E., and Myers, G.D., "A Channel Detection Technique Using the Pulsed Neutron Log," Paper SPE 20042, 60th SPE California Regional Meeting, Ventura, California, April, 1990.

CHAPTER 17

FLUID MOVEMENT: BULK FLOW RATE MEASUREMENT

TYPES OF FLOWMETERS

Spinners and Other Types of Flowmeters

Nearly all flow measurements in producing wells are made with spinner type flowmeters. Other types, mainly acoustic, are discussed briefly later in this chapter. Spinner flowmeters are preferred for vertical production wells where one, two, or three phases of fluid may be present, since spinners respond to the bulk flow rate regardless of phase makeup. Spinners are also used in injection wells where the use of tracers is not common. Spinners may be classified into two basic types. The continuous spinners, both small diameter and full bore, are run continuously across the producing interval and entries are inferred from changes in the spinner rotational response.[1-4] Diverter spinners are logged in a stationary position and the flow is channeled through the tool housing for measurement. These basic spinner types are shown on Figures 17.1a,[5, 6] 17.1b,[17] and 17.1c.[17] Many varieties exist and specific tools, while functionally identical, may look different from that pictured.

Continuous Flowmeters

Continuous spinners are best suited for vertical wells with one or more flowing phases. Continuous spinners are also useful in high flow rate deviated wells where the flowing phases are well mixed. Full bore types are preferable over smaller diameter spinners since they sample a much larger fraction of the wellbore fluid mixture. If only a single phase is present downhole, continuous spinners are suitable for wells at any angle of deviation.

The small diameter continuous spinners are usually made of plastic or a light metal, and are mounted on jeweled bearings. Small diameter tools are useful for measurement of flow in the production casing and in the tubing. Such spinners are mounted either midway along or on the bottom of the tool string. Small diameter spinners should always be run centralized. Motor driven bow springs have been used in pairs to move a small diameter spinner located at the end of the tool string vertically across the wellbore in multiphase flow.

The full bore type spinners fold up for passage through tubing and do not rotate when in this position (see Figure 17.1b). When emerging from the bottom of tubing, they spring open to the full bore position and sample a large (50–75%) fraction of the borehole. The blade assemblies are available in various sizes to maximize the sampling area in various casing diameters. Some tools utilize the spinner centralizer cage as a caliper while others require a separate bow spring caliper sonde on the tool string. A caliper should always be run with a spinner in an open hole and in cased hole where scale or other buildup is expected on the inside casing wall.

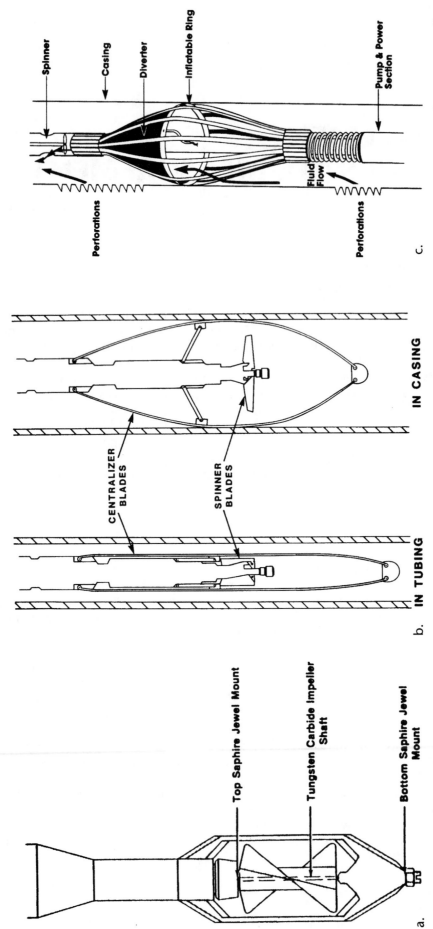

Figure 17.1 a. Spinner flowmeters: Small diameter continuous flowmeter (Reproduced with permission of Halliburton Co.), b. Spinner flowmeters: Full bore type continuous flowmeter (Courtesy Schlumberger, Ref. 17), and c. Spinner flowmeters: Diverter type flowmeter (Courtesy Schlumberger, Ref. 17)

Diverting Spinner Flowmeters

Diverting spinners are available in a number of configurations. The most common is a motorized centralizer cage with metal petals attached to the centralizer arms. When in the producing interval where measurements are to be made, the centralizer is opened and the petals form an inverted funnel to channel the flow through the tool which houses the spinner. This configuration does not seal perfectly and tends to leak, especially when gas is present. Other versions form the inverted funnel with a fabric and sealing lip (see Figure 17.1c) or may include an inflatable rubber packer bag as the primary diverting element. The fabric and packer versions are superior in term of sealing, but appear to be somewhat more susceptible to damage downhole.

Diverting spinners are run stationary at depths of interest. They are retracted and moved to another position for the next measurement. Diverting spinners are best suited for multiphase flow in vertical and deviated wells up to flowrates of about 3,000 B/D (475 m³/D) at downhole conditions, depending on the tool size and vendor. Horizontal wells pose special problems and are discussed in a later section.

SPINNER RESPONSE

Spinner Response under Ideal Conditions

Spinner responses to real flows downhole can best be understood by considering first their response under ideal downhole conditions. In this context, "ideal" means that the wellbore is filled with a fluid having no viscosity, the spinner bearings have no mechanical friction, and the tool has no cross sectional area. The response expected is that shown for zero flow in Figure 17.2. The vertical axis is scaled in revolutions per second (RPS) or hertz (hz), with the positive corresponding to, say, clockwise spinner rotation, and negative to counterclockwise rotation. The horizontal is scaled in tool velocity units, here shown in ft/minute (fpm). Tool speeds in the down direction are plotted to the right of the origin while up runs are plotted to the left.

The response of such an ideal spinner logging tool for zero flow (static conditions) downhole is a straight line passing through the origin. The slope, typically between .03 and .06 RPS/fpm (.1 and .2 RPS/[m/min]), is determined primarily by the angle of the blades with respect to the oncoming flow. In this case, the slope is .040 RPS/fpm (.131 RPS/mpm). When an upflow of 50 fpm (15.24 mpm) is encountered by the logging tool, its spinner response curve is shifted 50 fpm to the left. If an upflow of 100 fpm (30.48 mpm) is

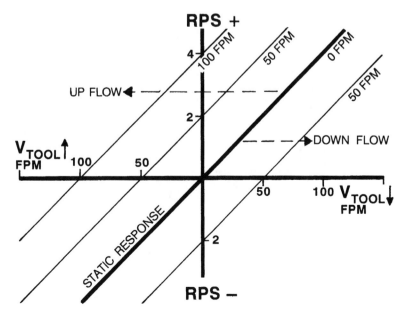

Figure 17.2 Ideal spinner response with effects of up and down flow

encountered, the response curve is shifted 100 fpm to the left. Similarly, a down flow causes a comparable shift of the response curve to the right of the static condition.

A rule is now apparent. It is that the effect of an upflow on the static response curve is to shift that response curve to the left an amount equal to the upflow velocity, and the effect of a down flow is to shift the response curve to the right an amount equal to the down flow velocity. This rule is assumed to hold true for all real flows, regardless of what the real static response curve looks like.

Spinner Response under Real Conditions

To most easily understand the real response curve of a spinner, consider what happens when each of the idealizations of the previous section is removed to create a new model response. When the fluid has a viscosity, a viscous drag occurs when the blades rotate, causing a reduction of spinner rotational speed. The greater the viscosity, the greater the reduction. However, as the rotational velocity of the blades gets higher, the response curve approaches a line parallel to the ideal static response. Compare the dashed (viscous) response with the straight line asymptote (solid) at higher speeds as shown on Figure 17.3. Note that in practice some changes in slope may be observed, especially when comparing gas and liquid flows.

Due to mechanical friction, the blades do not rotate at the slightest flow. Instead, a flow velocity of about 5 to 10 fpm (1.5 to 3 mpm) relative to the spinner is required to provide enough torque to overcome mechanical friction. This is the spinner rotational threshold. Any data gathered when the spinner is not rotating has an uncertainty in velocity of plus or minus this threshold velocity and is therefore usually not usable. The dotted curve of Figure 17.3 shows the combined effects of mechanical friction and viscosity.

Lastly, when a real tool inhabits the region of flow, the flow seen by the tool logging down is slightly different from that logging up. In the down direction, the flow is by and large unaffected by the tool string. In the up direction, the flow approaches the spinner in a more annular way. As a result, the response slope and intercept are slightly different for the up and down logging directions.

In summary, if the spinner response is approximated by a straight line, the effect of viscosity is to create an "offset" of the straight line up and down log response relative to the origin. This offset is sometimes called a "threshold velocity." While commonly used, such a "threshold velocity" is simply an offset of the linear response and not a mechanical friction threshold. This offset is not necessarily equal for the up and down logging runs, and the response slopes may be different in the up and down directions. The straight line static response curve in real conditions is shown as the two heavy lines labeled A-A and offset around the origin on Figure 17.4.[7]

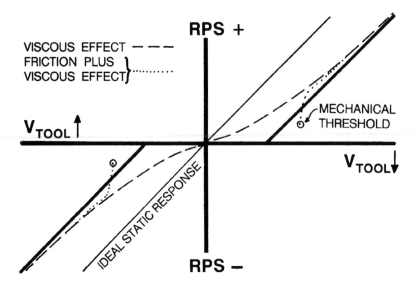

Figure 17.3 Effects of viscosity and mechanical friction on the ideal response

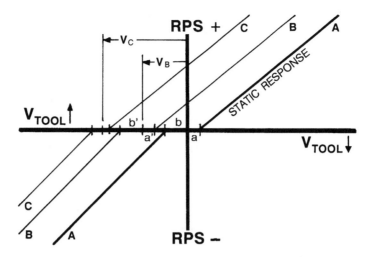

Figure 17.4 Real spinner response and effects of viscosity changes in the fluid flow

Effect of Downhole Flows on Response Curves

Even with real response curves, the rule that the fluid velocity downhole simply displaces the static response to the left or right still holds true. If the static condition response is labeled A-A, then the response for a spinner encountering a flow of velocity V_B is shown by the response B-B. Response B-B is simply displaced to the left an amount equal to VV. Notice that the separation and therefore viscosity have remained unchanged.

When the velocity and viscosity simultaneously change, the static response is displaced due to velocity and the separation of the up and down run intercepts is modified due to viscosity as shown by response C-C. To measure the velocity of the fluid with a new viscosity, the displacement of the static response must be measured more carefully.

One method to evaluate the velocity when a viscous change occurs is to locate a reference point between the static response intercepts, and assume the effect of viscosity is proportionally the same on each side. A convenient reference point is the origin. In Figure 17.4, the intercepts on the static response are displaced distances a and b from the origin. If $a'/b' = a/b$, then the origin reference point can be relocated distances a' and b' from the intercepts of C-C as shown in Figure 17.4. V_C is then equal to the displacement of the C-C reference point from the origin.[7]

An alternative method is to measure the displacements of the up, V_{uC}, and down, V_{dC}, runs separately and use the average as the fluid velocity, i.e.,

$$V_C = (V_{uC} + V_{dC})/2$$

Another alternative method is to perform a linear regression on the data points comprising A-A, B-B, and C-C. The shift of the x intercept relative to the static response is the velocity. This regression may be calculated with the following equations:[8]

$$V_i = \frac{\Sigma X \Sigma Y^2 - \Sigma Y \Sigma XY}{n \Sigma Y^2 - (\Sigma Y)^2}$$

$$m = \frac{n \Sigma XY - \Sigma Y \Sigma X}{n \Sigma Y^2 - (\Sigma Y)^2}$$

where

V_i = Velocity intercept

m = Slope of response curve, velocity units/RPS

X = Tool speed

Y = RPS

n = Number of points used for up and down runs

COMPUTING DOWNHOLE FLOW VELOCITY FROM LOG DATA

Multipass Technique

The multipass technique is the most common spinner procedure. It does not require any surface calibration, but is calibrated downhole by use of multiple up and down log passes across the production (injection) intervals of interest. The minimal program is three up and three down logging runs, with each run at a different but constant cable speed. Stationary measurements may also be taken and are used as an indicator of flow type and stability.[1,2]

A minimal suite of logs is shown on Figure 17.5. The well shown in this schematic is producing from two zones, and the rathole (A) is static. Spinner surveys are run to determine the velocity between the perforated zones and their bulk volume contributions. The logs are scaled in RPS or Hz, from 0 to 10. The up logging passes are at 40, 90, and 150 fpm while the down passes are at 50, 100, and 140 fpm. The velocity of the up runs and the down runs need not be equal to each other, and all runs should be at significantly different cable speeds. Note that some logging equipment may present spinner response in terms of counts. Since counts are linear with respect to RPS, counts can be treated just like RPS for analysis. The schematic of Figure 17.5 assumes that the equipment does not indicate the direction of spinner rotation, but only its absolute value.[7]

For purposes of quality control, it is good practice to display all of the up and down spinner surveys each on a single page. The runs in one direction should all look alike, except that they are displaced from each other due to different logging speeds. In Figure 17.5, the down runs are closely similar. The up runs, however, show a dissimilar response for the 40 fpm run. This is caused by a spinner reversal due to the flow moving up faster

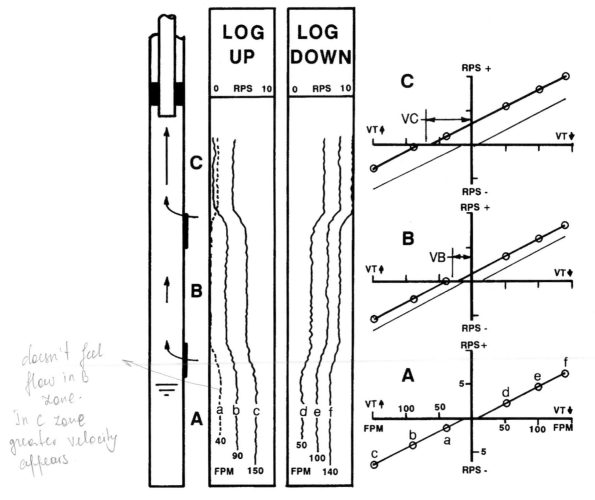

doesn't feel flow in B zone. In C zone greater velocity appears.

Figure 17.5 Multipass plot technique for flow velocity evaluation from up and down spinner surveys (After Schlumberger, Ref. 7)

than the tool. Other reasons for dissimilarity could be due to well or flow instability, inadvertent changes in cable speed, cavitation of the flow behind the spinner, grit in the spinner bearings, and the like.[9]

The response of the static rathole, interval A, is made from stable readings from each logging run. The down run readings are indicated as d, e, f, and the up run readings as a, b, and c. These data points confirm the likelihood of zero flow in the rathole due to the fact that there are offsets on each side of the origin. Choosing the origin as a reference point for velocity determination, it is located midway between the up and down intercepts. Note that spinner resolution is usually too poor to confidently confirm zero flow in the rathole. A high density or completion fluid in the rathole would confirm zero flow.

Interval B is evaluated by plotting stable readings and constructing the spinner response curves. By locating the midpoint between the B-B intercepts, the interval B response curve is displaced V_B to the left, indicating an upflow of that velocity. Note that if V_A is not zero, the velocity in interval B is $V_B + V_A$. A similar discussion can be made for velocity V_C as the velocity in interval C. Note in this case that the 40 fpm log data plots on the same curve as the down runs. This is due to the fact that the spinner has undergone a reversal and the flow is up relative to the spinner blades.

Multipass Technique in High Velocity Flowing Wells

When dealing with a high velocity flowing well, the spinner up runs may be too slow to pull through the flow, and the fluid is moving up with respect to the spinner during most of the logging run. An example of such an effect was shown on the up run at 40 fpm on Figure 17.5 in interval C. When the fluid velocity exceeded the logging tool speed, the spinner underwent a reversal. Sometimes, only down runs will be recorded in such circumstances.

Data recorded in a high velocity well is shown plotted in Figure 17.6. In this example, the static interval is labeled A, and there are three flowing intervals: B, C, and D. The measured response curves are shown for all four intervals. The velocities of each interval are

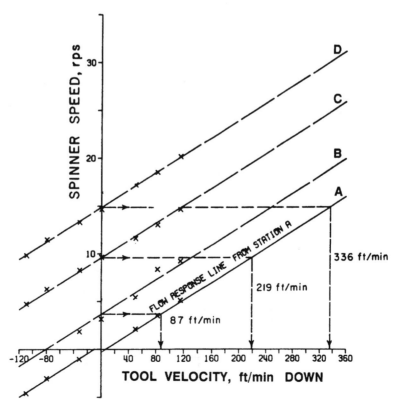

Figure 17.6 Multipass plot technique when only down logging runs are available (Courtesy Schlumberger, Ref. 1)

calculated by applying the rule developed earlier, i.e., measuring the displacement of the plotted response curves from the static response. In this case, the measurement uses only the data from the down run static curve. If velocities are low, this technique may introduce a significant error since it ignores any change in the viscous offsets between the up and down runs. However, where velocities are large, this effect is presumed to be small relative to the velocities measured.[10-12]

Two Pass Overlay Technique

The two pass overlay technique is a very easy to use technique when up and down runs are available. Since the up and down runs should each all look alike, except for being off-set due to cable speed, each up and down run contains the information necessary to compute a velocity profile. It is, however, necessary to know the spinner response curve slope for the up and down runs, B_u and B_d respectively. This is the slope of the response curves shown to the right on Figure 17.5. Typically, the slopes are computed from the response in the lowest interval, but it is good practice to examine the remaining intervals, since the lowest interval may contain both a non-representative fluid and response curve.

Figure 17.7 illustrates the use of the two pass technique.[7] First, the user must select the most representative up and down run. This could be an average or could be one that may be more stable or thought to be more representative than the others. In the schematic, the runs are overlaid in the lowest interval. The up and down runs overlaid must be at the same absolute RPS scale. While the lowest interval is usually static, it may have a non-zero flow, V_A. The velocity in any other interval, i, is then equal to

$$V_i = V_A + \frac{\Delta RPS\,i}{(B_u + B_d)}$$

where ΔRPS_i is the separation of the overlayed up and down runs in interval i.

If the fluid viscosity changes along the wellbore, the up and down runs do not plot symmetrically around a center line passing through the interval A. Suppose that the entry at

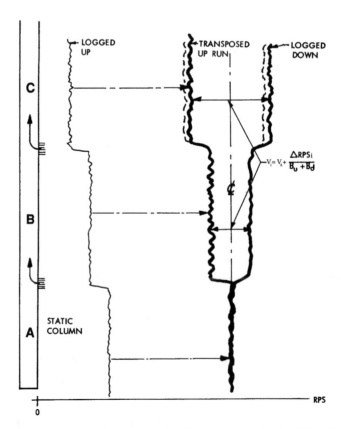

Figure 17.7 Two pass overlay technique for flowmeter analysis (After Schlumberger, Ref. 7)

the upper set of perforations of Figure 17.7 causes an increased viscosity in interval C. The effect is a reduction in the spinner rotational speed and changes the log response to that indicated by the dashed line. While such non-symmetry indicates a new fluid in the well-bore, the separation is essentially unchanged due to the viscosity effect.[7,13]

Interpretation with Only One Pass

When only one pass is available, an accurate interpretation may be made if the flow velocities are high and viscous effects are relatively small. If the response slope is not known, then the total production (injection) just below tubing must be assumed to be 100% of the surface flow rate. Figure 17.8 shows such a situation.[1] In this case, only one down run is available. 100% of the flow is assumed at the top above the top perforations. 0% is assumed below the lowest interval in the rathole. In this example, the rathole is too short to obtain an accurate zero. Instead, stationary measurements, x, are made at various depths. The difference between the moving and stationary readings is the effect of cable speed only, and this difference corresponds to the zero flow response in RPS. Notice that this treatment neglects the effects of viscosity. 100% is taken as 60 MMcf/D, and the flow is scaled linearly between that value and the zero line.[1]

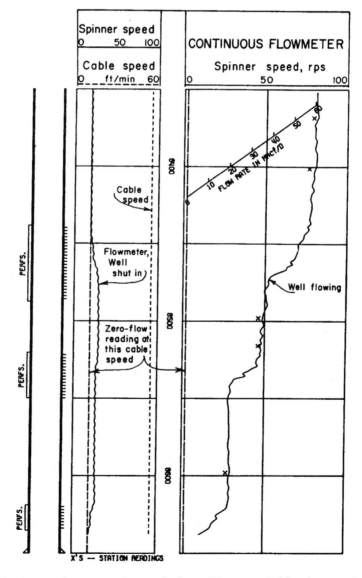

Figure 17.8 Single pass interpretation technique (Courtesy Schlumberger, Ref. 1)

COMPUTATION OF BULK FLOW RATE FROM THE SPINNER VELOCITY

Spinner Flow Profile Correction Factor

The spinner, if centralized, measures the flow velocity in the central region of the borehole, as shown in Figure 17.9. This velocity is too high and must be corrected back to provide a true average velocity. This correction depends on the flow Reynolds number, the spinner diameter, the casing diameter, fluid viscosity, and other parameters. This correction may be calculated in software packages available for production log interpretation. For purposes of computation by hand, a flow profile correction factor of C = .83 has been satisfactory in most cases.

Computation of Bulk Flow Rate

Assuming that the spinner has responded to the bulk flow rate, the bulk flow rate in an interval i can be computed from the velocity measured by the spinner using the following equation:[14]

$$Q_i \text{ (in B/D)} = 1.40 \times C \times V_i \times d^2$$

$$Q_i \text{ (in cf/D)} = 7.85 \times C \times V_i \times d^2$$

where

C = Velocity correction factor (about .83)

V_i = Flow velocity measured by spinner, ft/min

d = Internal diameter of casing, inches

or,

$$Q_i \text{ (in m}^3\text{/D)} = .1131 \times C \times V_i \times d^2$$

where

C = Velocity correction factor (about .83)

V_i = Flow velocity measured by spinner, m/min

d = Internal diameter of casing, cm

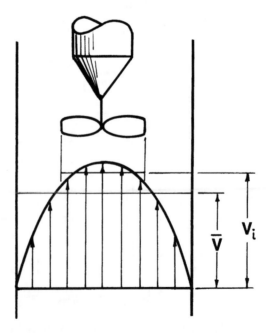

Figure 17.9   Spinner flow velocity correction

After such computations, a bulk flowing profile and entry profile can be constructed. The cumulative bulk flow profile for a well with three contributing sets of perforations is shown on Figure 17.10. The individual contributions of each set of perforations is the difference between the flows in each interval: A, B, C, and D. These contributions are shown in a bar chart representation labeled entry profile.

SPINNER LOGGING IN MULTIPHASE INCLINED FLOW

Effect of Flow Distribution

When two or more phases of differing densities are flowing in a deviated wellbore, the heavy and light phases gravity segregate. As a result, the light phase flows along the high side of the pipe and the heavy phase flows along the low side of the pipe. This situation is shown in Figure 17.11, part d, where a spinner is shown in a deviated wellbore flowing two phases.[15] The actual bulk flow profile is as shown in Figure 17.11, part a. Continuous spinner logs, however, would respond like that shown on Figure 17.11, parts b and c. In the interval just above the lowest perforations, the bubbles migrate to the top of the pipe. The light phase moves quickly along the top, carrying heavy phase uphole with it. If there is little or no heavy phase production from the lowest perforations, then the heavy phase will "fall back," causing the velocity profile across the wellbore to be as indicated. The spinner senses only the envelope through which the spinner passes. As a result, the spinner senses only downflow, even when there is a net upflow.[15]

As the flowrates get larger and the phases become better mixed, this effect diminishes. However, this effect can be severe even for downhole flowrates as high as a 1,000 B/D (150 m³/D) and higher, especially if one of the phases is gas. As a result, continuous spinners, especially those of small diameter, are usually not well suited to such flow. Better suited is the diverter flowmeter pictured on Figure 17.1c. Diverter flowmeters may be put on the same tool string with continuous flowmeters. In this way their data can be merged, with the diverter making accurate low volume measurements and the continuous or full bore flowmeter used for the higher more mixed flows.

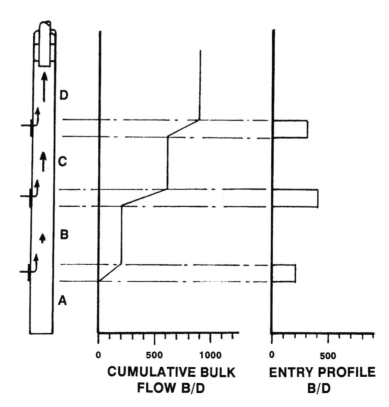

Figure 17.10 Bulk flow and entry profile presentation from spinner survey

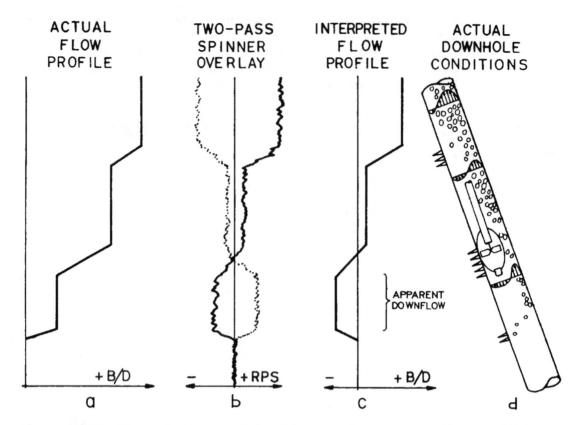

Figure 17.11 Effects of two-phase deviated flow on spinner response (Courtesy SPWLA, Ref. 15)

Effect of Well Inclination and Tool Centralization

Studies have shown that this effect of flow segregation and fallback can be severe at even small well deviation angles. Figure 17.12 shows the effects of wellbore deviation on a small continuous spinner (1.5 in. or 3.8 cm dia) held stationary at various angles of inclination in 8.5 in. (21.6 cm) pipe.[16] The flowrate is gas and oil at a rather large flowrate of 850 B/D (134 m³/D) oil plus 7.9 Mcfd (224 m³/D) gas. The response of the centralized tool holds up rather well up to about 10 to 15 degrees inclination. However, if the tool is not centralized, the response indicates a downflow, even at only a 2 degree angle wellbore! Full bore flowmeters are better due to their centralization and larger cross sectional area sampled, but they are still subject to this effect. Best are the diverter flowmeters which sample nearly all of the flow.[16]

Response of Diverter Flowmeters to Multiphase Inclined Flow

Diverter flowmeters divert the flow through a test section within the tool. There the flow moves much more rapidly, is well mixed, and its flowrate is measured by a spinner impeller. The effects of wellbore deviation are minimized as indicated by the comparison of a single and two-phase water/kerosene flow at a 25 degree inclination as shown in Figure 17.13a. This data reflects the response of a fabric basket type tool with a sealing lip which diverts 100% of the flow through the test section. This data is from the Schlumberger Inflatable Diverter Flowmeter*, IDF*, (* indicates mark of Schlumberger shown in Figure 17.1, part a. Metal petal baskets, while much better than continuous flowmeters, do not approach such linearity. The response of the fabric diverter for gas and various gas mixtures is shown on Figure 17.13b.[17] Response curves such as those of Figures 17.13a and 17.13b are available for most diverter tools. To compute bulk flow rate, it is simply a matter of setting the tool, measuring the RPS response in relatively known flowing conditions, and picking the corresponding bulk flow rate from the appropriate chart.[17-19]

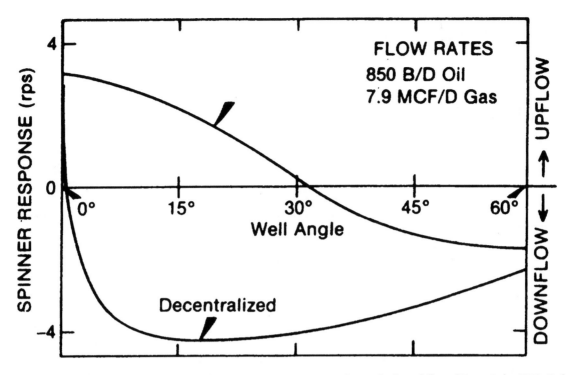

Figure 17.12   Actual small diameter spinner response to two-phase deviated flow (Copyright SPE, Ref. 16)

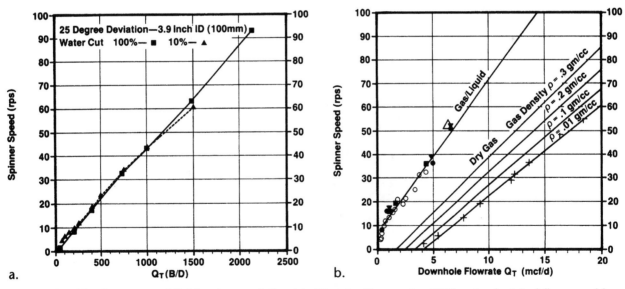

Figure 17.13   Response of Schlumberger Inflatable Diverter Flowmeter (IDF) a. to deviated flows and b. to gas flows (Courtesy Schlumberger, Ref. 17)

The diverting flowmeters, while much of an improvement, are somewhat less robust mechanically than their continuous counterparts. Depending on the tool design, size, and service company, diverting flowmeters are limited to no more than about 3,000 B/D (475 m³/D).

OTHER FLOWMETER TECHNIQUES

Horizontal Spinner

The horizontal spinner is unusual and not run by major companies. A schematic of the horizontal spinner is shown on Figure 17.14.[20] It is run continuously across the perforated interval and is not designed to respond to vertical changes in flow. Instead, it responds to lateral jetting such as gas from a perforation. This jet impinges onto the tool housing and is diverted into a spinner which jerks to a high response momentarily at the point of entry and quickly dies away after the tool passes. This tool is useful to locate active perforations, and a log example is shown on Figure 17.15.[20]

Acoustic Flowmeter

The acoustic flowmeter was developed through the Gas Research Institute and is useful to determine the gas velocity in low pressure and low flowrate gas wells. This tool measures the acoustic velocity of the gas medium and the time of flight of an acoustic signal parallel to the wellbore axis, from which a flow velocity profile can be constructed. This tool will not work when two or more phases of fluid are present.[21]

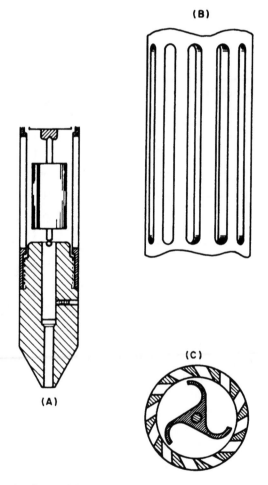

Figure 17.14   Horizontal spinner (Courtesy Southwestern Texas Short Course, Texas Tech University, Ref. 20)

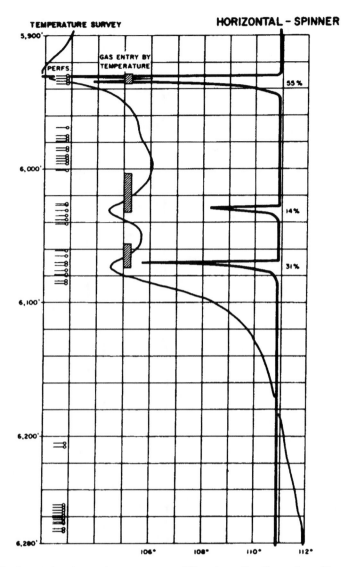

Figure 17.15 Horizontal spinner log response (Courtesy Southwestern Texas Short Course, Texas Tech University, Ref. 20)

Ultrasonic Two-Phase Measurements

Research is currently under way to assess the utility of ultrasonic signals for two-phase flow evaluation. Reflections of lateral bursts of an acoustic beam are detected to map out the distribution of bubbles around the tool. Reflections of ultrasonic signals parallel to the tool yield a distribution of velocities by a Doppler shift technique. This is a brute force approach to measuring two-phase flow and is in only the early stages of research.[22]

LOG EXAMPLES

Repeatability and Quality Control

The suite of three down spinner runs on Figure 17.16 shows good repeatability.[14] Such runs should always be shown side by side as a quality control measure. In this case notice that the separation between runs 1 and 2 is greater than between runs 2 and 3 below about 12,750 ft, but that above this depth the separations are about equal. This indicates that the second perforated interval, from 12,710 to 12,750 ft is not stable, contributing relatively more for run 3 than run 2.[14]

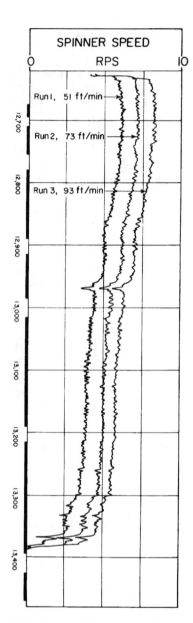

Figure 17.16   Spinner repeatability for quality control (Courtesy Schlumberger, Ref. 14)

Continuous and Diverter Spinner in a Water Cut Gas Well

The condition of gas with water cut is probably the worst possible condition to log with a flowmeter in a vertical or deviated well. Such a condition is shown on the log of Figure 17.17.[17] This well was making about 1.6 MMscfd gas and 640 B/D water. The continuous flowmeter is a full bore flowmeter in a 7 inch, 20 degree deviated borehole. The up and down spinner runs are overlaid in the lowest (static) interval. The spinner runs exhibit a high variability with the indication of downflow appearing in some cases between the lowest and middle set of perforations. Results from a fabric diverter are also shown.

Two-Pass Overlay with Downflow

The log of Figure 17.18 shows a two-pass overlay of up and down runs.[23] In this case, the logs cross over and only show the correct relationship for upflow at and above the perforations at D. Where the negative separation occurs, the overlay indicates downflow. The rathole, below perforation I, is static.[23]

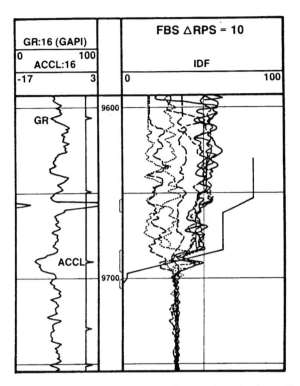

Figure 17.17 Diverter and continuous spinners in deviated multiphase flow (Courtesy Schlumberger, Ref. 17)

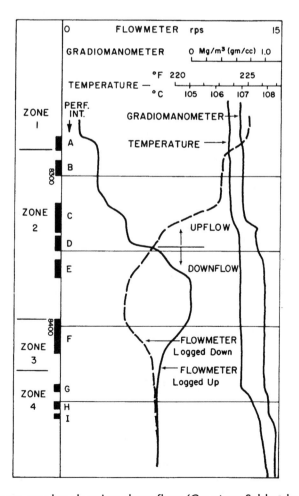

Figure 17.18 Two pass overlay showing downflow (Courtesy Schlumberger, Ref. 23)

POINTS TO REMEMBER

- The well should be stable when the logs are run.
- Determine the downhole volumetric flow rates to select tool.
- Small diameter continuous spinners are suitable for vertical flows in tubing and casing.
- Small diameter spinner must be centralized.
- Full bore spinner best for multi-phase well mixed flows in vertical and to a lesser extent inclined flows.
- Diverter flowmeters are best for inclined multi-phase flows up to about 3,000 B/D (475 m³/D).
- Diverter flowmeters take stationary measurements.
- Continuous flowmeters are suitable for all ranges of flow velocities.
- Spinners measure bulk flow volume, regardless of phase makeup.
- Continuous spinners are not well suited for low flowrate inclined flows.
- Three up and three down runs are typical for continuous spinners.
- Cable speed must be recorded.
- The effect of an up flow is to shift the static response curve to the left an amount equal to the flow velocity.
- The effect of a down flow is to move the static response curve to the right an amount equal to the flow velocity.
- All of the up runs and all of the down runs should look alike.
- Service company spinner response and flowrate calibration curves are necessary to interpret a diverting flowmeter.
- Techniques which make interpretations using only down log runs neglect the effects of viscosity and are suitable only for high flowrate or single-phase wells.
- A caliper must be run with continuous spinners in open holes or in cased holes where there may be scale or other buildup on the pipe walls, or swelling of the casing over the perforated interval.
- Calipers are not required for basket flowmeters if the basket opens all the way to the wall.
- Diverter and continuous full bore spinner data may be merged in highly deviated multi-phase flow wells.

References

1. Schlumberger, "Production Log Interpretation," Document No. C-11811, Houston, Texas, 1973.
2. Atlas Wireline Services, "Interpretive Methods for Production Well Logs," Document 9635, Houston, Texas, 1991.
3. McKinley, R.M., "Production Logging," Paper SPE 10035, International Petroleum Exhibition and Technical Symposium, Beijing, China, March, 1982.
4. Hill, A.D., "Production Logging—Theoretical and Interpretive Elements," SPE Monograph Volume 14, Henry L. Doherty Series, Richardson, Texas, 1990.
5. Piers, G.E., Perkins, J., and Escott, D., "A New Flowmeter for Production Logging and Well Testing," Paper SPE 16819, 62nd Annual Technical Conference, Dallas, Texas, September, 1987.
6. Gearhart (now Halliburton Energy Services), "Production Logging Seminar," Document No. WS-774, Fort Worth, Texas, 1977.
7. Peebler, R., "Multipass Interpretation of the Full Bore Spinner," Document SMP-1061, Schlumberger Well Services, Houston, Texas, 1978.
8. Gearhart (now Halliburton Energy Services), "Production Logging," Document G-1975 R1-87, Fort Worth, Texas, 1975.
9. Carlson, N.R., and Barnette, J.C., "Judging the Quality of Continuous Spinner Flowmeter Logs," 31st SPWLA Annual Symposium, June, 1990.
10. Meunier, D., Tixier, M.P., and Bonnet, J.L., "The Production Combination Tool—A New System for Production Monitoring," Paper SPE 2957, 45th Annual Fall SPE Meeting, Houston, Texas, October, 1970.
11. Wade, R.T., Cantrell, R.C., Poupon, A., and Moulin, J., "Production Logging—The Key to Optimum Well Performance," *JPT*, pp. 137-144, February, 1965.
12. Davarzani, J., and Sloan, M., "Analysis of Geothermal Wells Using Simultaneous Logging Instruments," Paper SPE 16818, 62nd Annual Technical Conference, Dallas, Texas, September, 1987.
13. Pande, K.N., "An Approach to Evaluate Flow and Amount of Fluid (Oil) in Different Abnormal Conditions Using One Up and One Down Run of Flowmeter Log," 26th Annual SPWLA Symposium, June, 1985.

14. Leach, B.C., Jameson, J.B., Smolen, J.J., and Nicolas, Y., "The Full Bore Flowmeter," Paper SPE 5089, 49th Annual SPE Fall Meeting, Houston, Texas, October, 1974.
15. Smolen, J.J., "Cased Hole Logging: A Perspective," *The Log Analyst*, pp. 165-174, March-June, 1987.
16. Hill, A.D., and Oolman, T., "Production Logging Tool Behavior in Two-Phase Inclined Flow," *JPT*, pp. 2,432-2,440, October, 1982.
17. Piers, G.E., "Inflatable Diverter Flowmeter," 30th Annual SPWLA Symposium, June, 1989.
18. Roesner, R.E., LeBlanc, A.J., and Davarzani, M.J., "Effects of Flow Regimes on Production Logging Instruments' Responses," Paper SPE 18206, 63rd Annual SPE Technical Conference, Houston, Texas, October, 1988.
19. Carlson, N.R., Barnette, J.C., and Davarzani, M.J., "Using Downhole Calibrations for Improved Production Profiles in Three-Phase Flows," 29th Annual SPWLA Symposium, June, 1988.
20. Kading, H.W., "Horizontal-Spinner, A New Production Logging Technique," Southwestern Petroleum Short Course Association, Texas Tech University, Lubbock, Texas, April, 1975.
21. McBane, R.A., Campbell, R.L. Jr., and DiBello, E.G., "Acoustic Flowmeter Field Test Results," Paper SPE 17722, SPE Gas Technology Symposium, Dallas, Texas, June, 1988.
22. Morris, S.L., and Hill, A.D., "Ultrasonic Imaging and Velocimetry in Two-Phase Pipe Flow," ASME Energy Sources Technology Conference, New Orleans, Louisiana, January, 1990.
23. Anderson, R.A., Smolen, J.J., Laverdiere, L., and Davis, J.A., "A Production Logging Tool With Simultaneous Measurements," Paper SPE 7447, 53rd Annual Fall Conference, Houston, Texas, October, 1978.

CHAPTER 18

FLUID MOVEMENT: FLUID IDENTIFICATION AND MULTIPHASE FLOW

DOWNHOLE FLOW CONDITIONS

Bulk Flowrate

In producing wells, it is quite common to encounter two- and even three-phase flow downhole. Ideally, the flowmeter measures the bulk total flowrate regardless of the number and distribution of phases present. As well inclination angle becomes significant in multiphase flow, some types of flowmeters deviate from this ideal. However, an appropriate mix of flowmeter tools can usually provide the necessary bulk flowrate information necessary. The measurement of the bulk flowrate profile indicates the size of the fluid entries, but not the type of fluid. In order to evaluate the phase entering, e.g, water, oil, or gas entries, another measurement is needed. That measurement is the fraction by volume of the fluid phases present in the wellbore. This is the phase holdup. Fluid identification devices (FIDs) are used for this purpose.

Flow Patterns Downhole

A wide variety of flow patterns may exist downhole. Figure 18.1 shows a map of such patterns for a liquid/gas mixture in vertical flow.[1] Needless to say, these become quite different in deviated and horizontal flow conditions.[1,2] The map is scaled in terms of superficial velocity of the liquid, V_{LS}, and the superficial velocity of the gas, V_{GS}. Superficial velocity is defined as the volumetric flowrate of either the liquid or gas phase divided by the cross sectional area of the pipe through which the mixture is flowing. Therefore,

$$V_{LS} = Q_L/A$$
$$V_{GS} = Q_G/A$$

where

Q = Volumetric flowrate of phase L or G

A = Cross section area of the mixed flow

Comment on Slugging Flows

With gas/liquid systems, it is quite common to flow in slug, churn, or annular regimes. Note that these are not nearly as uniform as pictured in the figure. Instead, the flows are quite violent and cause spinner responses to be quite erratic. Liquid/liquid (water/oil) systems tend to be more like as pictured, with bubble flow common over a wider range. If slugging flow in a water/high viscosity oil system is encountered, it can easily be identified

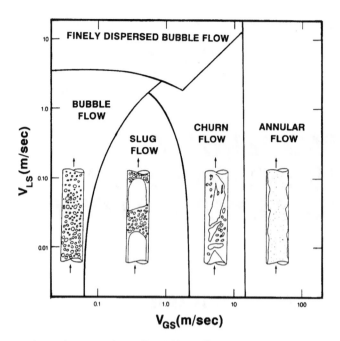

Figure 18.1 Flow regimes in two-phase liquid/gas flow (Courtesy Western Atlas, Copyright SPE, Ref. 1)

from the log. The spinner will respond with a periodic square wave response as the spinner encounters each individual slug or bullet. The FID log will produce a similar signature. If the bulk flow is upward, the square wave frequency will be higher when the log is run against the flow in the down direction.

Custom Characterization of Production Logging Tools

The accurate understanding and measurement of the parameters necessary for correct multiphase flow evaluation discussed in this and the previous chapter remain elusive. In deviated multiphasic conditions, especially with gas present, computed answers may not be much better than educated guesses. Service companies can simulate anticipated downhole conditions of multiphase deviated flow in test fixtures. Here the response of the spinners and FIDs can be studied under simulated downhole conditions. Such tool characterization is often the only way to have a reasonable assurance of good quality answers. Needless to say, such characterization is not economic for one or two logging jobs, but would be advisable if a single service company is to provide field-wide production logging.

PHASE HOLDUP AND CUT
Fluid Identification Devices Generally

Fluid identification devices are capable of making some measurement of the bulk flow. This measurement is useful to determine the fraction of volume, also called "holdup," of fluid phases present in the flowing stream. If only two phases are present, then a single FID, if properly selected, may be adequate to accurately evaluate the holdup of each phase. If three phases are present, then at least two FIDs are required to make a similar assessment. The main types of FIDs include devices which measure bulk fluid density. Certain devices also measure the dielectric constant of the fluid mixture. Previous chapters indicated that pulsed neutron capture (PNC) logs may measure a borehole sigma or perhaps an inelastic count ratio parameter such as RIN (PDK-100). Carbon/oxygen (C/O) logs, such as the Reservior Saturation Tool (RST), are capable of measuring the holdup based directly on C/O counts from the near and far detectors. At present, it would appear that holdup, at least in two–phase flow, can be measured accurately and with confidence.

Understanding How FIDs Work

Figure 18.2 is useful to understand how FIDs are useful to locate entries.[3] On the left, the well sketch shows five entries and six intervals including the rathole. An analysis of flowmeter data provides just this information. On the right is an FID bulk density measurment of the wellbore fluids. In this example, three phases are indicated with densities of water, oil, and gas of 1.0, .7, and .2 gm/cc, respectively. A density reading of 1.0 gm/cc in interval A-B (between entries A and B) indicates that entry A is 100% water. A similar conclusion can be drawn for entry B. In interval C-D, the density is .7 gm/cc. Clearly, this cannot be only an oil entry. There is no way to mix a finite amount of oil at .7 gm/cc with water flowing up at 1.0 gm/cc and come up with a mixture density of .7 gm/cc. The entry at C must contain gas and may contain oil and water. Above C, three phases may be flowing, and a second FID is necessary to solve the holdups of each phase. Entry D can be 100% oil, or it can be water and gas, or it can be all three phases. The entry at E must clearly contain gas, but again may contain other phases in appropriate proportions.

Phase Holdup

The holdup of a fluid phase is the fraction of volume of that phase present in the flowing stream. It is critical that the difference between "holdup" and "cut" be understood. Consider the schematic of Figure 18.3. At the left is a vertical pipe containing water and oil. A valve is located at the bottom of the pipe. Suppose that the valve is opened and that oil is pumped into the pipe as shown on the right side of the figure. After a while, when the interface moves up and stabilizes at a higher level, what phase is pouring out the top? The oil bubbles up through the water and coalesces with the oil column at the interface. Oil is being produced with no water. If this was a well, it is producing 100% oil cut and 0% water cut.

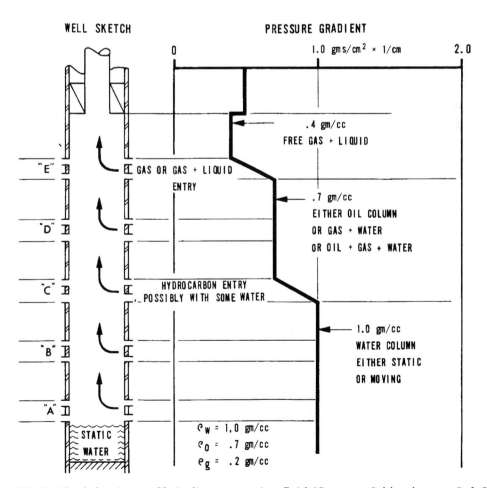

Figure 18.2 Fluid density profile indicates entering fluid (Courtesy Schlumberger, Ref. 3)

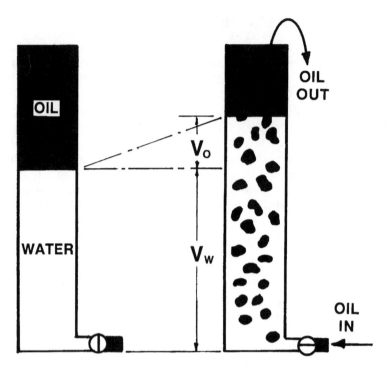

Figure 18.3 Fixture to illustrate difference between "holdup" and "cut"

Now, consider what a logging tool sees. If the tool responds to the amount of each phase present by volume, the tool clearly sees both water and oil if placed below the oil/water interface in the bubble flow. The distance times the pipe area to the original interface position is the water volume, V_w. The distance that the interface moved up in the pipe times the pipe area corresponds to the oil bubble volume, V_o. The water holdup, Y_w, is

$$Y_w = V_w/(V_w + V_o)$$

and the oil holdup, Y_o, is

$$Y_o = V_o/(V_w + V_o)$$

and

$$Y_w + Y_o = 1$$

Judging from the schematic, the FID sees about 30% oil holdup and 70% water holdup, even though the flow is 100% oil cut.

Why the difference? The difference is due to the fact that each phase moves at a different velocity. While the water moves at an average of zero velocity, the oil phase moves up at a velocity called the "slippage velocity" or "slip velocity." Another way of looking at holdup is to imagine that each phase occupies a fraction of the mixture flowing cross section equal to its holdup. This is sort of like two separate flows in tubes of differing sizes.

Relation Between Holdup and Cut

Figure 18.4 shows how cut and holdup might interact in a well.[3] This chart is for a specific set of fluids indicated on the figure and for flow in a six inch (15.24 cm) hole. Note that 1,000 B/D = 159 m³/D. When flowrates are high, say over 10,000 B/D, the difference between the phase velocities (slip velocity) is not large relative to the bulk flow velocity. As a result, holdup and cut approach each other. However, when the flow is low, say 200 B/D, then even only a 1% water cut well can have a holdup of water downhole of nearly 70%! As a result, in most lower flowrate wells, large holdups of water are often detected downhole. This is more common in oil producing wells since gas wells tend to have relatively much larger light phase volumes downhole.

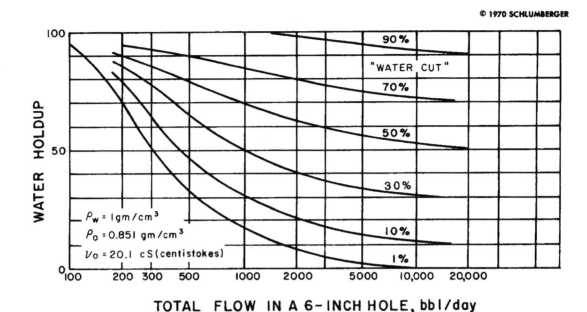

Figure 18.4 Relation between hold and cut over a range of flow rates (Courtesy Schlumberger, Ref. 3)

FLUID IDENTIFICATION DEVICES

Pressure Differential Fluid Density Devices

The bulk fluid density in the wellbore is measured by a number of different techniques. These include both pressure differential devices and nuclear devices which measure gamma ray attenuation between a source and detector. If three phases are present, the bulk density is a linear function of the density of each phase and the holdup, i.e.,

$$\rho_B = \rho_O Y_O + \rho_W Y_W + \rho_G Y_G$$

where

ρ_B = Bulk density of fluid mixture

ρ_W = Density of water phase

ρ_O = Density of oil phase

ρ_G = Density of gas phase

Y = Holdup, fraction by volume, of phase W, O, or G

The earliest pressure differential device for measuring bulk density is the Gradiomanometer* (* mark of Schlumberger), shown on Figure 18.5.[3] This device measures a pressure differential by means of pressure sensitive bellows spaced over about a 2 foot (60 cm) interval. The response is linear and the tool is calibrated on the surface in air (0.0 gm/cc) and fresh water (1.0 gm/cc). Newer pressure differential tools utilize a quartz crystal pressure differential gage. The gage accesses the flowing stream by means of an oil filled capillary tube both above and below, with such access ports spaced about 2 feet apart.[3-5]

Note that the pressure differential seen by such tool is a function of the bulk density plus a pressure drop along the pipe length due to friction. This friction effect may become significant when flow rates become large. For example, a correction of about 5% is required in 5 1/2 in. (14 cm) casing at flowrates of 14,000 B/D (2225 m³/D).[6] Consult the service company for a correction appropriate to the tool used if flow rates warrant it.

Pressure differential tools must be adjusted for wellbore inclination. The gradiomanometer required a simple correction by multiplying the reading by the cosine of the well angular deviation from vertical. However, the newer capillary tube type tools must

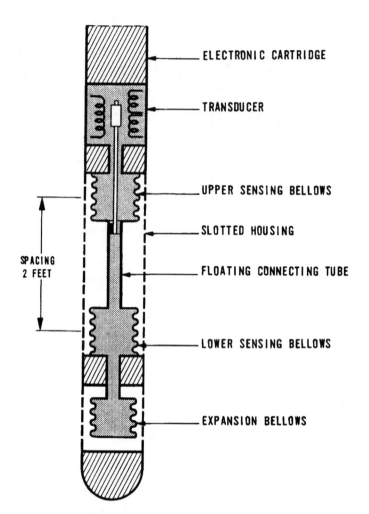

ELECTRONIC CARTRIDGE

TRANSDUCER

UPPER SENSING BELLOWS

SLOTTED HOUSING

SPACING
2 FEET

FLOATING CONNECTING TUBE

LOWER SENSING BELLOWS

EXPANSION BELLOWS

Figure 18.5   Schlumberger gradiomanometer (Courtesy Schlumberger, Ref. 3)

be corrected using a somewhat more complex equation which takes into account the density of the fluid in the capillary tubes. Consult the service company for proper correction in inclined wells. When the well is highly deviated or horizontal, tools of this type are no longer useful.

Nuclear Densimeters

Density is also measured using a nuclear densimeter type of tool. There are two basic types, the focused or collimated, and the non-focused. A focused nuclear density device is shown on Figure 18.6.[7] A cesium 137 gamma ray source is set up to focus a beam of gamma ray across a gap of about 4 inches (10 cm). This gap is open to the bulk fluid flow and hopefully representative of it. The counts at the detector are calibrated in terms of bulk density, and the output on the log is a gram/cc density of the bulk fluid. The greater the density, the fewer the counts at the detector.[7, 8]

The non-focused type also uses a cesium 137 source, but emits gamma rays in the region around the tool and depends on scattering of gamma rays back to the detector. Imagine these tools investigating an ellipsoidal shaped region between the source and detector. These tools are most often used for gravel pack evaluation. However, by adjusting the spacing between the source and detector, the tools can be "tuned" to respond mainly to borehole fluid density. In highly deviated and horizontal wells, the focused tools are of little use since they detect only the fluids cutting the gamma ray beam. If the flow is segregated, this is the heavy phase. The non-focused tools are the only density devices which can be used in highly deviated and horizontal wells.

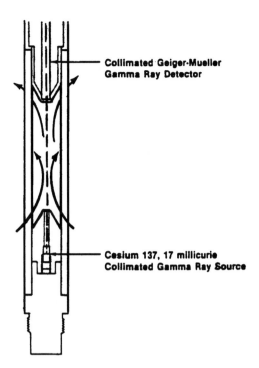

Collimated Geiger-Mueller
Gamma Ray Detector

Cesium 137, 17 millicurie
Collimated Gamma Ray Source

Figure 18.6 Focused nuclear fluid density tool (Reproduced with permission of Halliburton Co., Ref. 7)

Fluid Capacitance Devices

The fluid capacitance instrument is shown on Figure 18.7.[7] This embodiment is a short tool, about 6 inches (15 cm) in length along the test section. The holes at the ends of the tool test section allow the flowing mixture to enter and pass through the tool. The test section is essentially an annular capacitor, with the core and outer jacket serving as plates of the capacitor. The output of the instrument is frequency, with lower frequencies indicating water, higher frequencies hydrocarbons.

The basic response of the fluid capacitance is described by the following equation:

$$C = Y_O D_O + Y_W D_W + Y_G D_G$$

where, at room temperature,

$$D_W = \text{Dielectric constant for water, 78}$$

$$D_O = \text{Dielectric constant for oil, 4}$$

$$D_G = \text{Dielectric constant for gas, 1}$$

At downhole conditions, the spread between water and the hydrocarbon phases becomes less, but is still significant.

Actual response of the capacitance tool to flows of water and kerosene in a surface flow loop are shown in Figure 18.8.[8] Over the range of water holdups from zero to about 40%, the capacitance tool shows a good dynamic response. Above 40% water holdup, the tool has poor resolution with a very narrow response range. This tool is well suited to higher flowrate wells where the phases are well mixed, but poorly suited to measure holdup in lower flow wells.

The capacitance device pictured relies on the fluid mixture ramming its way into the annular space and providing a representative sample. Other devices exist which are longer in length and which rely on pressure differential differences between the inside and outside of the tool to cause the well fluids to circulate through the annular region. Such circulation continues until the inside and outside pressure differentials match. One problem with such tools is that when three phases are present, there is not a unique combination of three phases needed to provide a specific pressure drop, and hence this tool may yield

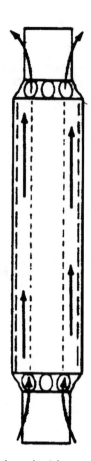

Figure 18.7 Capacitance tool (Reproduced with permission of Halliburton Co., Ref. 7)

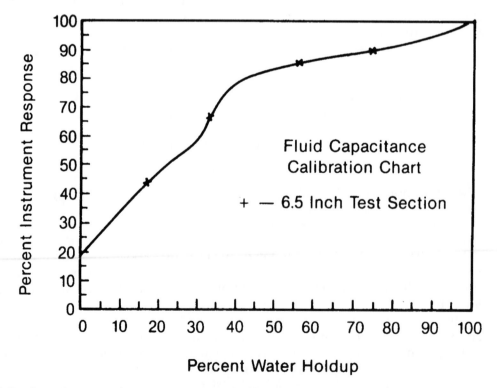

Figure 18.8 Capacitance tool response to water holdup (Courtesy Western Atlas, Ref. 8)

questionable results. In highly deviated and horizontal wells, it would appear that all capacitance devices have a problem in that they do not have access to all flowing phases, since the light phase will move along the high side of the well while the capacitance device is immersed in the heavy phase.[7-12]

Gas Holdup Tool

Halliburton is currently introducing the Gas Holdup Tool*, (GHT*) (* mark of Halliburton). The GHT configuration is similar to a non-focussed nuclear fluid density tool, except that it uses a low energy Cobalt 57 gamma ray source (122Kev). The source detector spacing is only about 1 inch (2.5 cm). The tool measures scattering of the low energy gamma rays. Unlike the nuclear density devices, the greater scattering and hence count rate is associated with higher density. The tool has very little sensitivity to cement or formation since such low energy gamma rays, if they pass outside of the pipe, do not have adequate energy to return. Furthermore, this tool is primarily responsive to the contrast in electron density. Since water and oil are similar in this respect, the tool can discriminate between liquids and gas, hence the name Gas Holdup Tool. Halliburton also has indicated that this tool provides accurate results in horizontal wells.[13]

PNC and C/O Tool Measurements

As discussed in the PNC and C/O sections and earlier in this chapter, the PNC and C/O measurements may prove useful to production logging, both in terms of the holdup measurements as well as water velocity using oxygen activation. As a result, the PNC tools have become combinable with the production logging tool string. If the borehole water is salty, then the borehole sigma can be used as a holdup indicator. Assuming that holdup is linear with respect to capture cross section, the equation would have the form

$$\Sigma_{BHB} = \Sigma_o Y_o + \Sigma_w Y_w + \Sigma_G Y_G + \Sigma_{BACK}$$

where

Σ_{BHB} = Bulk borehole capture cross section

Σ_w = Capture cross section of the flowing water phase

Σ_o = Capture cross section of the oil phase

Σ_G = Capture cross section of the gas phase

Σ_{BACK} = Background capture cross section, a constant

Similarly, measurements of RIN, the ratio of inelastic count rates in the near and far detectors, have been found to be responsive to gas and may be used as a holdup indicator (Atlas PDK-100). The Schlumberger RST-B makes a highly successful measurement of water holdup in two-phase flow using a cross plot of the near and far counts. There are probably many approaches that can be taken with such pulsed neutron tools to provide other methods to measure the wellbore phases.

Wellbore Scanners

Some companies are developing tools which utilize resistance, capacitance, or other sensors on arms to scan the wellbore. If the flow is stratified, as possibly expected in horizontal flows, these tools would easily detect such a condition. If the light phase is in a bubble or slug form, this can also be detected. One approach incorporates an array of four or six arms with sensors at the ends. Another is to use a single sensor on the end of an arm which rotates downhole. Such scanners can provide maps of the multiphasic flows downhole. They may even be able to measure the local velocity of the fluids and/or bubbles around the borehole. This is a brute force approach to the problem and appears to have promise. Schlumberger has recently introduced a four arm tool of this type designated the Digital Entry Fluid Imaging Tool*, (DEFT*) (* mark of Shlumberger).

COMPUTING TWO-PHASE HOLDUP, SLIP, AND FLOW RATES

Computations Necessary for Two-Phase Flow

This section shows the necessary steps to perform the computation of individual-phase volumetric flow rates and how to determine which fluids are entering and from where. The bulk flow rates, Q_t, are known for each interval from spinner flowmeter computations discussed in a previous section. The following steps are taken:

1. Compute holdups from the bulk density measurement.
2. Determine the slip velocity between the phases.
3. Calculate the flow rates of each phase in interval i.
4. Establish the phase and volume of each entry.

Determining the Phase Holdups from the Bulk Density

To determine the holdup of each phase, their downhole densities must be known. These may come from local experience in a particular reservoir, PVT data, or fluid conversion charts. An alternative method is to shut in the well after the production logs have been run, wait a short while, and run the FID through the static condition—hopefully to detect an interface and measure the phase density above and below that interface.

If the fluids are called the heavy and light phase, then the holdup of the heavy phase is given by the equation

$$Y_H = \frac{\rho_B - \rho_L}{\rho_H - \rho_L}$$

where

$$\rho_B = \text{Bulk denisty}$$

$$\rho_L = \text{Density of light phase}$$

$$\rho_H = \text{Density of heavy phase}$$

$$Y_H = \text{Heavy phase holdup}$$

Since $Y_L + Y_H$, the holdup of the light phase equals

$$Y_L = 1 - Y_H$$

Determination of Slip Velocity

The slip velocity is probably the weakest link in the two-phase flow computation. Separate correlations are used for liquid/liquid (water/oil) and gas/liquid systems.

An often used model for water/oil flow is illustrated in Figure 18.9.[6] This is sometimes called the standard slip velocity chart. The horizontal scale is the density difference between the water and oil phases. A number of curves are presented for various water holdups. If the water density is 1 gm/cc and the oil is .75 gm/cc, the density difference is .25 gm/cc. If the water holdup in an interval is .60, then the slip velocity, V_s, is 15 ft/min (4.57 meters/min). The oil phase is moving up at a rate of 15 ft/min relative to the water in this interval. This chart applies for vertical intervals.[14-16]

The following equation describes this chart and adds a correction for well deviation. This equation applies for oil having a density greater than .5 gm/cc, and water holdup greater than .3. While not stated, this chart may be limited to deviations less than about 60 degrees. Note in this equation that θ is expressed in degrees, and density, ρ, in gm/cc. For V_{SLIP} in units of ft/min

$$V_{SLIP} = 39.414(\rho_H - \rho_L)^{.25} \exp\left\{-.788\ln\left(\frac{1.85}{\rho_H - \rho_L}\right) \times (1 - Y_H)\right\}(1 + .04\theta)$$

[handwritten annotation: θ deviation angle in degrees]

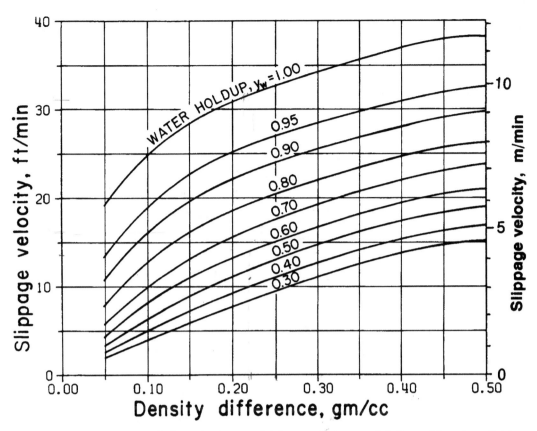

Figure 18.9 Conventional slip velocity model for vertical water/oil flows (Courtesy Schlumberger, Ref. 6)

For V_{SLIP} in units of m/min

$$V_{SLIP} = 12.013(\rho_H - \rho_L)^{.25} \exp\left\{-.788\ln\left(\frac{1.85}{\rho_H - \rho_L}\right) \times (1 - Y_H)\right\}(1 + .04\theta)$$

For gas/liquid systems, a constant value of 60 ft/min (18.3 meters/min) has been used with some success. However, the following equation probably provides a significantly better estimate of slippage velocity of gases. Note that θ is the wellbore deviation from vertical and is expressed in degrees.

  A. $\rho_{GAS} < .5$ gm/cc and $Y_{LIQUID} < .35$

  $V_{SLIP} = 0$ (mist flow)

  B. $\rho_{GAS} < .5$ gm/cc and $Y_{LIQUID} > .35$

For V_{SLIP} in ft/min,

$$V_{SLIP} = \{60\sqrt{.95 - (1 - Y_{LIQ})^2} + 1.50\}(1 + .04\theta)$$

For V_{SLIP} in m/min

$$V_{SLIP} = \{18.288\sqrt{.95 - (1 - Y_{LIQ})^2} + .457\}(1 + .04\theta)$$

An alternative and simple to use model is illustrated in Figure 18.10.[17] This chart provides a different correlation for slippage velocity and includes a chart for methane gas at downhole conditions (downhole density of .05 to .15 gm/cc). It also provides a correction for wellbore deviation.[17–19]

Another alternative is to back calculate the slip velocity in the highest interval from the total bulk and individual phase production detected at the surface. Then, use that slip velocity for all logged intervals regardless of holdup.

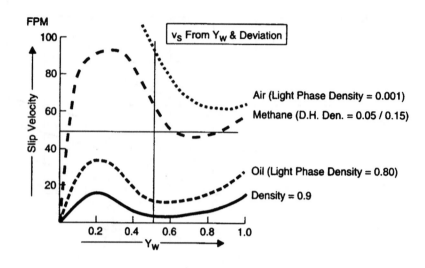

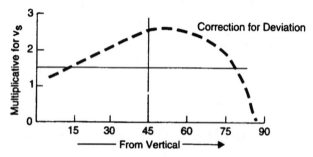

Figure 18.10 Slip velocity model with correction for deviation (Courtesy Western Atlas and SPWLA, Ref. 17)

Compute the Phase Flowrates in an Interval Downhole

The computation of flowrates in an interval i downhole in two-phase flow is a simple matter once the following are known:

$$Q_i = \text{Bulk flowrate determined from the flowmeter}$$

$$Y_{Hi} = \text{Heavy phase holdup determined from the FID}$$

$$V_{SLIPi} = \text{Slip velocity}$$

For each interval i, the individual flowrates of the heavy and light phases are given by

$$Q_{Hi} = Y_{Hi}\{Q_i - 1.4(ID^2 - d^2)V_{SLIPi}(1 - Y_{Hi})\}$$

and

$$Q_{Li} = Q_i - Q_{Hi}$$

where, for an interval i

$$Q_i = \text{Bulk flowrate, B/D}$$

$$Q_{Hi} = \text{Heavy phase flowrate, B/D}$$

$$Q_{Li} = \text{Light phase flowrate, B/D}$$

$$Y_{Hi} = \text{Heavy phase holdup}$$

$$V_{SLIPi} = \text{Slip velocity, ft/min}$$

$$ID = \text{Internal diameter of flowing mixture, inches}$$

$$d = \text{Tool diameter, inches}$$

And in metric units,

$$Q_{Hi} = Y_{Hi}\{Q_i - .1131(ID^2 - d^2)V_{SLIPi}(1 - Y_{Hi})\}$$

and

$$Q_{Li} = Q_i - Q_{Hi}$$

where (see definitions above)

$$Q_i, Q_{Hi}, Q_{Li} = \text{Flowrate, m}^3\text{/D}$$

$$V_{Si} = \text{meters/min}$$

$$Y_{Hi} = \text{no units}$$

$$ID, d = \text{cm}$$

Determining Entering Phases

The difference in the flowrates between the intervals is an indication of the entering flow. Occasionally, the computed answer may be perplexing in that it indicates one phase entering and another leaving the wellbore flow at a set of perforations. Obviously, this cannot be, and the analyst must look at the big picture to assess what is happening. Remember, this computation requires a number of separate inputs, few of which have a high degree of accuracy.

THREE PHASE FLOW

The Equations for Three-Phase Flow

For three-phase flow, two FIDs are required which respond to the total flowing mixture. Typically the two most often used FIDs are the nuclear density log and the capacitance device. The borehole capture cross section from PNC logs may also apply.

The basic equations for holdup are as follows:

Density device:

$$\rho_O Y_O + \rho_W Y_W + \rho_G Y_G = \rho_B$$

Capacitance device:

$$D_O Y_O + D_W Y_W + D_G Y_G = C$$

Borehole sigma from PNC:

$$\Sigma_{BACK} + \Sigma_O Y_O + \Sigma_W Y_W + \Sigma_G Y_G = \Sigma_{BHB}$$

Holdup equation:

$$Y_O + Y_W + Y_G = 1$$

From any two tools and the holdup equation, the holdup of each of three separate phases can be calculated. A graphical presentation of this concept is shown on Figure 18.11.[20] This figure plots the responses of the Schlumberger's Gradiomanometer and capacitance device called the HoldUp Meter*, (HUM*), (* indicates a marks of Schlumberger). The area of the map comprises the possible data points when both tools are responding to the bulk fluid properly. A data point, i.e., a reading of the gradiomanometer and a reading of the capacitance device is shown, and the corresponding oil, water, and gas holdups are indicated. This type of chart can be drawn for any two types of sensors. The only question is the linearity of the tool response to holdup.[20, 21]

The slip velocity is determined from the charts and equations used for two-phase flow. Since the density of the gas phase is so low relative to the liquid phases, the oil and water mixture is treated as a single liquid phase, and the gas slippage velocity is calculated. It is assumed that the water and oil are well mixed and the slip between the liquid phases is zero.

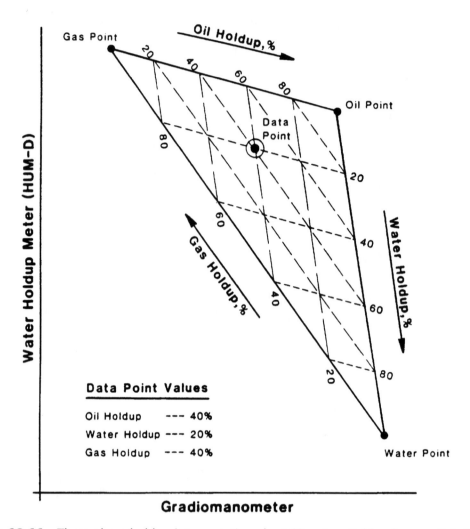

Figure 18.11 Three-phase holdup interpretation chart (Courtesy Schlumberger, Ref. 20)

To compute the individual phase production, the same equations as used in two-phase flow are used, with the heavy phase being the mix of oil and water. The individual flowrates of oil and water are then determined from the heavy-phase flow in proportion to their holdup, i.e.,

$$Q_{OIL} = \left\{ \frac{Y_O}{Y_O + Y_W} \right\} \times Q_{LIQ}$$

and

$$Q_{WATER} = \left\{ \frac{Y_W}{Y_O + Y_W} \right\} \times Q_{LIQ}$$

EXAMPLES OF MULTIPHASE LOGS

Fluid Density and Capacitance Comparison

The log of Figure 18.12 shows a capacitance and nuclear fluid density log run over a high flowrate gas well (300 MMcfd) producing 200 B/D water.[22] Both devices show a similar entry profile. The water entry appears to be coming from the perforations at 11,746. Notice that the density increases slightly above that depth and the capacitance also shows a sudden increase of water. The dynamic range of the capacitance device appears quite good at the low water holdups higher in the well.[22]

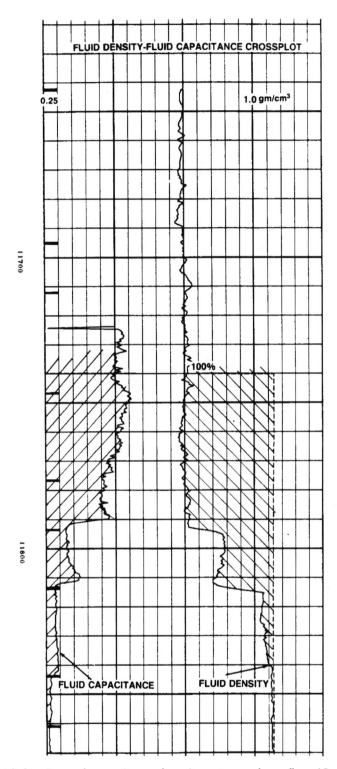

Figure 18.12 Fluid density and capacitance logs in water and gas flow (Courtesy Western Atlas and SPWLA, Ref. 22)

Low Gravity Oil and Water Production

In the example of Figure 18.13, 22° API (.92 gm/cc) oil is flowing with water downhole.[22] There are two sets of perforations. With the small density contrast between the water and the oil, the density log shows only the slightest indication of the oil entries. The capacitance log, however, shows the wellbore contents quite well since there is a large dielectric

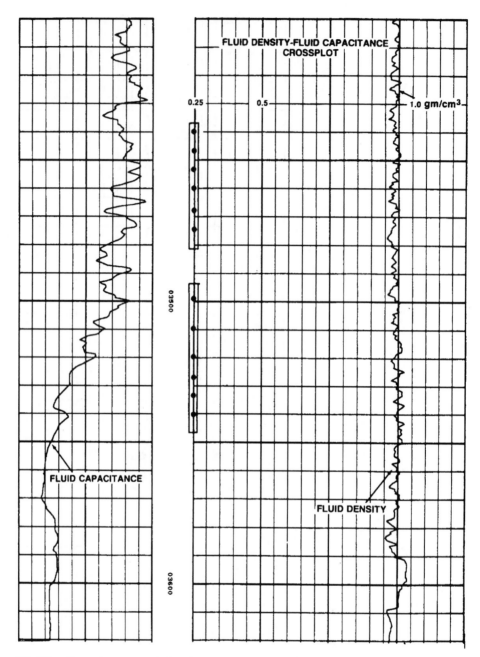

Figure 18.13 Capacitance and density comparison in low gravity oil and water flow (Courtesy Western Atlas and SPWLA, Ref. 22)

contrast between the oil and water. Entries from both the upper and lower perforations are clearly indicated from the capacitance device.[22]

Two-Phase Production

Refer to the log suite of Figure 17.18. The flowmeter shows a downflow below the perforations labeled D, and upflow above that point. The gradiomanometer shows the downflow to be essentially water, with lighter phases moving uphole. The interpretation of these logs is shown on Figure 18.14.[23] This curious well is acting like a downhole separator due to the slip velocity effect. The water flows downhole, and the oil, along with a small amount of water, flows uphole. An entry profile is also shown.

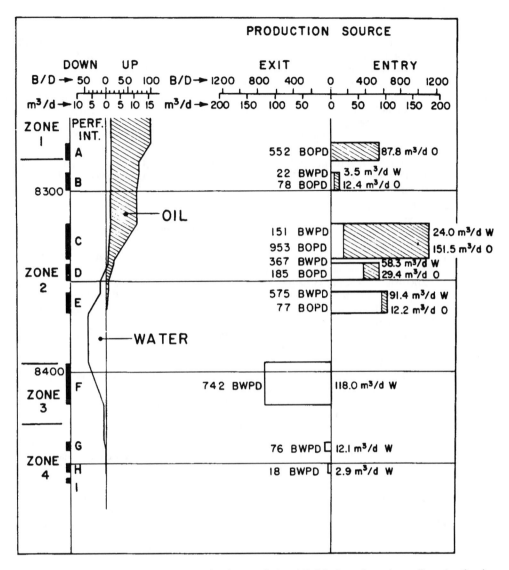

Figure 18.14 Production profile for the logs of Fig. 17.18 showing downflow in the lower part of the well (Courtesy Schlumberger, Ref. 23)

POINTS TO REMEMBER

- Bulk fluid density and capacitance are the most common FIDs.
- Fluid density is based on pressure differential or gamma ray absorption.
- Fluid density measures phase holdup in two-phase flow.
- Phase densities at downhole conditions must be accurately known to evaluate phase holdup based on bulk fluid density.
- Capacitance measurements are sensitive to water and are most quantitative at water holdups less than 40%.
- Holdup refers to the fraction of each phase present by volume in a flowing mixture.
- Cut refers to the individual phase flow rate through the pipe.
- Cut and holdup are different due to the effects of "slip."
- If any water is produced to the surface, especially in a low flowrate well, large water holdups can be expected downhole.
- A friction pressure drop correction may be required for pressure differential density tools at higher flowrates.
- Slip velocities increase significantly when the wellbore is deviated.

References

1. Roesner, R.E., LeBlanc, A.J., and Davarzani, M.J., "Effect of Flow Regimes on Production Logging Instruments' Responses," Paper SPE 18206, 63rd Annual Technical Conference, Houston, Texas, October, 1988.

2. Ding, Z.K., Ullah, K., and Huang, Y., "A Comparison of Predictive Oil/Water Holdup Models for Production Log Interpretation in Vertical and Deviated Wellbores," 35th Annual SPWLA Symposium, June, 1994.

3. Schlumberger, "Production Log Interpretation," Schlumberger Limited, Houston, Texas, 1970.

4. Curtis, M.R., "Flow Analysis in Producing Wells," Paper SPE 1908, 42nd Annual Fall SPE Meeting, Houston, Texas, October, 1967.

5. Curtis, M.R., "Flow Analysis with the Gradiomanometer and Flowmeter," Document C-11715, Schlumberger Well Service, Houston, Texas, 1966.

6. Schlumberger, "Production Log Interpretation," Document C-11811, Houston, Texas, 1973.

7. Gearhart (now Halliburton Energyy Services), "Production Logging Seminar," Document WS-774, Fort Worth, Texas, 1977.

8. Atlas Wireline, "Interpretive Methods for Production Well Logs," Third Edition, Document 9441, Houston, Texas, 1986.

9. Carlson, N.R., and Johnson, M., "Importance of Production Logging Suites in Multi-Phase Flows," 24th Annual SPWLA Symposium, June, 1983.

10. Carlson, N.R., and Davarzani, M.J., "New Method for Accurate Determination of Water Cuts in Oil-Water Flows,"30th Annual SPWLA Symposium, June, 1989.

11. Zhenwu, J., Haimin, G., and Xiling, W., "New Response Equations of Fluid Capacitance Instrument and Applications in Multiphase Flows," 31st Annual SPWLA Symposium, June, 1990.

12. Knight, B.L., "Flow-Loop Evaluation of Production-Logging Holdup Meter," *The Log Analyst*, pp. 421-430, July-August, 1992.

13. Kessler, C. and Frisch, G., "New Fullbore Production Logging Sensor Improves the Evaluation of Production in Deviated and Horizontal Wells," Paper SPE 29815, SPE Middle East Oil Technical Conference, Manama, Bahrain, March, 1995.

14. Nicolas, Y., and Witterholt, E.J., "Measurements of Multiphase Fluid Flow," Paper SPE 4023, 47th Annual SPE Fall Meeting, San Antonio, Texas, October, 1972.

15. Davarzani, M.J., and Miller, A.A., "Investigation of the Flow of Oil and Water Mixtures in Large Diameter Vertical Pipes," 24th Annual SPWLA Symposium, June, 1983.

16. Guo, H.H., Zhou, C.D., and Jin, Z.W., "The Determination of Key Parameters in Production Log Interpretation for Two Phase (Oil/Water) Flows," 33rd Annual SPWLA Symposium, June, 1992.

17. Ding, Z.X., Flecker, M.J., and Anderson, C., "Improved Multiphase Flow Analysis Using an Expert System for Slip Velocity Determination," 34th Annual SPWLA Symposium, June,1993.

18. Davarzani, J., Sloan, M.L. and Roesner, R.E., "Research on Simultaneous Production Logging Instruments in Multiphase Flow Loop," Paper SPE 14431, 60th Annual Technical Conference, Las Vegas, Nevada, September, 1985.

19. Chenais, P.V., and Hulin, J.P., "Liquid-Liquid Flows in an Inclined Pipe," *AICHE* Journal, pp. 781-789, May, 1988.

20. Schnorr, D., "Production Logging," Production Logging School, Schlumberger Offshore Services, Anchorage, Alaska, 1990.

21. Carlson, N.R., Barnette, J.C., and Davarzani, M.J., "Using Downhole Calibrations for Improved Production Profiles on Three-Phase Flows," 29th Annual SPWLA Symposium, June, 1988.

22. Carlson, N.R., and Roesner, R.E., "Water-Oil Flow Surveys With Basket Fluid Capacitance Tool," 23rd Annual SPWLA Symposium, July, 1982.

23. Anderson, R.A., Smolen, J.J., Laverdiere, L., and Davis, J.A., "A Production Logging Tool With Simultaneous Measurements," Paper SPE 7447, 53rd Annual SPE Technical Conference, Houston, Texas, October, 1978.

CHAPTER 19

FLUID MOVEMENT: PRODUCTION LOGGING IN HORIZONTAL WELLS

THE TOOLS AND EQUIPMENT OF HORIZONTAL WELLS

Logging Tools

As of this time, the logging tools used in horizontal wells are the same as those used in vertical and deviated holes. These tools have been designed to evaluate dispersed vertical flows, and many tools are known to have problems in inclined flow conditions. As a result, in horizontal wells, their responses are often unusual and sometimes bizarre. As will be discussed later, each tool has a diameter of investigation around the center line of the tool. If that diameter of investigation does not correspond to the flow cross section, strange results appear due to the tools sampling only part of the flow. Furthermore, horizontal flows can be stratified or slugging, depending on the flowrate and well angle. Applying the traditional interpretation principles can lead to erroneous conclusions.

Logging Operations

Logging operations were discussed in an earlier chapter. Most production logs are run at the end of coiled tubing.[1, 2] In most cases there is an electric wireline inside of the coiled tubing, and so the logs can be pushed and pulled across the logging interval and data recorded at the surface in real time much like is done in vertical holes. In some instances, mainly due to short radius wells, the tools are mounted directly onto a conventional coiled tubing and data is recorded downhole. Such "slickline" type tools are usually shorter and simpler. Tool sections are connected together with flexible couplings to accommodate the curvature of the wellbore between the vertical and horizontal sections.

Nuclear tools can be run in a pump down mode.[3] Conventional tubing joints are run to the bottom of the well, and the tool assembly is set up with a pump down sub at the lead end. The tool is connected to the surface unit with an electric wireline. When the pump down sub reaches bottom, it opens a port to allow flow through the unit. The nuclear tools, such as a pulsed neutron capture (PNC) or oxygen activation tool, are turned on and the tool pulled back across the interval of interest.

CONFIGURATION OF A HORIZONTAL WELL

What Is a Horizontal Well?

The term "horizontal well" usually means any well inclined over 75 to 80 degrees from the vertical. Such a classification may have meaning for drilling or financial considerations, but is virtually meaningless for logging fluid movement. Consider the well configurations, all of which can be classified as "horizontal," shown in Figure 19.1.

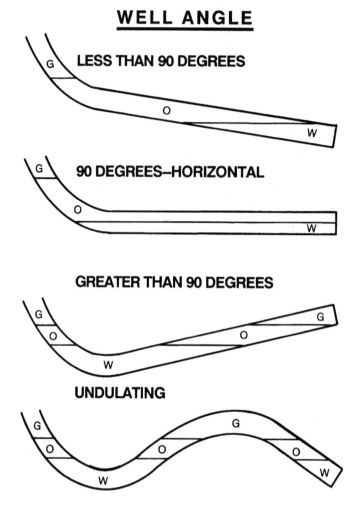

Figure 19.1 Horizontal wells

If a well produces and contains only one phase of fluid, then any of the configurations of Figure 19.1 are equivalent from a logging standpoint. However, if more than one phase is present, gravity segregation occurs. The water phase (W) puddles at the low points of the well and the gas (G) is trapped at the high points. Oil, if present, is sandwiched in between. These puddles and gas traps remain even when the well is flowing, unless the flow rates are large enough to thoroughly mix the flow. If the well angle is less than 90 degrees, then water and completion fluid accumulate at the bottom of the well. The perfectly horizontal well is probably a rarity. If the deviation is greater than 90 degrees, the water puddle is at the low point of the bend and gas accumulates toward the end of the well. Most common is the undulating well, where there are likely to be numerous gas traps and puddles along the well length. The presence of these traps and puddles plays havoc on spinner and fluid identification device response. It would appear that accurate directional information is necessary over the full length of the horizontal section if the production logs are to be understood.[4]

Types of Completions

A number of different completion types common to horizontal wells is shown on Figure 19.2.[5, 6] The conventional completion includes both the cemented and perforated casing as well as the open hole configurations. The common factor in such completions is that the flow is confined to a single cross section. There is no flow outside of the casing string unless it is inadvertent channeling or the like. The slotted liner and prepacked gravel completions would be classed as conventional if the formation collapsed onto the outside of

TYPES OF COMPLETIONS

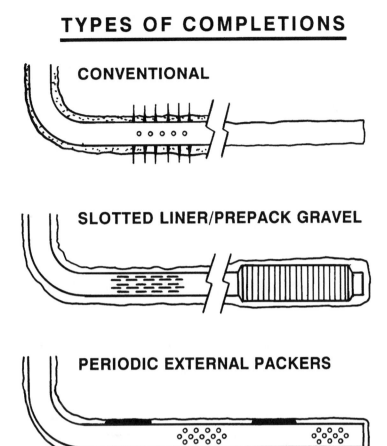

Figure 19.2 Types of completions

the screen or pack. However, if this does not occur, it is possible for fluid to flow both inside and outside of the casing. Flow outside of slotted casings is commonly observed, with logging tools often erroneously indicating all of the flow entering at the top of the slotted interval.[7] The use of periodic external packers offers some aspects of both types of completions. While annular flow outside of the casing is possible, data gathered at the packers is all in-bore and could be more easily interpreted for an accurate flow profile.

Flow Patterns for Completion Types

For conventional completions, the flow patterns are in-bore and shown on Figure 19.3. For single-phase flow, the profile is just the same as in a vertical well, and as a result, most tools would have no problem in such a horizontal well. Multiphase flow, however, is stratified over a wide range of flowrates.[8] Even as flowrates get higher, the slugging flow remains at the high side of the borehole. For mixtures of kerosene and water (densities of .83 and 1.00 gm/cc, respectively), at least 1,000 B/D (160 m³/D) is required to achieve a non-stratified flow in a 4.0 inch (10.2 cm) ID pipe.[9] A slight deviation causing the flow to move upward causes the heavy phase to accumulate, causing its holdup to increase and the flow to become violently slugging. A slight downward deviation produces a waterfall effect, causing the heavy phase holdup to be reduced and reinforcing the stratification of the flow. Clearly, this is a complex flow and accurate well inclination data is important to a sensible interpretation.

Concurrent flows, both in bore and annular, may exist as shown between packers. In this case, the annular flow can be either in the same or opposite direction to the main in-bore flow. The annular flow can be independent of, moderately independent of, or not independent of the in-bore flow. Depending on the nature of the measurement, the possibility

FLOW PATTERNS

SINGLE PHASE

IN-BORE FLOW

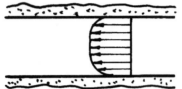

MULTIPHASE

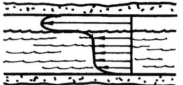

CONCURRENT FLOW—IN-BORE AND ANNULAR

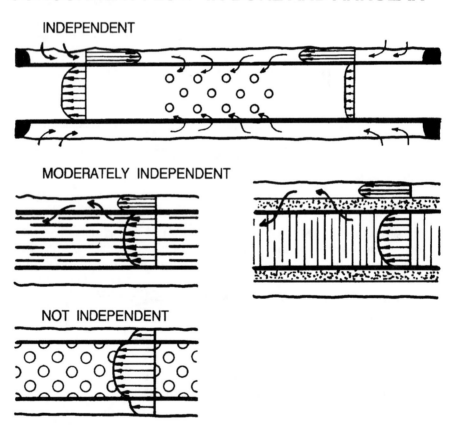

Figure 19.3 Flow patterns in horizontal wells

of concurrent flows must be considered, especially if two or more phases are flowing. In such cases, the light phase often flows high in the annular region and is not detected until diverted into the main borehole region.

If the formation outside of the slotted or gravel packed liners has collapsed onto the liner, then the flow is essentially confined to the in-bore condition. If the formation has closed in around the casing at some places and not at others, then the flow may be diverted into the borehole at the points of cave-in and cause erroneous indications of fluid entry. It would appear advisable to assess this possibility with gravel pack type logs. This log can identify regions of possible single or concurrent flows.[4]

PRODUCTION LOGGING TOOL RESPONSE IN HORIZONTAL FLOWS

Spinner Flowmeters

Figure 19.4 shows a schematic of horizontal section having a two-phase, in-bore flow. This well is undulating with one gas (G) and two water (W) entry points. This same well configuration is used for all logging tools and their expected responses to be discussed in the following sections. This well section is of a moderate flowrate and the flow is not well mixed. The changes in the velocity profile and water holdup are apparent from the figure.

At the top of the figure is shown the full bore and small diameter spinners. To their right, the casing is shown centralized in an open hole, simulating an open annulus all around the casing. The arrows emanating from the tool centerline extend to a dashed curve which describes the envelope of investigation of the tool.

At the bottom of Figure 19.4 is the expected response of the spinners to the flow in the well schematic. Note that both detect the water fallback downstream of the gas entry. As the flow waterfalls past the high point, the small spinner response increases dramatically since it is on the low side of the hole. Past the trough, the small spinner sees water fallback while the larger spinner sees some unknown mix of flows.[8–11]

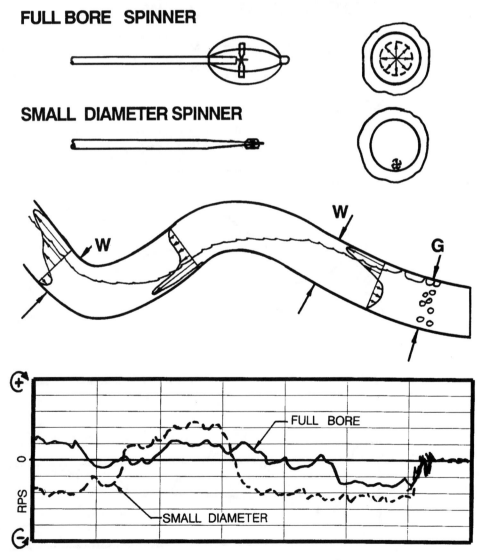

Figure 19.4 Spinner flowmeter envelope of investigation and response

Stationary Flow Measurements

Two basic types of stationary flow measurements are indicated on Figure 19.5. The basket flowmeter, which measures the total of the in-bore flow and the radioactive tracer. Tracers may be used which carry two phases of radioactive (RA) tracer downhole. If a light phase RA tracer is ejected, the tracer moves with the velocity of the light phase. If RA water tracer is used, the tracer travels with the velocity of the water.

The results are quite interesting. Theoretically, the basket measures the total flow and should provide a good measurement as indicated in the log schematic.[8–10] However, in practice, horizontal wells may be loaded with debris and diverting flowmeters very likely will clog up quickly. Furthermore, some basket equipment has had difficulty sealing in the horizontal environment, and so they must be supported by centralizers both above and below the diverter section. The tracer has made a good measurement of the water phase velocity. With a tracer tool having two ejectors, both the water and hydrocarbon phase may be measured in the same trip.[12] Notice that the tracer can make measurements even in the annulus, provided that the annulus flow is not too independent of the main bore-hole flow.

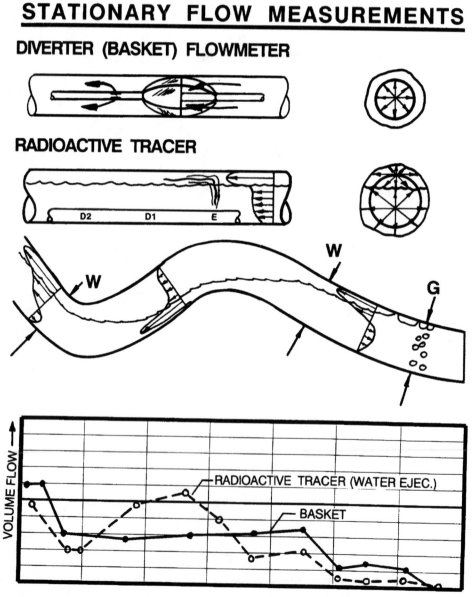

Figure 19.5 Stationary flow measurement techniques in horizontal wells

Fluid Identification Devices

Both the focused nuclear fluid density and the capacitance tools examine only a very small region of the wellbore as shown on Figure 19.6. The gravity segregation of the phases clearly limit their usefulness to more vertical flow conditions. Pressure differential devices are also useless in the horizontal logging environment. The non-focused nuclear fluid density, however, can examine the holdup of in-bore flow or even the holdup of concurrent flows, since its diameter of investigation encompasses these regions.[13, 14] It appears that by varying the spacing between the source and detector the diameter of investigation can be tailored to a particular casing or flow diameter.

The log response shows that the focused density and capacitance tools respond only to the water and totally miss the presence of the gas. The non-focused tool mimics the true holdup very well.

Temperature and Noise

Temperature logs in horizontal wells do not have much character, but are useful. They do examine the total region and can show entries. The key is the temperature scale, which is usually set to about one degree over the log track. Entries may cause anomalies of a 10th of a degree or less. The temperature log for this well schematic shows the gas entry as a

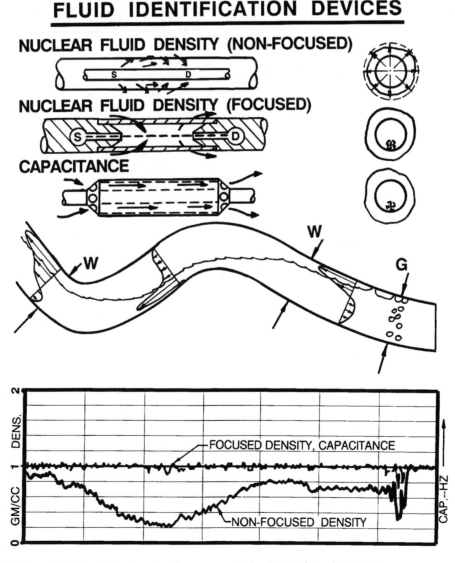

Figure 19.6 Fluid identification tool response in horizontal environment

small cooling anomaly. The water entries are shown heating up the flow, which has been cooled by the upstream gas entry. Water entries may be indicated by a small warming anomaly since the water often comes from deeper in the reservoir.[13]

The noise log also examines the whole region around the wellbore, including annular flow. At the point of gas entry, bubbles cross the wellbore and cause a noisy two-phase response. The stratified flow downstream is fairly quiet. When the well turns up, the liquid may choke off the gas stream, causing it to surge and slug at the low point. Hence, there may be an increase in noise level at the low point, but possibly not at the water entry. The temperature and noise logs are shown on Figure 19.7.

Pulsed Neutron Logs

Both the inelastic counts and the borehole sigma measurements, as shown in Figure 19.8, have large diameters of investigation and see the full flow, either in-bore or concurrent. The ratio of the inelastic near and far counts for gas entry detection has been used in horizontal wells. This response is called RIN for the Atlas Wireline PDK-100, from which this technique comes. This ratio is responsive to the amount of gas in the wellbore, and therefore is an indicator of gas holdup.[15] The borehole sigma measurement, if the produced water is adequately saline, is very similar. The borehole sigma measurement is essentially a measure of the water holdup.

TEMPERATURE AND NOISE

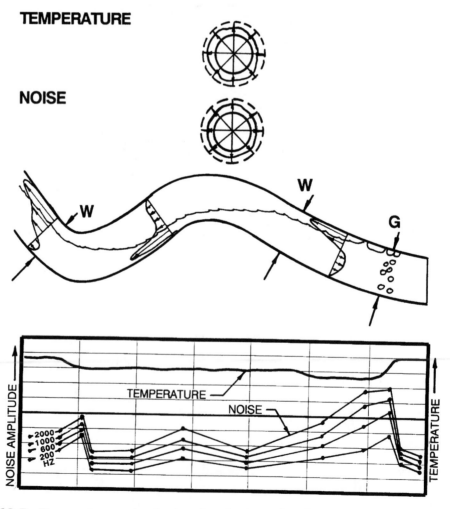

Figure 19.7 Temperature and noise logs in a horizontal well

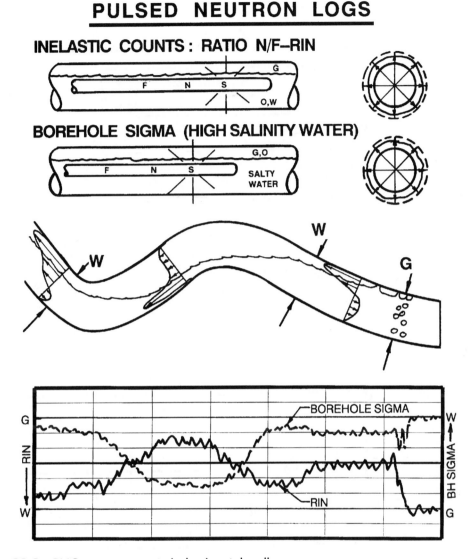

Figure 19.8 PNC measurements in horizontal wells

Oxygen Activation

The oxygen activation measurement, shown on Figure 19.9, can measure both single in-bore and concurrent flows. Its limitation, however, is that it only measures the water phase. The oxygen activation measurement can be made two ways. The traditional way with PNC logs is to record the background counts. If water flow exists which moves past the tool, the activated water causes an increase in the background counts. Such a detection of the water flow in the horizontal well is shown. This technique fails to detect slow water flows which do not flow faster than the tool and hence reads zero over some parts of the flowing interval. The station oxygen activation techniques, the Water Flow Log of Schlumberger, or the Hydrolog of Atlas Wireline Services are capable of detecting the water velocity. With holdup information, the water flow profile can be calculated.[16, 17]

CONCLUSION

Running Production Logs in Horizontal Wells

To successfully run production logs in horizontal wells, the individual tools of the tool string should be considered based on their region of investigation and the type of

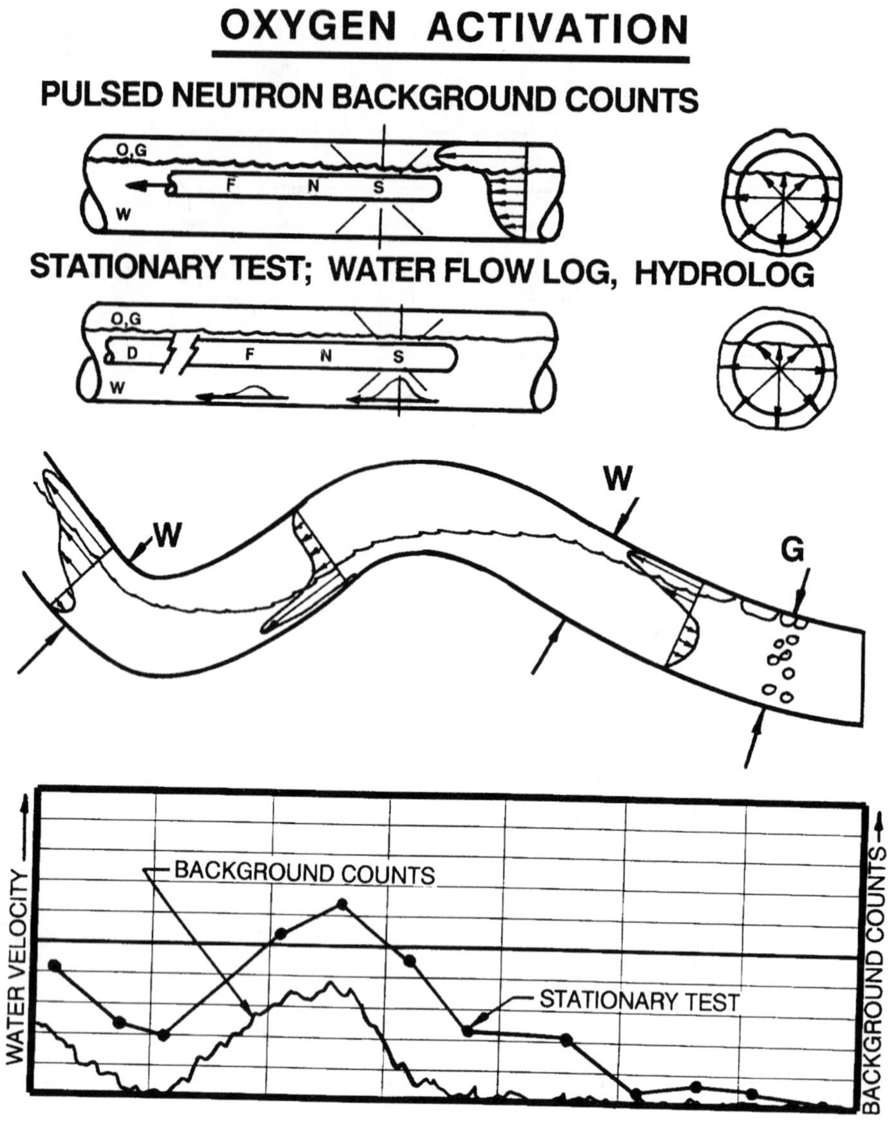
OXYGEN ACTIVATION
PULSED NEUTRON BACKGROUND COUNTS
O,G
F
N
S
W
STATIONARY TEST; WATER FLOW LOG, HYDROLOG
O,G
D
F
N
S
W
W
W
G
BACKGROUND COUNTS
STATIONARY TEST
WATER VELOCITY
BACKGROUND COUNTS

Figure 19.9 Oxygen activation applications in horizontal wells

completion. Serious consideration should be given to what kind of flows are to be encountered, i.e., whether in-bore or concurrent, and whether the flow is likely to be mixed or stratified. A mix of tools having different regions of investigation can help assess the presence of the flow stratification.

Directional surveys are extremely helpful to determine the cause of changes in holdup. Are these caused by entries or by wellbore undulations? Upflow and downflow are also quite likely to have very different flow patterns, even with the volumetric flowrate of the phases unchanged. Upflows will tend toward unstable slugging flows with higher heavy phase holdup while downflows tend to more stratified flows with lower heavy phase holdup.

Open hole caliper and gravel pack logs are very useful when concurrent flows may be present. The open hole caliper indicates the true diameter of the bulk flow and is very helpful in interpreting flow velocity changes. The gravel pack log is used to examine whether the annulus is open to flow or whether the flow is being diverted into the liner. It maps regions of possible in-bore and concurrent flows.

It has also been recommended that the well be shut in after taking the flowing production log data. The contrast between the non-flowing and flowing condition is extremely helpful in understanding the fluid identification logs and flowmeters. At such shut in conditions, the well can also serve as a downhole calibration tank, since the locations of the various phases, if present, will be known.

POINTS TO REMEMBER

- Most production logging tools run in horizontal wells are designed for vertical well mixed flow.
- Coil tubing is the typical conveyance method for production logging tools in a horizontal well.
- Nuclear tools may be pumped down through tubing installed to total depth.
- Horizontal may mean slightly more or slightly less than 90 degrees, a big difference when the flow regime is considered.
- Most horizontal wells will undulate, causing water puddles and gas traps.
- Completions may have either in-bore or concurrent flows in the casing and annulus.
- Concurrent flows are much more difficult to interpret since the light phase may flow in the annulus and the heavy phase in the wellbore.
- When selecting logging tools, pay particular attention to the tool's diameter of investigation and its size and location relative to the flow to be measured.
- Direction surveys are helpful to find high and low points in a well and locate points of expected puddles or gas traps.
- Open hole calipers are helpful when interpreting flow in slotted liners or other configurations with possible annular flow.
- Gravel pack logs are helpful to locate intervals where the formation collapsed onto the pipe and would divert annular flow into the casing.
- A shut-in survey after taking the flowing data is helpful.

References

1. Lohuis, G., Flexhaug, L., Huber, P., and Baker, C., "Coil Tubing/Production Logging in Highly Deviated and Horizontal Wellbores," Canadian Fracmaster, Calgary, Alberta, Canada, 1991.
2. Copoulos, A.E., Costall, D., and Nice, S.B., "Planning a Coiled Tubing Conveyed Logging Job in a Horizontal Well," Paper SPE 26090, SPE Western Regional Meeting, Anchorage, Alaska, May, 1993.
3. Noblett, B.R., Gallagher, M.G., "Utilizing Pumpdown Pulsed Neutron Logs in Horizontal Wellbores to Evaluate Fractured Carbonate Reservoirs," 34th Annual SPWLA Symposium, June, 1993.
4. Nice, S.B., "Production Logging in Horizontal Wellbores," 1992 World Oil Horizontal Well Conference, Houston, Texas, 1992.
5. McLamore, R.T., "Advances in Horizontal Completion Technology," World Oil's 5th International Conference on Horizontal Well Technology, Amsterdam, Netherlands, July, 1993.
6. Schlumberger, "Horizontal Wells: Evaluation, Completion and Stimulation," Document SMP 8551, Houston, Texas, 1992.
7. Catterall, S.J.A., and Yaliz, A., "Case Study of a Horizontal Well in the Ravenspurn North Field: A Stratified Rotliegendes Gas Reservoir," Paper SPE 25046, European Petroleum Conference, Cannes, France, November, 1992.
8. Branagan, P., Knight, B.L., Aslakson, J., and Middlebrook, M.L., "Tests Show Production Logging Problems in Horizontal Gas Wells," Oil and Gas Journal, pp. 41-45, January 10, 1994.
9. Carlson, N.R., and Davarzani, M.J., "Profiling Horizontal Oil-Water Production," Paper SPE 20591, 65th Annual SPE Technical Conference, New Orleans, Louisiana, September, 1990.
10. Knight, B.L., "An Experimental Comparison of Production Logging Tool Response in Vertical and Deviated Wet Gas Wells," Topical Report GRI-92/0386, Gas Research Institute, Chicago, Illinois, September, 1992.
11. Schnorr, D., "Production Logging Horizontal Wells," Schlumberger Well Services, Canada Alaska Unit, Calgary, Alberta, Canada, 1991.
12. Chauvel, Y.L., "Production Logging in Horizontal Wells: Applications and Experience to Date," Paper SPE 21094, SPE Latin American Petroleum Engineering Conference, Rio de Janeiro, Brazil, October, 1990.
13. Robertshaw, S.E., and Peach, S.C., "Well Illustrates Challenges of Horizontal Production Logging," Oil & Gas Journal, pp. 33-38, June 15, 1992.
14. Kessler, C., and Frisch, G., "New Fullbore Production Logging Sensor Improves the Evaluation of Production in Deviated and Horizontal Wells," Paper SPE 29815, SPE Middle East Oil Technical Conference, Manama, Bahrain, March, 1995.
15. Barnette, J.C., Copoulos, A.E., and Biswas, P.B., "Acquiring Production Logging Data with Pulsed Neutron Logs from Highly Deviated or Non-Conventional Production Wells with Multi-phase Flow in Prudhoe Bay, Alaska," Paper SPE 24089, SPE Western Regional Meeting, Bakersfield, California, March, 1992.
16. McKeon, D.C., Scott, H.D., and Patton, G.L., "Interpretation of Oxygen Activation Logs for Detecting Water Flow in Producing and Injection Wells," 32nd Annual SPWLA Symposium, June, 1991.
17. Chauvel, Y., and Clayton, F., "Quantitative Three-Phase Profiling and Flow Regime Characterization in a Horizontal Well," Paper SPE 26520, 68th Annual SPE Technical Conference, Houston, Texas, October, 1993.

Index

W

Water
  fingering, 22, 23
  movement detection using oxygen activation,
    295
  N-F background response to activated, 296–297
  oil contact (WOC), 22
  producing perforation, 37
  See also Carbon/oxygen log
Wave properties, 125
  propagation and refraction, 127–128
  types of, 126–127

Well
  fluid content and noise, 261
  integrity, 20
  integrity acoustic cement, 161–190
  integrity downhole casing, 215–233
  natural completions, 6
  profile, 5–6
  pumping, 6–7
  rod pumped, 6–7
  set-up, 9
  temperature deviation, 236
  ultrasonic pulse echo cement evaluation, 193–212
Wellbore integrity of cased hole logs, 2

Wellhead, 5
  connection, 10
  rigging up, 8–9
  set up coil tubing, 11–14
  set up pump down, 14
  typical, 10
Windows method for c/o, 100–103
WOC. *See* Water, Oil contact

Z

Zones A-D, 17–19

Biography

JAMES J. SMOLEN has more than 25 years of oil industry experience in cased hole well logging and applications, related research, and training. For 10 years, he designed, developed, and field tested cased hole logging equipment and established new log evaluation techniques for Schlumberger. Since 1980, Smolen has been an independent consultant, working with a variety of petroleum and service companies. He has conducted seminars on cased hole logging, wireline formation testing and related subjects worldwide. During this period, he also was an officer and director of Petroleum Computing, Inc., a developer and marketer of PC-based log evaluation software. Smolen has numerous technical publications to his credit. He was a Distinguished Lecturer for both the SPE and the SPWLA during 1988.

Smolen holds a B.S. from Northwestern University and M.S. and Ph.D. degrees from the University of California at Berkeley.